The PACIFIC HALIBUT

The PACIFIC HALIBUT

The Resource and the Fishery

F. Heward Bell

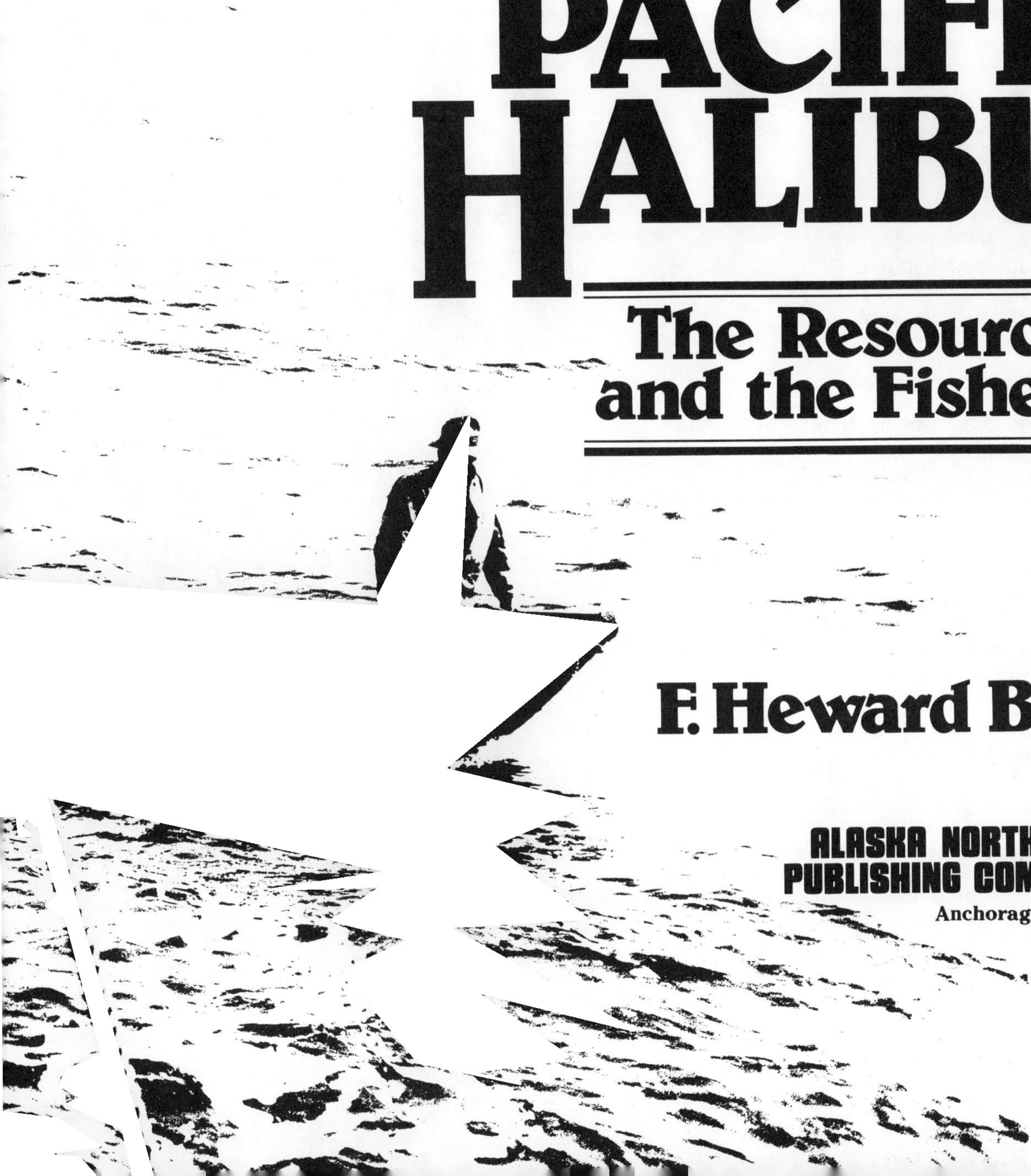

ALASKA NORTHWEST PUBLISHING COMPANY
Anchorage, Alaska

Library of Congress cataloging in publication data:

Bell, Frederick Heward, 1902-
The Pacific halibut, the resource, and the fishery.

Bibliography: p.
Includes index.
1. Halibut fisheries—Northwest, Pacific—History. 2. Pacific halibut. 3. Fishing ports—Northwest, Pacific—History. 4. Fishery management, International —Northwest, Pacific—History. I. Title.
SH351.H2B44 333.95'6 80-29218
ISBN 0-88240-141-6 Softbound
ISBN 0-88240-158-0 Hardbound

COVER PHOTO: The crew of the *Leviathan* struggles to haul a halibut over the rail. Shown are James, Eric and David Olsen. Their sister Christine Olsen, who also longlines on board the *Leviathan* and has for the past eight years, shot this photograph in the Gulf of Alaska in 1977.

TITLE PAGE PHOTO: International Pacific Halibut Commission

DESIGN BY VAL PAUL TAYLOR

ALASKA NORTHWEST PUBLISHING COMPANY
BOX 4-EEE, ANCHORAGE, ALASKA 99509

PRINTED IN U.S.A.

Dedication

To Edith,
my deceased wife who contributed innumerable editorial suggestions and was a constant source of encouragement during my 50-year association with the halibut fishery.

TABLE OF CONTENTS

PREFACE

This book is best described as an encyclopedic documentary on the Pacific halibut and its fishery. It covers many facets of those two related subjects but does not profess to be an exhaustive treatise on any one aspect of the resource or of the fishery.

Items are included regarding Washington, British Columbia and Alaska maritime and fishery history of the past 100 years that have not been previously recorded. There are also first-hand accounts of what transpired in the Pacific halibut fishery and in its management over the past one-half century.

ABOUT THE AUTHOR

William Brothers

The Pacific halibut fishery was less than 40 years old when F. Heward Bell began his 45 years of service with the newly established "Halibut Commission" of Canada and the United States. He knew and interviewed many individuals who had long been a part of the industry even from its beginning in the 1880s.

Beginning with investigations in 1925 the author was involved with all phases of the fishery, first as chief biologist, then as assistant director and later as director until retiring in 1970. In all capacities close contact was maintained with the dealers, the resource and the fleets. Many months were spent at sea on board halibut vessels of all types, which were either chartered to the Halibut Commission or being used commercially.

The author was awarded the 1953 Coronation Medal of Elizabeth Regina II for services rendered Canada with respect to the Pacific halibut fishery. He was the first person to receive the Golden Halibut Award presented annually by the Halibut Fishermen's Wives' Association and was awarded a Golden Halibut by the Halibut Association of North America for services rendered the industry with respect to the misuse of the name "Greenland Halibut" in Canada and United States. In 1962 he received a dedicated service award from the Maritime Press Association.

Upon graduation from the University of British Columbia in 1924 with honors in biology he was appointed instructor in biology for that school. In 1925 he served as research assistant, tagging salmon on the west coast of Vancouver Island, for the Fisheries Research Board of Canada. While on leave from the Halibut Commission from 1940 to 1941 he was assistant director of the International Pacific (Sockeye) Salmon Fisheries Commission of Canada and United States. He was Special Lecturer in Fisheries at the University of Washington from 1930 to 1970, a founding fellow of the American Institute of Fishery Research Biologists and is a member of other scientific societies.

ACKNOWLEDGMENTS

International Pacific Halibut Commission

This book owes much to the many who have been a part of the Pacific halibut fishery during the past century. Hundreds of persons have been interviewed over the years, and the scientific reports and records of the Halibut Commission, newspaper files of Pacific Northwest cities and industrial journals such as the *Canadian Fisherman* and *Pacific Fisherman* were major sources for materials and for figures used in this volume.

Scientific staffs of the Halibut Commission including that during my tenure as director (pictured at left) are to be given full credit for their contributions to halibut research. Alonzo T. Pruter, Deputy Director, Northwest Fisheries Center, United States National Marine Fisheries Service; Dr. Richard Van Cleve, Emeritus Dean, College of Fisheries, University of Washington, and the editorial staff of Alaska Northwest Publishing Co. provided editorial review and contributed many useful suggestions.

Of the line figures in the book, many are the work of Kenneth W. Exelby.

1 HALIBUT AND ITS NAMES

Some views are presented that are at variance to what is generally held regarding the word "halibut" and the species it represents. It will be shown that the etymology of the word halibut as recorded in dictionaries of the English language should be reexamined by lexicographers; that the technical name *Hippoglossus stenolepis* (Schmidt, 1930), by which the Pacific halibut is currently known in the scientific community, should revert back to the earlier-accepted *Hippoglossus hippoglossus,* (L.); the continued use of the word *halibut* in commerce for flounders other than of the genus *Hippoglossus* offends the rules of good usage. Such poor usage as in the case of Greenland halibut is also contrary to recent determinations by the governments of Canada and of the United States as well as being in substantial conflict with judicial decisions in the highest courts of the United States.

ETYMOLOGY

Many myths have arisen and most persist about the name halibut. Such myths seem to stem from the fact that dictionaries generally have ascribed a defective origin to the "hali" portion of the word. Some current writers have fantasized that since halibut were relatively rare in "Earlie Englanda" they became a delicacy and the "festive fare on Holy Days." Others, less poetic but probably as incorrect, state that it was the high esteem in which halibut had been long held in England during the Middle Ages that resulted in its being reserved for consumption on holy days. Thus halibut became known as the Holy Butte (butte-flounder).

Most lexicographers regard the word halibut as being derived from the Middle English *haly butte.* This word has been spelled variously according to the *Oxford English Dictionary.*

In English literature the word halibut seems to occur first in the *Boke of Nurture,* written by John Russel between 1430 and 1440 A.D. (Furnivall, F.J., 1868). This document may be located in the Harleian MS No. 4011, British Museum, vol. 171. John Russel was Lord Chamberlain to the Duke of Gloucester, and his writings provide a most detailed description of the lifestyle, including all the services and duties of a major-domo of a princely establishment in England during the Middle Ages. Some of the accounts deal with most delicate matters as pertained to the royal bedchamber, including the connubial affairs therein.

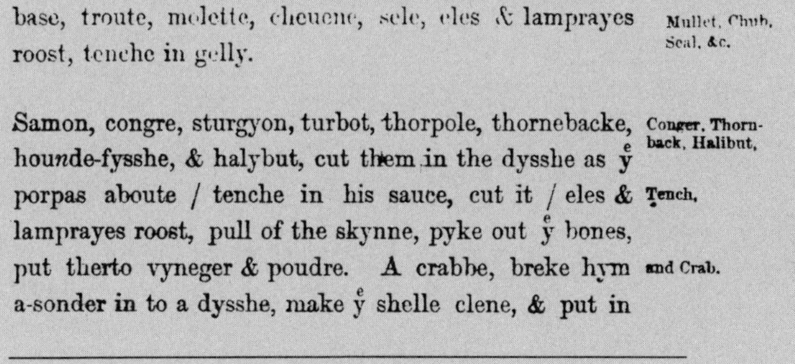

¶ The seconde course. — *Second Course:*

¶ Gelly whyte and rede, dates in confetes, congre, samon, dorrey, brytte, turbot, halybut / for standarde, base, troute, molette, cheuene, sele, eles & lamprayes roost, tenche in gelly. — Jelly, dates, &c. For a standard. Mullet, Chub, Seal, &c.

Samon, congre, sturgyon, turbot, thorpole, thornebacke, hounde-fysshe, & halybut, cut them in the dysshe as ye porpas aboute / tenche in his sauce, cut it / eles & lamprayes roost, pull of the skynne, pyke out ye bones, put therto vyneger & poudre. A crabbe, breke hym a-sonder in to a dysshe, make ye shelle clene, & put in — Conger, Thornback, Halibut, Tench, and Crab.

FIGURE 1. One of the first records of the compound word "halybut" in English literature.
F.J. Funivall

The word *halybut* first occurs on page 41 of the *Boke of Nurture,* verse 584:

"Halybut, to hym put hathe heele alle pese cut
in be dische as youre lord etethe at meale."

It also appears on page 51, verse 735 of the same treatise.

The word halybut is again found in Russel's even more specific account regarding the use and serving of fish, namely his treatise on *Kervynge of Fische* (page 166, line 12 and page 167, line 10) also written between 1430 and 1440 A.D.

Halybut may go back to the Old English *halig* (holy) and possibly the Germanic *but* (flounder). It is noteworthy, however, that *but* or *butt* or *butte* is not recorded for Old English. It is absent, for example, in the dialogue between the teacher and the fisherman in Alfric's *Colloquy*, c. 1020 A.D. While Alfric wrote in Latin, someone interlined the Old English equivalents. In *Colloquy* the Old English for flounder was given as *floc*.

The word butt first occurs in the Middle English metrical romance, *Havelock the Dane,* verses 749-759, written about 1300 A.D. It is glossed *flounder* by recent editors of the romances, Hale and French. While it is quite possible that the word but or butt may go back to Old English, it is evident that the very early history of the word halibut is defective in its two parts, both *haly* and *but.* It was not until the fifteenth century, 1430 to 1440 A.D. that the compound word halibut has been observed to occur in English literature.

From a phonetic point of view, some have believed the word halibut might be borrowed from Scandinavian or other Germanic languages with the stronger probability that Scandinavia was the source.

In the United States G. Brown Goode (1887), a scholarly fishery scientist of the nineteeth century, asserted that the true derivation of the name halibut is found in Scandinavia where the prefix *halle* in the Swedish halleflundre refers to the holes "hallir" or "haller," or deep places at sea where halibut tend to be found. G. Brown Goode (p 309) stated as follows:

> In studying these names it should be borne in mind that "But" or "Bott" is another word for a flounder or flat fish, and that the English, Dutch, German, and Scandinavian prefixes to either this word, or the equivalent word Flounder, are presumably of the same meaning. A false derivation has been imagined for the name, which is exemplified in the German word "Heiligbutt" just mentioned, and also in the English orthography, which is sometimes encountered "Holybut." This is without foundation, for the Halibut has never been mentioned more than any other species of flat fish, and the derivation is as fanciful as the New England one of "Haul-a-boat," which our fishermen have frequently assured me is the proper name, referring to the size and strength of the fish. The true derivation of the word is best understood by a study of its Scandinavian names, from which it appears that the prefix has reference simply to the holes or deep places at sea in which the fish is found, and that the name simply means, "a deep-sea fish," or "a deep-sea flounder."

Smitt (1893) noted that the Swedish name for halibut is written, pronounced and explained in different ways. The Icelanders call it flydra- or heilig-fiske (holy fish), *flydra* being the origin of the Swedish helgeflundra and *heilig* of the German heiligbutt.

Much earlier, Qvensel (1806) derived the Swedish helgeflundra from Helgeland in Norway. The Norwegians call the species qveite (kveite), or halleflyndre. Nilsson (1855) stated that "halibut haunt those spots in deep water where springs of freshwater occur. These spots, which are discovered by the Norwegian fishermen by the common method of taking bearings, are known to them as hallir (holes), whence came the Norwegian name halleflyndre and the Swedish halleflundra." In Bohuslan (the Baltic Sea) the halibut is generally called hallefisk according to Malm (1877).

Another writer many years later, William E. Meehan (1918), reiterated the views of Goode and others, in the following:

> The name Halibut is of Scandinavian origin, and it is made up of two words meaning deep-sea flounder or flatfish. Curiously and very much out of the ordinary rule, the common name is the same almost everywhere with trifling exceptions. In England, it is often called (and spelled) Holibut; in Germany it is Heilbutt; in Holland, Heilbot; and in Sweden it is Halleflundra.

All modern English dictionaries, such as the Oxford, Webster's, Funk & Wagnall's, Random House and others, give *holibut* as an alternative spelling. In the nineteenth century the holibut version was used almost to the exclusion of that of halibut. The *Century Dictionary and Cyclopedia* (1913) states that holibut is etymologically more correct than halibut, as it relates better to "holiday." That 1913 source also states that the sense seems to have been lost, and that the forms have suffered corruption. This conclusion in the *Century Dictionary* may stem from an attempt to subscribe to the unclear and defective "holy" origin rather than to the more sensible "hole" lineage from Scandinavia. It is noteworthy, however, that the name of the species is still frequently pronounced holibut by some of the older fishermen of New England, Nova Scotian and Newfoundland origins and even occasionally by some of Norwegian backgrounds. However, it has rarely been seen spelled holibut by this writer in the literature of this, the twentieth century.

The apparent unanimity* of modern lexicographers in subscribing to the "holy" origin of the word, probably arises from the fact that they tend to follow almost verbatim the *Oxford English Dictionary*, the most prestigious lexicon of the English language. However, there is some reason to believe that the abridged edition rather than the unabridged first edition was used as their guide. In the latter it was stated that the "holy" origin stemmed from the belief that halibut "was supposed to be socalled from being commonly eaten on holy days." In the abridged edition, the phrase "supposed to be" was omitted. Such omission could be critical to the issue, if not noted by later lexicographers.

It seems almost certain that inclusion of the phrase "supposed to be" in the first edition was not accidental. It suggests that the many hundreds of distinguished scholars who contributed to the *Oxford English Dictionary* had not found any record or usage that would allow any more definitive statement to be made about the presumed "holy" origin. It might be said that later lexicographers, by not noting the phrase "supposed to be," have probably led to the "holy" origin theory becoming ingrained in all modern dictionaries.

There is considerable evidence, though circumstantial, that supports the contrary views of Goode and this author. If the species had in fact been used specifically or exclusively for holy days, it seems extremely unlikely that anyone like the Lord Chamberlain John Russel would have repeatedly discussed the use and the serving of halibut in a royal household without making some reference to the religious constraints on its use. Instead, Russel, who so meticulously described the most intimate features of the life and living in the royal household, speaks of halibut repeatedly without any reference to it's holy status. It is further unlikely that he would so ignore any proscriptions or limitations about its use and thereby hazard the wrath of a then all-powerful church, particularly when the royal household itself included many churchmen.

Also, the fact that there are so very few references to the word halibut in Middle English literature generally does not give much weight to the theory that the species was a holy flounder for use on holy days, having been singled out from other fish and flounders for such distinctive use. Certainly such distinction should have led to the word being more visible in English literature during the time of Church dominance with its multiplicity of holy days.

However, if it is desirable to find both Old English or Middle English origins for both parts of the word holibut as it was known in earlier times, one only has to look at the Middle English as well as the Old English word *hol* meaning, amongst other things, a hole or hollow. Thus, halibut becomes the *hole but* of both Old and Middle English, i.e., the flounder caught in deeper water. Such a simple solution conforms to the Norwegian halleflyndre and the Swedish halleflundra theories.

The name holey or deep water flounder is also most descriptive of the species bathymetric or depth distribution during its adult phases. The relationship to holes or hollows thus does not depend upon some rather tenuous and undocumented religious requirements of the Middle Ages as in the case for the holy flounder proponents. Furthermore, the adult sizes which would be sought after for food are usually found in water deeper than where most flounders congregate. In Norway, with the most plentiful European supply of halibut and where the name probably originated, the halibut at certain seasons of the year are in fact found in very deep water or "holes" far up the fjords.

Despite the fact that the German *heiligbutt* or Icelandic *heiligfiske* with their "holy" prefixes continue to pose some problems for the deep water, "holey" flounder theory, there remains many more reasons to accept the "holey" rather than the "holy" argument. For one, the German "heil" could well be mere impropriety.

It is unfortunate that Kohler (1906), in his classic study of the genealogy of the names of many important European species of fish, did not discuss the word halibut. However, if the halibut had been the holy fish, prescribed for eating on holy days, it is likely more would have been said about it in the literature of the Middle Ages when such religious considerations were matters of great moment, and Kohler probably would then have included it in his list of fish names studied.

One might summarize as follows:

1. The origin of halibut is defective with the "but" portion of the word not recorded for Old English.

2. Phonetically, the word halibut might have been borrowed from Scandinavia.

3. Halibut was always more abundant and available for use in Scandinavia where the name most likely had its origin.

4. The prefix halle in halleflyndre and halleflundra, the Norwegian and Swedish names respectively for halibut actually mean holes (hallir or haller).

*In some editions of Webster the *perhaps* is added to the word "so-called" suggesting some doubt about the "holy" origin theory.

5. Halibut in fact tend to congregate in holes, i.e., in deeper water than do other flounders.

6. Probably due to failure to find any reference in the early literature to the "holy" origin of the word, the many hundreds of scholars contributing to the first edition of the *Oxford English Dictionary* found it necessary to insert the phrase "supposed to be" in describing any such origin.

7. Such explicit chroniclers as the Lord Chamberlain John Russel (1430) failed to mention the holy aspect of the halibut, despite the dominance of the Church and the frequency of holy days.

8. One finds in both Middle English and Old English the word hol meaning, among other things, hole, and thus providing a direct and simple link to the word holibut, the spelling in common usage in the recent past.

Although there are many reasons to support the "holey" rather than the "holy" origin, subscribing to the "holy" origin does confer upon the species an unearned aura of importance that is lacking in the rather mundane "holey" designation.

A NEW BUT OLD SCIENTIFIC NAME

The systematic or scientific classification and naming of living organisms is of great academic interest as well as value in bringing some order to the multiplicity and endless complexities of living forms on planet Earth. Where an animal or plant has become the object of trade and commerce its scientific identification takes on added importance in the competitive economic world. Also, in the current trend toward consumerism, a stable and generally accepted product identity, source and content is a necessity. The *caveat emptor* principle seems to have fallen into disrepute. Society now goes to great lengths to protect the buyer from his ignorance as well as from his own cupidity.

The classification and naming of halibut has been subjected to a number of versions since the time of Carolus Linnaeus who developed the binomial system of nomenclature in the mid-1700s that is still in general use throughout the scientific world. This binomial system provides each organism with two types of names, genus and species. The genera are further classified into various larger groupings such as subfamilies, families, orders, classes and phyla. Species are sometimes divided into subspecies or varieties.

One system of classifying the genera of flatfishes is that of Jordan and Evermann *(Fishes of Middle and North America,* 1896). These authorities classified the

FIGURE 2. The Pacific halibut with the dark topside showing (A) and with the white underside (B). A medium-sized halibut flails its strong body after being pulled over the rail (C).
Insert photographs (A) and (B) courtesy of the Northwest & Alaska Fisheries Center. Photograph (C) courtesy of Christine Olsen.

flatfish as the suborder, Heterosomata, and divided it into the following subgroups:

- Family Soleidae (the soles)
- Family Pleuronectidae (the flounders)
  - Subfamily Pleuronectinae (flounder tribe—including 14 genera)
  - Subfamily Psettinae (turbot tribe—including 10 genera)
  - Subfamily Hippoglossinae (halibut tribe—16 genera including *Hippoglossus, Paralichthys, Reinhardtius* and 13 others).

The halibut tribe, Hippoglossinae, are large-mouthed with ventral fins symmetrical, while the flounder tribe, the Pleuronectinae are small-mouthed. The Psettinae or turbot tribe are large-mouthed but with asymmetrical ventral fins.

Among the 16 genera placed in the halibut tribe by Jordan and Evermann is the genus *Hippoglossus,* which is the true halibut, the genus *Paralichthys*, the so-called "bastard" halibuts that include a large number of species throughout the world, and the genus *Reinhardtius*, comprising but one species, the turbot. The turbot is usually and improperly called Greenland halibut by the scientific community.

As knowledge of the anatomy and the evolutionary relationships of the flatfish increased, other ichthyologists modified the foregoing classification.

In 1920, Regan presented the following grouping which elevates the flatfish to the order Heterosomata rather than the suborder rank of Jordan and Evermann.

- Order Heterosomata
  - Family Psettodidae
  - Family Bothidae
    - Subfamily Paralicthyinae
    - Subfamily Platophrinae
    - Subfamily Bothinae
  - Family Pleuronectidae
    - Subfamily Pleuronectinae
    - Subfamily Samarinae
    - Subfamily Rhombosoleinae
    - Subfamily Paralichthodinae
  - Family Soleidae
  - Family Cynoglossidae

Later Jordan (1923) subscribed to the above grouping of Regan in general but raised most of the subfamilies to the rank of family. Norman (1934), a more recent authority on the flatfish of the world followed the Regan scheme, which he regarded as a "perfectly natural one." He did add one subfamily to the Pleuronectidae, namely the Poecilopsettinae.

In Regan's system the genus *Hippoglossus,* the true halibut, is in the family Pleuronectidae, but the genus *Paralichthys*, the bastard halibut, is further removed from the true halibut by being transferred to another family, the Bothidae. *Reinhardtius*, the turbot, remains in the Pleuronectidae.

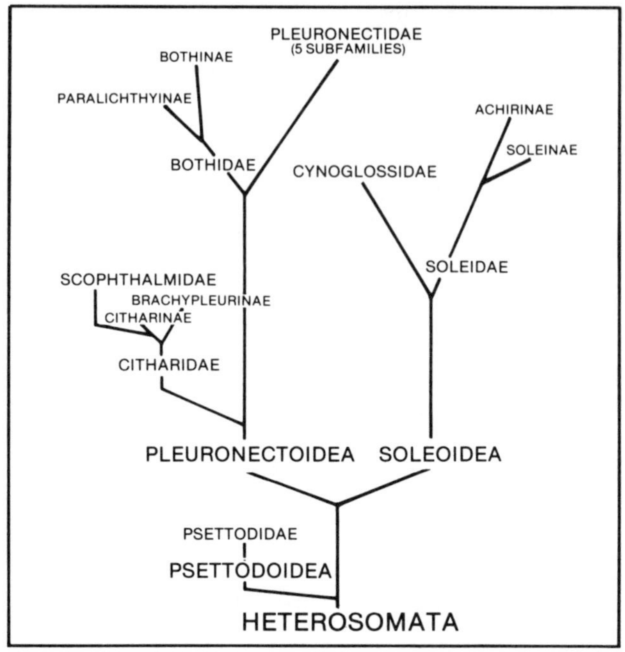

FIGURE 3. Phylogeny of the Heterosomata.
C.L. Hubbs

A later structure of the Heterosomata is that of Hubbs (1945) who added one more family, the Citharidae, to Regan's five. Hubbs also reviewed the phylogeny of the flatfishes and presented an up-to-date family tree of the group (Figure 3).

In the field of taxonomy or classification there is a great amount of flexibility and change. Some stem from expanding knowledge in the fields of genetics, palaeontology and phylogenetics generally. Also, some of the changes may result from whether the leading workers in the field at the time belong to the school of "splitters," those who wish to divide organisms into the smallest taxonomic groups, or the "groupers," who are inclined to combine organisms into larger and sometimes simpler divisions. However, much of the foregoing is of academic concern only. In the field of commerce, while scientific names are important, taxonomic status should be interpreted with discretion.

Whether two species belong to the same genus, family or subfamily or tribe should not necessarily be used to claim similarity or contrast in the market-place. A peanut is clearly a peanut and to involve it with a sweet pea in commercial considerations merely because both belong to the same taxonomic family Leguminosae is to confuse the issue to absurd levels. Similarly with fish, merely because two forms may be closely related from a taxonomic standpoint, they may possess very divergent attributes as far as their use by man is concerned. This should be remembered when some of the problems are discussed that have arisen with respect to three of the genera of the flatfish or Heterosomata, namely *Hippoglossus, Paralichthys,* and *Reinhardtius.*

There has been unanimity throughout the years that the genus *Hippoglossus,* the true halibut, belonged in the family Pleuronectidae. However, the generic and species names for the true halibut have been subjected to many versions. Most present authorities on classification or taxonomy presently designate both the Asiatic and North American Pacific halibut as *Hippoglossus stenolepis* Schmidt and the Atlantic halibut as *Hippoglossus hippoglossus.* Government agencies and the scientific community generally now subscribe to the latter designation. This is a desirable unanimity but such concurrence without independent research adds no strength to the validity of the separation of the Pacific and Atlantic halibut into two species.

The true halibut of the Atlantic was first described as *Pleuronectes hippoglossus* by Linnaeus (1758), and in 1813 Pallas used the same name for the halibut of Kamchatka. Cuvier (1817) established the genus *Hippoglossus,* and the halibut became *Hippoglossus hippoglossus.* In 1828 the name *Hippoglossus vulgaris* was introduced by Fleming and was accepted both for the Atlantic and Pacific forms for the next 60 years. Jordan reestablished *Hippoglossus hippoglossus* in 1885 for both Atlantic and Pacific halibut.

In 1904 P.J. Schmidt described the halibut taken off Sakhalin Island as a separate species, *Hippoglossus stenolepis.* In 1930, this Asiatic form was compared by Schmidt with that of the Pacific coast of North America, and it was concluded that they were identical to each other but different from the Atlantic halibut.

In 1931, Rendahl compared a Kamchatka specimen with Atlantic specimens from Sweden and regarded them as intermediate between H. *hippoglossus* and H. *stenolepis* and named it H. *hippoglossus camchaticus.*

Schmidt (1930) based his distinctions between the Atlantic and Pacific form chiefly on apparent differences in scale proportions, ray counts and numbers of supplementary scales using only seven Pacific and three Atlantic individuals in his comparisons. These are unacceptably small samples to be used for such a purpose. (The writer in 1929 filled Schmidt's written request for the Pacific material.)

Vernidub (1939) with sample sizes slightly larger than Schmidt's, made an exhaustive study of the morphology including the characters used by Schmidt. She concluded that the differences observed between the Pacific and Atlantic halibut were of varietal *not* species rank and that the North Pacific form should be designated H. *Hippoglossus stenolepis* Schmidt, a variety of the Atlantic H. *Hippoglossus.*

With respect to Schmidt's most definitive character, namely the shape of the scales from which the name stenolepis (*steno*-narrow and *lepis*-scale) was derived, additional and more extensive data do not manifest the sharp difference in the width of the scales that Schmidt (1930) reported for the two oceans.

In Table 1 the scale ratios of both Atlantic and Pacific samples are shown.

The differences shown by Vernidub's data and those of Bell's five Northeastern Pacific samples with the respective Atlantic data are not statistically significant. It is also to be noted that the average width of the Goose Islands Pacific scales is *actually greater* than Vernidub's Atlantic material, limited in sample size as the latter may be.

Another difference noted by Schmidt was in dorsal and anal ray counts. The range of Schmidt's Atlantic dorsal count of 99-104 quoted from Smitt in 1893 did not even overlap Schmidt's Pacific count of 93-97, and the ranges of the anal counts of 73-79 Atlantic and 72-74 Pacific were also sharply different. However, the samples involved were extremely small for median ray counts in halibut which vary widely in number.

TABLE 1. Length/Width Ratios of Atlantic and Pacific Halibut Scales

	Pacific			Atlantic		
Source	**#Scales**	**Mean Ratio**	**L/W Range**	**#Scales**	**Mean Ratio**	**L/W Range**
Schmidt, 1930	25	2.57	1.90-3.44	8	1.74	1.57-2.01
Vernidub, 1936	32	2.06	1.44-3.50	24	1.99	1.30-2.51
Bell (Goose Is.)	214	1.94	1.30-2.80	216	1.81	1.25-2.90
Bell (Oregon coast)	56	2.12	1.50-2.90			

Note: 60-scale samples by Bell for other Northeast Pacific grounds, Chirikof Island, Yakutat, and Cape Addington showed L/W ratios of 1.91, 2.11, and 1.92 respectively.

It is possible to compile from the literature a number of individual dorsal and anal ray counts both for the Atlantic and Pacific halibut. Many of them do suggest a tendency for higher counts in the Atlantic. However, they resemble Schmidt's data in being based on very few individuals, but unlike Schmidt's material they show a great amount of overlap. Furthermore, it does also appear that some observers have, on occasion, overcounted the split rays at the posterior end of the fins.

Despite the absence of an extensive series of Atlantic ray counts, it can be shown that the available Atlantic material grades into the range of counts shown by some typical samples chosen from the several hundred such samples available from this writer for the Pacific area. It is likely that if more Atlantic material was available and from over a wider geographical range, there would be even more overlap between the samples from the two oceans.

In Table 2 the samples are arranged in descending order of dorsal ray counts.

TABLE 2. Dorsal and Anal Fin Ray Counts from Atlantic and Pacific Locations

Location	Mean Dorsal Count	Mean Anal Count
Banquereau, N.W. Atlantic (Bell, 1932)	100.94	75.81
Bering Sea, North Pacific (Bell, 1966)	100.30	74.86
Oregon Coast, Pacific (Bell, 1926)	100.12	74.67
Northeastern Atlantic (Vernidub, 1936)	100.00*	76.00
Kodiak Is., Pacific (Bell, 1963)	99.88	75.09
Icy Strait, Pacific (Bell, 1964)	97.75	73.50
Okhotsk and Bering Seas (Vernidub, 1936)	97.52	73.46

*Corrected for an anomalous count.

The "very striking difference" in the length of the pectoral between Atlantic and Pacific halibut noted by Schmidt seems to be true only for some individual fish. Even the opposite is the case in many instances. Indeed, the length of the pectoral seems to be a very variable character and observed differences give no support to Schmidt's conclusions.

While there has been no exhaustive study made of the frequency of supplementary scales in Atlantic and Pacific forms, any differences that seem to exist do not justify species separation.

Thus, in view of the scarcity of Schmidt's data and the contrary evidence of Rendahl's single observation coupled with the findings of Vernidub and the large amount of data collected by the writer for the International Pacific Halibut Commission, there is little basis for designating the Pacific halibut as *H. stenolepis,* the "narrow-scaled" halibut, as a separate species from the Atlantic halibut.

It also appears that the differences that do exist are probably between stocks rather than varieties. Consequently, under these circumstances, reversion to Jordan's (1885) original designation of *H. hippoglossus* for both the Atlantic and Pacific halibut is indicated.

With other Pleuronectidae evidencing a circumpolar distribution with the same or parallel forms in both oceans, halibut *(Hippoglossus)* would not be unique. Hubbs and Wilimovsky (1964) in an exhaustive study of *Reinhardtius hippoglossoides,* the so-called but falsely designated Greenland halibut, came to a similar conclusion about that species as follows:

> The preceding analysis of inter-ocean and intra-ocean variations of *Reinhardtius hippoglossoides,* in both morphometric and meristic characters, provide no confirmation of previous opinions that this species is divisible into species or subspecies. What average differences do exist, most notably in the number of vertebrae, are indicated to be on the race, less than subspecies, level. The Greenland halibut thus becomes another example of a species exhibiting an amphiboreal distribution pattern (i.e., occurring in the North Atlantic and North Pacific, but absent in the intervening Arctic Ocean.)

L.S. Berg pointed out the amphiboreal distribution of some species and groups of species in the Northern Hemisphere. Also, P.Y. Schmidt (1950) noted that of the 25 species of Pleuronectiformes in the Sea of Okhotsk (Pacific) eight have close relatives, if not subspecies in the North Atlantic. It is postulated that the interchange between oceans occurred during the Pliocene epoch, the spread being occasioned by the changing climatic conditions in the Arctic at that time. P.Y. Schmidt also identifies Pacific halibut as a subspecies of H. *hippoglossus,* namely H. *hippoglossus stenolepis* (Schmidt) and not as a separate species as did his namesake P.J. Schmidt.

From the above it is believed that the evidence supporting a reversion to the name H. *hippoglossus* for Pacific halibut is much stronger than that upon which the name H. *stenolepis* was established nearly 50 years ago by Schmidt.

HALIBUT—AN EXCLUSIVE COMMERCIAL NAME

It will be evident that the use of the name halibut in commerce should be restricted to fish of the genus *Hippoglossus.* However, despite the need for such exclusive use there are a number of instances where the name halibut continues to be misused and abused, usually to achieve some trading advantage. This has been to the detriment of the consuming public as well as the Pacific halibut fishing industry.

Deception of the consuming public in the past seems to have had some official sanction. In 1930 it was suggested officially in Canada that the consumption of a less desirable flounder such as turbot could be furthered by identifying it "on the fishmonger's slab" as halibut (Thompson, Harold, 1930). In more recent years, while officialdom no longer directly advocates such practices, it sometimes seems to condone them because of bureaucratic indifference or ignorance. Some distributors and importers alike have been relatively free for a long time to capitalize upon the superior qualities of halibut *(Hippoglossus)* by using its name to promote less desirable species for a larger profit.

One such species poses a continuing problem. The name halibut is used in describing various species of a flounder scientifically known by the generic name *Paralichthys.* One species of that genus is found off California and Mexico and is marketed in the United States as halibut or as California halibut. *Paralichthys* is a very large genus and is represented by many species throughout the world. It has been imported into the United States from such countries as Taiwan and Mexico under the United States Customs category of halibut. The Import and Export Corporation of Mainland China once tendered for export out of Tientsin, Dairen, and Kwantung the flatfish *Paralichthys olivaceous* under the designation of halibut. Even a federal fishery agency in the United States had regularly included species of *Paralichthys* in their published compilations of true halibut imports.

While the situation has only reached serious proportions on a few occasions in restricted marketing areas, it seems inadvisable for responsible public agencies to continue to countenance such deception. The disastrous results that were involved with *Reinhardtius,* the Greenland halibut, should have been sufficient warning. Unlike *Reinhardtius,* the genus *Paralichthys* does not even belong to the same systematic family as the true halibut *Hippoglossus.* Also, the American Fisheries Society in 1960 finally ceased to recognize the use of the name halibut as an acceptable common name for species of the genus *Paralichthys.*

Why the use of the name *halibut* should be restricted to fish of the genus *Hippoglossus* may best be understood by a brief review of what some called "The Great Greenland Halibut Masquerade" that occurred in the late 1960s. In planning the importation of *Reinhardtius hippoglossoides* from Newfoundland, a request was made to the United States authorities by an exporter for a common name for the species. In Newfoundland, the species had "always been known as turbot" (Templeman, 1966) and had been long indicated as such in official Canadian documents and publications. With the authorities in the United States, there had been some confusion regarding whether it should be labeled as flounder or Greenland halibut. The flounder designation that was chosen and officially approved in 1963 was probably regarded on second thought as an "inadvertance" and no doubt unsatisfactory to the importer.

The importer's further request for a determination by the United States for a common name produced the desired result. The experts of the federal government used the following sources:

1. *Check List of the Fishes and Fishlike Vertebrates of North and Middle America*—David Starr Jordan, B.W. Evermann and H.W. Clark, 1930.
2. *Yearbook of Fishery Statistics*—Food and Agriculture Organization (FAO), 1964.
3. *Common and Scientific Names of Fishes from the United States and Canada-American Fisheries Society,* 1960.

The conclusion from such a limited and unfortunate choice of references was inevitable. The common name for *Reinhardtius* was, of course, concluded to be Greenland halibut. Some of the experts consulted at the time stated "that they were unable to obtain *evidence* that would support a finding that turbot is the common or usual name for *Reinhardtius.*" The foregoing finding was made despite the fact that the common and usual name for *Reinhardtius without exception has been turbot in the United States and Canada for the previous 80 years!*

Also, regardless of the inadequacies of the three references used by the ichthyologists consulted, it is still difficult to understand why others in authority did not require a more exhaustive investigation.

Of the above three references sought to justify the deceptive common name Greenland halibut, two were clearly defective. What appears to be a 1930 source, namely, the *Check List* of Jordan, Evermann, and Clark, mentioned above, merely repeats without any amendments all the common names used by

Jordan and Gilbert in 1882, 48 years previously. Over 70% of those common names had already been disqualified by the American Fisheries Society in its published list of common names in 1960. Furthermore, in the flounders or Heterosomata alone, 17 of the 29 common names of Jordan and Evermann had been rejected by the American Fisheries Society by 1960.

The FAO reference also does not qualify as an independent confirmatory source for English common names. It is specifically stated in the FAO yearbook that the common names of fishes are based upon what the respective national or regional offices use. It makes no independent studies.

As for the 1960 list of common names of the American Fisheries Society more must be said. First, it must be recognized that there was very little known about *Reinhardtius* by the scientific community in the United States. *Reinhardtius* was not even listed in the 1948 compilation of common names by the society and the 1960 edition incorrectly indicated *Reinhardtius* as occurring only on the Atlantic coast, when it also occurs quite widely on the Pacific coast. Under the circumstances, the American Fisheries Society publication would hardly qualify as an authority regarding the common or usual name for *Reinhardtius.* Furthermore, there had been no commercial landings of *Reinhardtius* in the United States for many decades, and only seven specimens were known to exist in the ichthyological museums in the United States (Hubbs and Wilimovsky, 1964). Consequently, *even in the scientific community,* it could not be said that any name could possibly "reflect broad current usage" as the society maintained. As for the general public or layman for whom common names have their greatest, if not only relevance, it is even more remote that there would be any significant knowledge regarding the species or certainly its common name! Thus the recent claim by the American Fisheries Society that the name Greenland halibut is "widely, if not universally accepted as the common name and is obviously stable since it has endured for so many years," has no basis in logic nor in fact. Even in papers of the *Transactions of the Society* the word *turbot* is used in describing *Reinhardtius.*

Also, the prefix *Greenland* is, quoting one American Fisheries Society criterion for a common name, "a poorly descriptive and misleading geographical characterization" for a species that is found throughout the boreal or cooler waters of the entire Northern Hemisphere, Atlantic and Pacific alike. Waters about Greenland are probably the origin for less than two percent of the present world production. It is interesting to note that *Reinhardtius* has the dubious reputation of being the only species recorded in the 1960 list of common names of the American Fisheries Society for which a geographic designation is *completely inconsistent* with the known distribution of the species.

It was claimed, however, that the name Greenland halibut is "deeply entrenched in the literature and has been in existence for at least since 1881." Superficially, this would appear to be true except for one overriding and incontestable fact. The so-called common name was only "deeply entrenched" in the restricted literature of the savant. It was not in the lay literature nor in use by the lay community or body public. Admittedly, some recently constituted international agencies such as the FAO tend to list Greenland halibut, but they, too, also function largely in the scientific arena.

In the nonscientific world, the world of the common people, the common name for *Reinhardtius* has always been turbot, both in the United States and Canada. This lineage goes back unbroken well into the nineteenth century, when United States fishing vessels fished *Reinhardtius* off Greenland for the East Coast United States market. In those earlier years the word Greenland was sometimes appended to turbot but never to the companion word halibut.

Even the official statistical publications of both Canada and United States up to most recent times have recorded a preference for the use of the name turbot as the common name for *Reinhardtius.*

It has been contended by the American Fisheries Society that if halibut is to be assigned to no other fish than those of the genus *Hippoglossus,* then the word salmon should be restricted to *Salmo salar* and similarly for other generic types such as bass, perch, cod. This is, of course, illogical. Nevertheless it seems the American Fisheries Society does presently hold the view *"that the term halibut is to be applied broadly to a number of species, just the same as salmon and bass are used!"** To so broaden the use of the word halibut is to revert to the generally rejected tribal groupings of Jordan of over 80 years ago with the attendant taxonomic and commercial chaos.

Such a viewpoint is obviously an arbitrary one. In addition, it is contrary to the principles pursued to date by the society itself in attempting to achieve order in the common and scientific names of fish. It would even require reversal of one of the society's own recent decisions. The society in the 1960 list of common names officially designated *Atheresthes stomias* as the arrowtooth flounder despite the so-called "deep entrenchment" of the name Arrowtooth

*Author's italics.

Halibut over the past 75 years including its acceptance as such in the society's 1948 list of common names.

It is improbable that the American Fisheries Society will alter its recent decision on the common name of *Reinhardtius* in the foreseeable future, regardless of the merits of any request for reversal. It required about 22 years of intensive effort on the part of a group of Pacific coast ichthyologists, of which the writer was one, to have the society change the name of the *California sardine* to the more descriptive *Pacific sardine.*

Furthermore, the recent regressive statement mentioned above that it is their "view that the term halibut is to be applied broadly to a number of species, just as the names salmon and bass are used," augurs very poorly for any rational treatment of the Greenland halibut name by the American Fisheries Society. One member of the society involved in its decision stated to me that some of my objections to the term Greenland halibut were dictated solely by commercial considerations and the society would not accede to such "pressures!"

Also, the society has stated that "if properly and accurately labeled, Atlantic halibut, Greenland halibut, and Pacific halibut should be easily distinguishable from each other by the consumer." This is a clear indication of a lack of understanding of the real world of merchandising fresh and frozen fish in United States and Canada.

The change will probably be made only when the name Greenland halibut is examined according to the society's three main criteria of 1960 for a uniform common name: namely, to (1) *avoid confusion* (2) reflect *broad* current usage and (3) be adaptable to *lay* uses. The continued recognition of the name Greenland halibut for *Reinhardtius* confounds each of these criteria.

The sanctioned use of such a deceptive name, along with the low cost of *Reinhardtius* in the countries of origin and the large potential supply of the species had a drastic effect on the North American halibut market. There was a precipitous decline in halibut prices in 1967, representing a \$7 million loss in revenue for the halibut fleet and 30% drop in crew and vessel earnings. *At least 80% of the decline was attributable to the "Greenland Halibut Masquerade."* The well-being of one of the most valuable single species fisheries in Canada and United States was thus threatened, the livelihood of many persons sharply reduced, and life savings of fishermen and others in the halibut fishing industry jeopardized.

Despite the rising crescendo of complaints from the Pacific Coast halibut industry provoked by the disastrous fall in halibut prices, federal officialdom in the United States continued to be both defensive and inert. During the passing of responsibility from one federal agency to another, one agency head notified a concerned member of the United States Congress that "Reinhardtius had always been known in the United States as Greenland halibut." How could such a statement be forthcoming when the official *Glossary of Names of Fishes* published by the same United States agency annually from 1930 to 1957 in its *Statistical Digest,* as well as in other of its publications back to 1880, indicated turbot to be the only or at least the preferred common name for *Reinhardtius* in the United States.

It was also contended by federal officials that the decline in halibut prices resulted from the removal of the papal ban against eating other than fish on Fridays. While this had a minor effect, it could not account for the fact that the proportionate decline in halibut prices was very much greater than for competing species despite the fact that supplies of halibut in the market had been at their lowest level in several years.

In Canada, there was some political ambivalence evident between the East and West coasts. Federal officialdom in Ottawa wished to placate the Pacific Coast halibut industry without jeopardizing the possible growth of a new fishery in economically-depressed Newfoundland on the Atlantic Coast. The British Columbia government on the other hand, strongly protested the situation to Ottawa. This apparently disturbed federal equilibrium and eventually constructive action was forthcoming at the federal level in Canada.

During the market chaos that prevailed during the *Reinhardtius* masquerade there were even some rather ludicrous situations. One large national diet-watching organization producing low calorie convenience fish dinners used the cheaper Greenland halibut, little knowing that they were using one of the fattest fish on the United States market. True halibut is a low fat, high protein species. Clients may well have wondered what was happening to their waistlines.

There were also some potentially serious aspects to the charade that was being played on the American consumer. In designing low fat, low sodium and high protein diets for cardiac patients, halibut *(Hippoglossus)* is regarded as a desirable source of animal protein. In contrast, *Reinhardtius* has about 25% more sodium and 20% less protein than *Hippoglossus.* Since extensive consumer studies have shown beyond the remotest doubt that it was practically

TABLE 3. Names Given *Reinhardtius hippoglossoides*

UNITED STATES

Five-Year Periods	Scientific Sources	Lay Sources
1875-1880	TURBOT (Goode & Bean, 1879)	GREENLAND TURBOT (Goode, 1880)
1881-1885	GREENLAND HALIBUT (Jordan & Gilbert, 1882)	GREENLAND TURBOT (Goode, 1884)
		NEWFOUNDLAND TURBOT (Collins, 1885)
1886-1890	GREENLAND HALIBUT (Jordan & Goss, 1889)	GREENLAND TURBOT (Goode, 1887)
1891-1895		
1896-1900	GREENLAND HALIBUT (Jordan & Evermann, 1898)	
1901-1905	GREENLAND HALIBUT (Jordan & Snyder, 1901) GREENLAND HALIBUT (Jordan, 1905)	
1906-1910 1911-1915		GREENLAND TURBOT (Bureau of Census, 1911)
1916-1920 1921-1925	GREENLAND HALIBUT (Bigelow & Welsh, 1925) GREENLAND HALIBUT (Jordan, 1925)	
1926-1930	GREENLAND HALIBUT (Breder, 1929) GREENLAND HALIBUT (Jordan, Evermann & Clark, 1930)	TURBOT (Fish. Ind. U.S., 1930)
1931-1935		TURBOT (Fish. Ind. U.S., 1931-1935)
1936-1940		TURBOT (Fish. Ind. U.S., 1936-1940)
1941-1945		TURBOT (Fish. Stat. U.S., 1942-1945)
1946-1950		TURBOT (Fish. Stat. U.S., 1946-1950)
1951-1955	GREENLAND HALIBUT (Bigelow & Schroeder, 1953) GREENLAND HALIBUT (Schroeder, 1955)	TURBOT (Fish. Stat. U.S., 1951-1955)
1956-1960	GREENLAND HALIBUT (Am. Fish. Soc., 1960)	TURBOT (Fish. Stat. U.S., 1956-1959)
1961-1965	GREENLAND HALIBUT	

CANADA

Five-Year Periods	Scientific Sources	Lay Sources
1906-1910	TURBOT (Cornish, 1907)	
1911-1915 1916-1920		TURBOT (Huntsman, 1918) TURBOT-GREENLAND TURBOT GREENLAND HALIBUT (Meehan, 1918) GREENLAND HALIBUT (Huntsman, 1920)
1921-1925 1926-1930	GREENLAND HALIBUT OR "TURBOT" (Thompson, 1930)	

TABLE 3, CANADA
Continued

Five-Year Periods	Scientific Sources	Lay Sources
1931-1935	GREENLAND HALIBUT AND NEWFOUNDLAND TURBOT (Anon., 1932 Rept., Newfoundland Fish. Res. Comm.)	TURBOT (Anon., 1931, Rules & Reg. Respecting Fish. of Newfoundland) TURBOT (Anon., 1931-1932 Annual Repts., Dept. Mar. & Fish. Newfoundland)
	GREENLAND HALIBUT AND NEWFOUNDLAND TURBOT (Anon., 1934 Rept., Newfoundland Fish. Res. Comm.) GREENLAND HALIBUT (Vladykov & McKenzie, 1935)	NEWFOUNDLAND TURBOT OR GREENLAND HALIBUT (Anon., 1934 Rept., Newfoundland Fish. Res. Comm.)
1936-1950		
1951-1955	GREENLAND HALIBUT (Dunbar & Hildebrand, 1952)	
1956-1960	GREENLAND HALIBUT (Stewart, 1956) GREENLAND HALIBUT (Templeman & Fleming, 1956) GREENLAND HALIBUT (Anon., 1956 Dept. Fish. & Fish. Res. Bd.) GREENLAND HALIBUT (Templeman, Squires & Fleming, 1957)	TURBOT (Anon., 1957 *Trade News*) TURBOT (Anon., 1958 *Trade News*)
1961-1965	GREENLAND HALIBUT (Hubbs & Wilimovsky, 1964)	TURBOT (Anon., 1964 Stat. for 1961-1965, Fish. Stat. of Canada) TURBOT (Martin, 1963)
1966-1970	GREENLAND HALIBUT (Leim & Scott, 1966)	TURBOT (Anon., 1966+ Fish. Stat. of Canada)
	GREENLAND HALIBUT OR TURBOT (Templeman, 1966)	GREENLAND HALIBUT OR NEWFOUNDLAND TURBOT (Templeman, 1966 *Trade News*)

impossible to prevent consumers and dietitians from being confused and deceived by the name Greenland halibut, the deception had the potential for serious health consequences. It is interesting to note that as late as 1978 some retail chains were still promoting the low calorie content of *Reinhardtius.*

Fortunately, much public interest in correcting the situation was engendered in Pacific Coast states. Alaska, Washington and Oregon passed legislation in a regional effort to thwart the fraud and deception that was being perpetrated on its citizens.

The Washington State Legislature in the closing hours of its 1967 session, cognizant of the urgency, unanimously suspended cloture on the introduction of new bills, in order to pass remedial legislation. By a unanimous senate vote of 45 to 0 and a 91 to 1 house vote, Senate Bill 642 was passed on April 16, 1967. It provided for the exclusive use of the word halibut as applied to *Hippoglossus.* Due to the emergency character of the act, it went into effect on its signing by the governor on April 25, 1967.

Oregon followed suit with similar legislation, and Alaska and Massachusetts had like legislation in hand. Massachusetts, an important fisheries state, was concerned not only with the immediate case, but with the chaos that would result if such market practices were permitted to become more general. The various states also memorialized the Congress of the United States and remedial legislation was introduced into the House of Representatives. There was a widespread sense of urgency and outrage. It became a "last straw" when *Reinhardtius* was listed as halibut on menus in restaurants of the United States House of Representatives!

In November, 1967, an extensive test of current public knowledge of *Reinhardtius hippoglossoides* was made by a national market research firm for the halibut industry. The firm had no information whatsoever about the purpose of the survey. The 2,751 respondents constituted a very large national probability sample of married women over 18 living with their husbands. The results indicated that only 1 out of 130, or 0.7%, knew there was any difference between *Reinhardtius* displayed as "fillets of

Greenland halibut" and *Hippoglossus* displayed as "steaks of halibut." In light of this and other responses, the research firm concluded that "in every respect, greater inability to make rational buying choices can hardly be imagined." The name Greenland halibut was considered by the U.S. Supreme Court in 1951 to thoroughly mislead the ordinary person "who is neither savant nor dolt, who lacks special competency with reference to the matter at hand but has, and exercised, a normal measure of the layman's common sense and judgment."

The passage of legislation by Washington and other states, precipitated counter legal action by the importers and others interested in assuring the continuation of the very lucrative masquerade. Suit was brought in federal court to enjoin Oregon from enforcing the law passed by the Oregon State Legislature that required the name halibut be used only with respect to fish of the genus *Hippoglossus.* On March 27, 1968, the United States District Court for the District of Oregon made the following unanimous findings of fact:

> Although the term "Greenland halibut" is accepted in the scientific community to describe *Reinhardtius,* it is *not a commonly understood name among the general public.*
>
> That "Greenland halibut" *is not a common name in the market place* and the consumer *confuses* it with halibut.
>
> That the addition of the word "Greenland" will not protect the consumer.
>
> That the Oregon law was designed to prevent *deception* of the consumer.*

Upon receipt of the unfavorable judgment, the plaintiffs forthwith appealed to the Supreme Court of the United States. During the interim, the State of Massachusetts on July 9, 1968, passed legislation similar to that in Oregon, Washington and Alaska. The director of that state's Food and Drug Division said the law was necessary "to prevent an apparent *fraud* on the consumer."

On November 12, 1968, the Supreme Court of the United States summarily disposed of the case and affirmed the judgment of the Oregon District Court both on the state supremacy clause as well as on the right of that state to protect its citizens from the *deception* inherent in the name, Greenland halibut.

In light of the evidence that had been brought to attention of the United States Food and Drug Administration, that agency promulgated on August 13, 1968, a statement of policy to the effect that labeling *Reinhardtius* as Greenland halibut would be considered misbranding. While it was agreed that the scientific community may have accepted the common name Greenland halibut, the ordinary public had not and would thus be confused by its use. The order was to take effect 90 days after publication on August 21, 1968, in the *Federal Register.*

It was also stipulated that *Reinhardtius* shall bear either the name flounder or, since the species is caught only in circumpolar waters, the name northern flounder. These names had been proposed by this writer whose research and documentation had demonstrated beyond any reasonable doubt to the courts that the name Greenland halibut failed to meet the basic requirements for common names laid down by the American Fisheries Society. Admittedly, the flounder, or northern flounder designations had their limitations but they were neutral and nonobjectionable to the consumer. Pricing was appropriate for the name, and they were scientifically precise. While the name turbot had been used heretofore, it had a twofold objection: It was used in some areas for designating several unmarketable species of flatfish, and it was the name for *Psetta maxima,* a high-priced European flatfish which was imported to United States in small amounts from Great Britain and France for the gourmet restaurant trade.

Needless to state, the use of the name flounder encountered considerable objection on the part of the flounder industry in New England, particularly in Bedford, Massachusetts. Public hearings were held and a Revised Formal Statement of Policy was published in the *Federal Register* on December 20, 1968, which provided that Greenland turbot was an acceptable name for *Reinhardtius.* Thus with this last finding, the recorded lineage of the true common name for *Reinhardtius* in United States remains unbroken over the past 100 years or more.

In Canada where the name turbot had been almost universally used except in the scientific community, the main problems seemed to be the placating of the Pacific Coast interests in *Hippoglossus* and Atlantic Coast interests in developing export markets for *Reinhardtius.* On January 13, 1969, the Federal Department of Fisheries announced that in Canada the name Greenland turbot, recently approved for sale in United States, was acceptable in Canada, but to the exclusion of the word halibut. It also contained a 1965 ruling against the use of the name flounder in labeling *Reinhardtius.* While to the purist, the use of the name turbot may be unacceptable, it does seem that it has been accepted by the North American public for over 100 years, even though not widely known. Furthermore, under that designation,

*Author's italics.

Reinhardtius now commands a national and profitable market in the United States with very large and increasing importations from Japan and Newfoundland. Exports from Canada in the 1970s were 4 to 5 million pounds annually and in 1977 appear to have reached over 9 million.

All the evidence of the past 100 years points to the fact that the name halibut should be used exclusively in referring to fish of the genus *Hippoglossus.* The remaining problem seems to be for the scientific community to decide whether common names are for use only by the lay public or *only* for the scientific community or savant. Clearly, there has been a failure to recognize the dichotomy that has existed in the lineage of the common name for *Reinhardtius* over the past 90 years and has required formal judicial action to resolve the matter and remove confusion in the mind of the public.

2 THE FISHERY

INDIAN FISHERY

Halibut fishing along the Pacific Coast began with the Indians of the region, and a discussion of the fishery would not be complete without at least a brief account of the Indian fishery.

Aboriginal Fishery

With halibut occurring in both inshore and offshore waters of the Northeastern Pacific it is not surprising that many tribes of coastal Indians made halibut an important part of the year-round food supply.

In northern California the coastal Yurok were known to fish halibut with lines 50 feet in length which were equipped with a gorge and thrown from the shore. There is also archaeological evidence that early-on the Tolowa had the equipment and dugout canoes to fish halibut outside the breakers off the mouth of California's Smith River. However, the coastal fishing by Indians south of Cape Flattery was relatively limited due to too few good harbors, poor watercraft for offshore fishing and the presence of more abundant and accessible anadromous species, particularly salmon, in nearly all coastal rivers.

Archaeological evidence indicates that the aboriginals of the western coasts of North America had consumed halibut from time immemorial. The early European explorers reported on its abundance and its extensive use by the native people encountered. Captain Cook in 1778 recorded the taking of 100 halibut in one hour by his crew while anchored near Shumagin Islands. La Perouse in 1786 noted the catching of halibut (fletan) by the Indians in Lituya Bay. Dixon (1789) observed that seven of his crew in a whaleboat fishing with handlines could not equal the catch made by two Indians fishing alongside with their native gear.

There are many good descriptions of the primitive but effective gear. One of them from Report No. 5 (1930) of the Halibut Commission included the following account:

> The hooks of the tribes in Southeastern Alaska were large and made of wood, two pieces being lashed together at an angle of about 15 degrees. They were sometimes much carved. To one of these pieces the line was attached, and at the end of the other a barb was lashed firmly by means of strong fibers. Another type of hook characteristic of the Cape Flattery Indians of Washington was made of splints from hemlock knots, steamed inside the bulb of the giant kelp, *Nereocystus,* shaped as desired, and allowed to cool. The barb was of bone, lashed with strips either of spruce cut thin like a ribbon, or of the bark of the wild cherry. A hook in the possession of the writer is of iron shaped in the typical Indian fashion. The bait was lashed back of the barb by fibers made of the sinews of animals, such as the whale. As the halibut mouth is vertical, it took the peculiarly shaped hook readily, and the upper part to which the line is attached operated as a spring to hold the barb in the flesh.
>
> The lines were made of twisted fiber of the cedar, of animal sinews or intestines, or of the giant kelp. The kelp lines were made by bleaching the kelp stems in fresh water, partially drying them in smoke, then stretching them repeatedly until they were the diameter of cod lines. They were brittle when dry, but exceedingly strong when wet. The twisted fiber or sinew lines were beautifully made by the squaws and had a uniform diameter and great strength. A line was made as long as 80 or 100 fathoms (Swan, 1870, p. 29), but this length was seldom used unless the black cod was fished for, as the latter was taken in greater depths.
>
> In use the line was attached to a stone sinker with the hook close to it and apparently often, if not usually, suspended just off the bottom. At the other end of the line was attached a buoy made of bladders or skin and to it a float or a flag by which the strike of the halibut was

FIGURE 4. Using halibut as a subsistence staple, Indians are shown as they prepare to butcher the fish on a beach near Cape Flattery, Washington, about 1910.
International Pacific Halibut Commission

signaled to the watching Indians. It was said that a canoe with two men could watch 10 or 15 such lines, but that if a larger canoe with more men was used, fishing was over the side. Fishing was commonly done in 10 to 20 fathoms of water, usually close to shore, but in the case of the Neah Bay Indians most often 15 or 20 miles west from Tatoosh Island, hence well out to sea.

Generally the same type of gear was used by all the Northwest native people including the Cape Flattery Makah near Neah Bay in northwest Washington, the Nootka of the west coast of Vancouver Island, the Kwagiutl of Queen Charlotte Sound, the Haida of the Queen Charlotte Islands and south end of Prince of Wales Island, the Tsimshian of the Hecate Strait coast of British Columbia and the Tlingit of Southeastern Alaska. Each group often incorporated some distinctive feature in their gear (Stewart, 1977).

With fishery products being so important in the lives of the coastal tribes both as a direct source of food and as trade items with inland groups, it is not surprising to find fish, including halibut, commemorated in the heraldry on totem poles. Halibut are also included in clan crests elaborately portrayed on the great ceremonial robes or painted on the crest boards that sometimes cover the entire front of the large clan or community houses. Indian folklore speaks of the Haida's supernatural halibut, the Invincible Crab of Nadan Harbor and many other symbolic characters.

FIGURE 5. Indian halibut float and line (A), and hook (B).
University of Washington Press

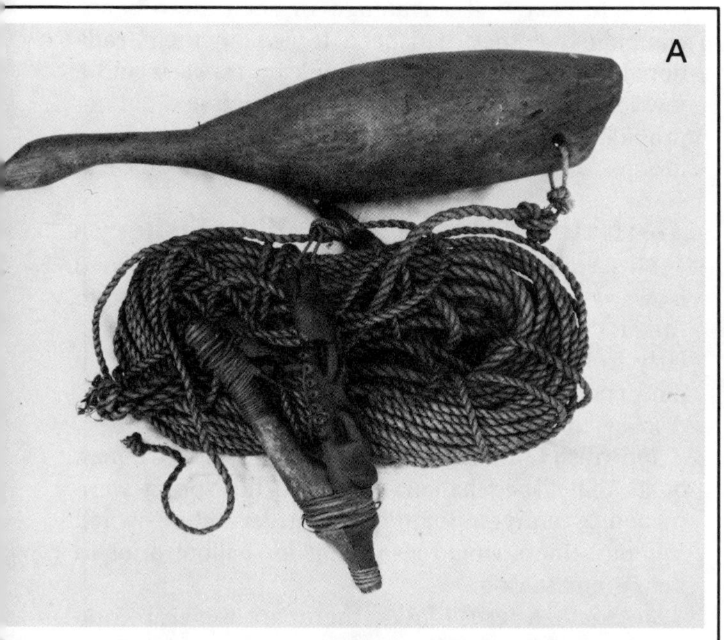

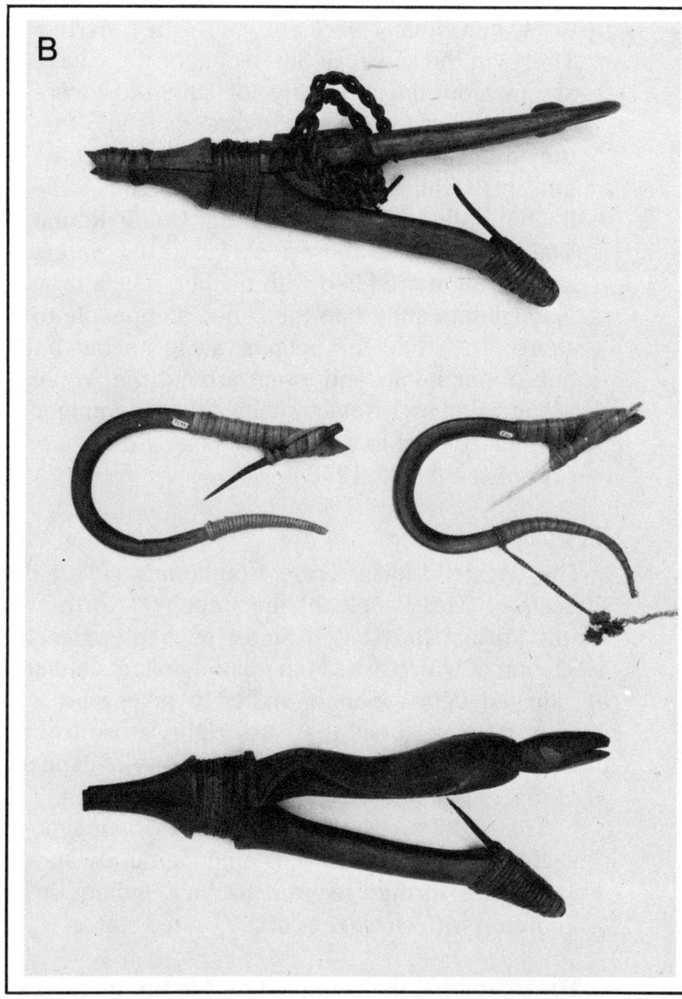

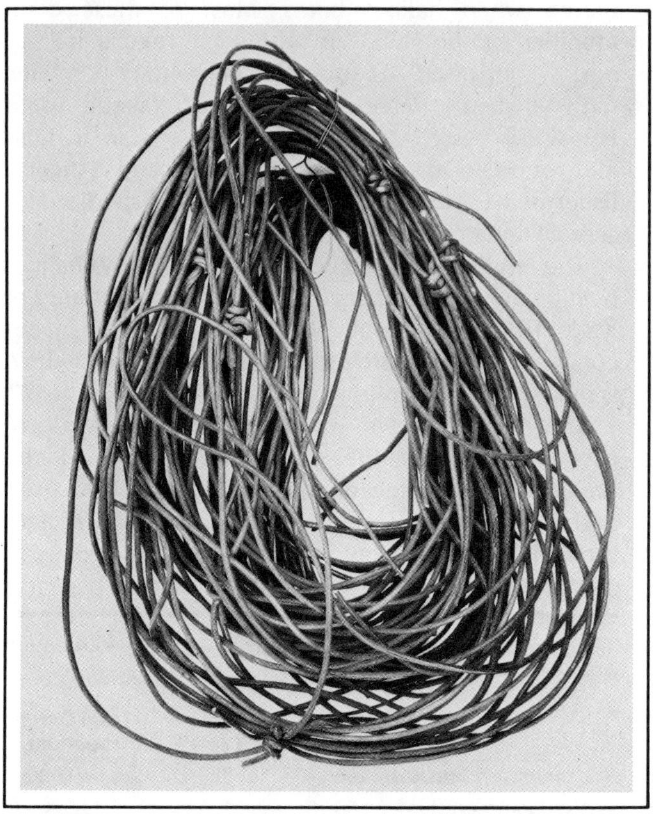

FIGURE 6. As part of their gear, Indians fishing for halibut often used kelp as fishing line.
International Pacific Halibut Commission

The story about the supernatural halibut has been recounted as follows:

> A tale of the Supernatural Halibut from Big Island west of Yan on Masset Inlet, Queen Charlottes. Two brothers went out fishing together every day. Only the elder ever caught any. He'd get up early and put on shaman robes and mask, go sing at water's edge to supernatural devilfish. The devilfish would bite his hook and end up as the day's bait. One morning, it finally happens. Younger brother gets curious. He follows his brother and watches the proceedings. Next morning rising early and peeking at his older brother still snoring in bed, he goes down to the beach, tries on the robes and mask. Sure enough, the devilfish rises to the song and is caught. Back to bed. Elder brother was somewhat amazed not to catch anything. They go out fishing. Soon Younger brother hooks a gigantic halibut.
>
> This halibut tows them around the entire North Island, around Cape Saint James, and

tows them finally back to where they started. There on the shore all the people of the village are in mourning, thinking for sure they were drowned, because it was several days ago that they disappeared. The canoe is stopped, so they pull in the line, heavy with the weight of the halibut called Never Appearing, Going Round The Island. When the fish surfaces, the waters around them are filled with halibut, which they load triumphantly into the canoe and paddle to shore. They told the people "a big halibut bit one of our hooks and swam around the whole island with us." Never again did the younger brother attempt to get the day's bait. *(Raincoats Chronicles,* p. 230, 1976.)

Treaty Claims

The Makah Indian Tribe, sometimes called the Classets or The People of the Cape, set forth two claims against the United States on September 15, 1952, one of which asked ten million dollars damages for alleged deprivation of rights to take seals and halibut. It was claimed that such rights arose from a passage in Article IV of the Makah Treaty of January 31, 1855, negotiated at Neah Bay:

> The right of taking fish and of whaling or sealing at usual and accustomed grounds and stations is further secured to such Indians in common with all citizens of the United States.

The plaintiff Makah claimed that if anything interfered with their treaty rights damages would be due. The United States claimed that the rights under the treaty were gratuitously given by the government and that they were merely equal in extent with the rights of other citizens and subject to control by necessary conservation measures.

The claims commission recognized that while the Makah claims were superior in some respects they were not so untouchable that they cannot be adjusted to meet changing circumstances when the government must take reasonable steps to conserve the natural resources which belong collectively to all the people. It was demonstrated that steps taken by the Halibut Commission to conserve the resource for all parties had been reasonable and effective and the plaintiff's case was dismissed on April 15, 1959, after nearly seven years of litigation (Docket No. 60).

In the late 1960s the Makah people developed a renewed interest in halibut as a result of securing, under governmental sponsorship, nine 10-ton fishing vessels. These vessels were adapted for two-man setline fishing, gill netting, trolling and other fishery pursuits.

One of the main concerns of the Makah in the 1950s had been the short halibut seasons which prevented them fishing into the summer. In 1969 the Makah petitioned the Halibut Commission for a special halibut season opening as early as March in order to take full advantage of their new boats, claiming that they had been forced off their traditional grounds on Swiftsure Bank by trawlers and a few larger setliners. The Halibut Commission did not provide for differential treatment of the various user groups.

The Native Commercial Halibut Fishery

Most of the coastal Indian tribes have carried forward their primeval interest in halibut fishing into the modern commercial fishery. The Makah, particularly from the 1940s to 1960s, maintained an active one- and two-man boat fishery out of Neah Bay and disposed of their daily catches to the Bay Fish Co.

In British Columbia a number of one- and two-man boats that fished halibut chiefly in the spring were owned by natives. Some of the larger native-owned salmon seiners would also outfit for halibut prior to the salmon season.

In Southeastern Alaska there are several communities which have been centers of activity for native-owned halibut boats. Most of these communities can be paired up with some regular halibut port which often becomes the operations point for native-owned vessels. Craig, Metlakatla and Hydaburg vessels tended to use Ketchikan to land and outfit. Kake used Petersburg, and Angoon fishermen used Sitka. Hoonah vessels fished out of Juneau and Pelican City.

The relative importance in the 1960s of the halibut fleet owned by members of the native communities of Kake, Hoonah, Angoon, Metlakatla and other places compared to the Southeastern Alaska fleet as a whole is indicated by the following data for 1963.

The larger crews on the native vessels are accounted for by the younger members of the families which come on board. In the recent past, native boats have accounted for as much as 20% of the halibut catch of Southeastern Alaska.

TABLE 4. Proportion of Native Vessels in the Southeastern Alaska Halibut Fleet, 1963

	Number of Vessels	Number of Crew	Total Catch (pounds)
Southeastern Alaska	184	575	9,049,000
Native Communities	30	156	1,148,000
Percentage Native	16%	27%	14%

THE HALIBUT FISHING FLEETS

The vessels that have participated in the Pacific halibut fishery can be grouped into a succession of overlapping types that include sailing vessels, steamers, "schooners," and combination or seine-boat-type vessels. Each such period of the fishery will be described in general terms with the activities of one or two noteworthy vessels of each type more closely chronicled.

The Period of Sail

While there had been sporadic attempts as early as 1870 on the part of sailing schooners to transport iced halibut from the Northwest to points as far south as San Francisco, such cargo vessels would not qualify as fishing vessels. Their cargoes included the fish caught by small boats on grounds in close proximity to various settlements, chiefly Victoria or Port Townsend. The duration of the period of sailing vessels actually fishing halibut was very short.

The Pacific halibut fishery was of relatively recent origin at the time when sail alone was used in the commercial fishery. In the 1880s a small group of two-man sailing sloops that could carry about 3,000 pounds fished halibut to supply the local markets of Victoria, Nanaimo, Vancouver, Port Townsend, and other Puget Sound points. The fish were caught in very close proximity to the places of landing.

The only vessels of a substantial size during the period of sail were the three that arrived on the coast in 1888 from New England, the *Oscar and Hattie*, *Edward E. Webster*, and the *Mollie Adams*. They were the typical Gloucester schooner ranging from 81 to 117 tons each with six dories carried in the waist of the vessel and a crew of about 14—the number depending upon the number of two-man dories carried. While these vessels experienced some success on trips where the fish was filleted and salted, a Pacific Coast fresh halibut fishery for the East Coast

FIGURE 7. Sloop *Jennie F. Decker* in 1901 with dories stowed amidships.
International Pacific Halibut Commission

market posed altogether too many problems for vessels depending exclusively upon sail. The high cost of ice and, sometimes the lack of any, high freight rates, long distances from the grounds to port and the difficulties of navigating the fjord-like rocky passages, along with strong initial resistance to Pacific halibut in the eastern markets, ended this episode of sail in its second season. Fortunately the vessels were also adapted to pelagic sealing and had been built primarily for that purpose. A few sloops and some small schooners using only sail continued in the fishery for several years, but after 1900 all had installed small auxiliary gas engines as they became available. In 1895 the *More* had been the first to install a gas engine.

The fleet of sloops and small schooners with their auxiliary power was an important factor in the industry during the first decade of the twentieth century. It supplemented the production of the growing fleet of large company-owned steamers. Most of the auxiliary-powered vessels were United States flag ships and their operations centered out of Seattle with a few vessels from other Puget Sound ports. Very few came from Canadian ports, but that fishery was dominated by the company-owned steamer.

During the summer most of the Puget Sound fleet fished Cape Flattery and the west coast of Vancouver

FIGURE 8. Power schooner *Knickerbocker*
Canadian Fisherman

FIGURE 9. M.V. *Pescawha*
Canadian Fisherman

Island. Going beyond Cape Scott at the north end was a rare event. In the fall and winter many went to Southeastern Alaska and fished the channels of the Alexander Archipelago. The catches were boxed in crushed glacier ice and shipped by common carrier to Puget Sound on consignment.

The auxiliary sloops and small schooners were later joined by several larger vessels that had been designed as sailing vessels on the East Coast. They were fitted with internal combustion engines as a primary source of power. They were the forerunner of the typical halibut schooner that was soon to dominate the fishery but they lacked the characteristic lines.

The *Knickerbocker* (Figure 8), a large vessel brought from the East Coast by the New England Fish Co. in 1914 had the lines and configuration of a downeast sailing ship made famous in the Bank fisheries of the Northwest Atlantic. This wooden vessel was built in Essex, Massachusetts, in 1913 and was 112.5 feet in length with a gross tonnage of 159 tons and a net of 101. On arrival to the Pacific Coast, it was refitted with two 100 h.p. engines. The vessel fished for New England Fish Co. from 1914 to 1917 with moderate success. Some years it carried a crew of 24 and usually fished 9 or 10 dories. Several names prominent in the fishery operated the *Knickerbocker* during its brief career. T.P. Quinn, who had discovered Yakutat Spit grounds in 1913 on the *Idaho*, ran the *Knickerbocker* in 1916. M.B. Scott, later captain of the S.S. *New England*, operated the *Knickerbocker* in 1917.

Of this small and hybrid group of vessels the *Carlotta G. Cox* and *Pescawha* were smaller than the *Knickerbocker* and of British registry. They fished for the Atlin Fish Co. of Prince Rupert, a subsidiary of New England Fish Co.

The *Pescawha* (Figure 9) had been built in Liverpool, Nova Scotia, in 1906 for the pelagic seal fishery of the North Pacific which operated out of Victoria, British Columbia. After the Tripartite Seal Treaty between Canada, United States and Japan terminated pelagic sealing in 1911, the *Pescawha* and several other sealers fished halibut. The *Pescawha* operated from 1913 to 1918 and later packed salmon. Initially the vessel had a 7 h.p. auxiliary engine but a 100 h.p. engine was installed for fishing halibut.

The *Carlotta G. Cox*, another converted sealer of 49 tons, slightly smaller than the *Pescawha*, also carried a full complement of sail. It installed a 130 h.p. internal combustion engine when it went halibut fishing in 1913. It left the halibut fishery in 1920. A third sealer, the *Emma H* also went halibut fishing then.

Like the larger *Knickerbocker* under United States flag, these converted Canadian sealers were only moderately successful. They were not well adapted to the needs of the fishery. The period belonged to the steamers and to the first of the halibut schooners that were now being built expressly for halibut fishing.

The Odyssey of the *Oscar and Hattie*

Many famous ships have "rounded the Horn" since the time of Ferdinand Magellan in 1520 and much has been said of their exploits. Almost unmentioned have been the innumerable small vessels which with their intrepid crews also braved the dangers of those almost always stormy waters. The *Oscar and Hattie* was one of the many such ships that made the hazardous passage.

Although she lacked the speed and grace of a *Flying Cloud* or the majesty of a *Royal Sovereign*, the *Oscar and Hattie* was a doughty vessel manned by a sturdy crew of 14. Constructed according to the New England clipper tradition, she was all of 85 gross tons, was 79 feet long, and carried six fishing dories.

Oscar and Hattie was to fulfill her destiny many thousands of miles from her home port of Swampscott, Maine. On October 10, 1887, she commenced a voyage that would take her to the other ends of the earth in search of halibut.

Dwindling supplies of fresh halibut from North Atlantic waters had made the accounts of the great abundance of halibut on the Northwest Coast a powerful lure to the adventurous seafarers of New England. They were also attracted by the promise of lucrative pelagic sealing in the same region.

Owned by Chris Johnson of Swampscott with Silas Calder, captain, the *Oscar and Hattie* sailed down the East Coast of the United States, across the Caribbean and, day after day, month after month, southward along the coast of South America. After rounding the Horn the vessel proceeded northward off Chile, Peru, and Ecuador, as more months passed and the seasons changed. Again across the equator and ever northward along the West Coast of North America until at long last came Cape Flattery and the end of a 15,000 mile voyage at Port Townsend, the new home port for the *Oscar and Hattie.*

Arriving too late in the spring for the pelagic sealing they proceeded to Rose Spit at the north end of Queen Charlotte Islands for halibut. It had made an unsuccessful set en route at Cape Scott at the north end of Vancouver Island. The crew fished on the north side of the "Spit" and dressed and fletched (to cut the fish lengthwise into fillets) when anchored under Tow Hill, which they named "Siwash" for Little Mountain. They set two skates a day from each of the six dories and usually loaded the dories five or six times from the two skates. They used 6-line, 28-pound groundline and 14-pound gangings. Manila beckets were set 12 feet apart. Little was realized from the sale of the fletches.

Undaunted by first failure, the *Oscar and Hattie* was outfitted for fresh halibut fishing at Cape Flattery where the *Mollie Adams* had been fishing. In turn the *Mollie Adams* then fitted out for fletching, as each crew thought the other had done better.

The *Oscar and Hattie* landed a fare of 50,000 pounds, and on September 20, 1888, a carload of fresh halibut left Tacoma on the Northern Pacific Railroad for the East Coast, fixing a date upon which the Pacific halibut fishery came into being. Very little was realized from the catch, however. Ice cost $22.50 per ton and rail freight rates were high. Nevertheless three more trips were made to the Cape Flattery grounds despite long delays because of the scarcity of ice. In the fall there was not a pound of ice in Puget Sound or in British Columbia. Calder went 800 miles north to the head of Glacier Bay in Icy Strait, Alaska, for natural ice. They broke a couple of booms and returned to Sitka for repairs. They had a very convivial time ashore. Sitka was still a busy place. Only 20 years previously it had been the capital of Russian Alaska.

After repairs were completed the *Oscar and Hattie* secured a trip of halibut fletches from Rose Spit. On their return to Seattle in November there was no demand for their trip, and they had to salt down the load on the dock. Thus ended their baptismal season in the halibut fishery for which they had come 15,000 miles.

Undaunted by the difficulties and setbacks encountered in the fall of 1888, Calder took *Oscar and Hattie* again in December on a fletching trip to Southeastern Alaska. They remained two months, January 1 to March 1 in the general vicinity of Sitka. They again experienced bad weather and caught few fish. They proceeded south to the Queen Charlotte Islands and in northern Hecate Strait off Porcher Island they managed to salt down about 140,000 pounds of fletches before the end of May. They arrived at Port Townsend in mid-June. Seattle was in ashes from the fire of June, 1889. Catastrophe was also to strike the *Oscar and Hattie.*

During their six months' absence the creditors had been busy and the ship was welcomed at the dock by a bailiff of the court. This was the last straw for these men who had struggled hard but in vain to open a new fishing frontier. Upon sale by the sheriff, the *Oscar and Hattie* went to new owners in San Francisco to pass into oblivion as far as the halibut fishery was concerned. Peter Nelson, mate, delivered the vessel to San Francisco and the crew recovered some of their share of the sale. So ended the halibut odyssey of the *Oscar and Hattie*—a brief but eventful struggle against adversity. The *Oscar and Hattie* was later observed engaging in pelagic sealing out of Victoria.

FIGURE 10. Unloading the *Oscar and Hattie* in Tacoma, Washington, September 20, 1888.
E.W. Wright, *Marine History of the Pacific Northwest*

One other calamity struck the vessel in 1892. While

off Yaquina Bay, Oregon, the captain and four of the crew drowned in the surf while going ashore in search of a missing crewman.

Oscar and Hattie also lived on in the halibut fishery through her crew who remained in the Pacific Northwest, although the vessel itself was finally abandoned and stricken from the register in 1922.

Peter Nelson, who had come out as mate on the *Mollie Adams* but had joined the *Oscar and Hattie* in the same capacity on its arrival in Port Townsend, was later the mate on the halibut vessel *Geo. H. Chance* out of Portland. This vessel also foundered on the shoals of financial disaster. After a season of salmon fishing on the Columbia River, Nelson became the first to fish halibut regularly out of Victoria. He started a partnership with Abbott in the sloop *Victoria* and supplied the Victoria market for two years from grounds in the Strait of Juan de Fuca. On occasion they sold any surplus in Nanaimo, already a busy coal mining center. The partners went into the Fraser River salmon fishery in 1893 and sold their sloop. They rowed from Vancouver to Victoria with the fishing dory and gear selling the latter to Captain Langley of the Steamer *Thistle* which was being outfitted for halibut fishing from Victoria, using a shore station near White Rocks in Hecate Strait. Nelson instructed Jim Anderson, the fish boss on the *Thistle* how to dress and ice halibut, for Anderson had never fished halibut. Nelson did not join the *Thistle* crew but went sealing out of Victoria. This ended his direct involvement in halibut fishing.

Captain John G. Daniels, who came out from the east as a crew member on the *Mollie Adams* also went with the *Oscar and Hattie* on her arrival in Port Townsend and remained until the fall of 1888. During the winter he fished on the three-man sloop *Alice* under Captain George Torry, the only vessel that fished regularly at Cape Flattery. Catches were sold to Ainsworth and Dunn in Seattle. Daniels sealed on the *Henry Dennis* in 1889 and went fletching halibut that fall on the *Edward E. Webster* in the Hecate Strait in the area of Bonilla Island and near Rose Spit. That winter Daniels worked on the *C.H. White*, about 84 tons and registered in San Francisco. It carried six dories and was chartered to Louch and Johnstone of Seattle to fletch and salt halibut for their smokehouse at what is now the foot of Pike Street. Louch and Johnstone later became Augustine and Kyer, the leading retail creditors, delivery provisioners and grocers in Seattle until their liquidation in the early 1930s. Their smokehouse was established in 1888 from which came the famous "88 Hams" that were an Augustine and Kyer specialty for the next 40 years. The *C.H. White* fished north of Bonilla Island on the "halves" basis, the same lay as on the vessels from New England. The *C.H. White* eventually returned to San Francisco. Daniels spent the next five years in San Francisco, and while there he observed the *Norman Sunde* and *Pilot*, both five-man sloops from Puget Sound, carrying two dories with signs in their rigging saying Fresh Northern Halibut. They had made a practice of selling some of their catch from the Hecate Strait in San Francisco.

In 1898 Daniels returned on the *Alceda* to fish for halibut out of Scow Bay near Petersburg. They fished from the first of October to the end of March in the Inside Passage of Southeastern Alaska. In 1900 Daniels bought the *Norman Sunde* (Figure 11). In the fall of 1912 Daniels ran the *Tom and Al*, which was the S.S. *Ragnhild* (Figure 12) converted to gasoline

FIGURE 11. The halibut schooner *Norman Sunde* in Petersburg, Alaska, September 9, 1907.
International Pacific Halibut Commission

power. It fished off Coronation Island along with such steamers as the *Weiding Bros.* and *Independent.* In the fall of 1913 when Captain T.P. Quinn of the *Idaho* discovered the famous Yakutat spawning grounds, Daniels sailed farther westward with the *Tom and Al* and discovered the equally famous "W" spawning ground 35 miles east of Cape St. Elias. In 1914 he operated the *Tyee*, later named the *Dorothy,* and fished such grounds as Marmot Island and Albatross off Kodiak Island. That same fall he again fished the "W" running a total of only 63, seven-line skates of nine-foot gear for a catch of 155,000 pounds.

In 1916 Daniels shipped as mate on the S.S. *Starr* owned by the San Juan Fishing and Packing Co. with Captain Pete Keough. They fished as far west as Shumagin Islands but spent most of the year on the northeast edge of Portlock Bank off Kodiak Island. In 1916 Daniels purchased the *Speculator* and in 1917 he acquired the *Jennie F. Decker* (Figure 7), which he owned until his retirement in 1923.

These rather detailed chronicles of two of the crew of the original *Oscar and Hattie* are indicative of how the halibut fishery developed in the years the two men spent in the fishery. They were an integral part of the legacy of the *Oscar and Hattie.*

There were two other vessels closely connected with the exploits of the *Oscar and Hattie*—the *Mollie Adams* and *Edward E. Webster*. Their crews had sought fortunes in the far off waters of the North-

FIGURE 12. The S.S. *Ragnhild,* which was later converted to the gas-engine powered *Tom and Al.*
International Pacific Halibut Commission

eastern Pacific from sealing and halibut fishing. While the *Adams* and the *Webster* participated in the early halibut fishery, they seem to have been primarily concerned with fur sealing. Since the story of the *Oscar and Hattie* would be incomplete without some account of these two other members of the trio that made the long journey to the Pacific Northwest, their contributions to the development of the halibut fishery are briefly recounted below.

The *Mollie Adams*

The *Mollie Adams,* owned by Captain Sol Jacobs and under the command of Captain Charles J. Johnson and with Peter Nelson as mate left Gloucester, Massachusetts, on October 2, 1887, for the Pacific Coast via Cape Horn, to engage in pelagic sealing and fishing. The crew also included Messrs. Crockett, Daniels, MacPherson, and Odson, each of whom was destined to play an important role in the later development of the Pacific halibut fishery and the maritime history of the Northeastern Pacific.

On arriving in Neah Bay on March 3, 1888, after a five-month trip with one stop for water at Good Success Bay at the tip of South America, the *Mollie Adams* went sealing until May 17, catching 700 seals. They secured $4.50 for each skin at Port Townsend but were much put out when they learned they could have received $5 each in Victoria. The *Mollie Adams* was outfitted for halibut fishing with eight two-man dories, each of which carried eight skates of gear.

The vessel made four trips to Swiftsure Bank for a total catch of 145,000 pounds of fresh halibut, fished about three days each trip, making two sets of four skates for each dory each day. The first catch, shipped east on consignment, netted $75 per man. The other two trips were "brokers," that is they failed to make expenses. Ice was difficult to secure and cost from $8 to $15 per ton. The vessel then secured a trip of 175,000 pounds of salted fletches from Rose Spit and was the first to fish the famous "Two Peaks" ground in that area. It was also the first fletching trip made on the Pacific Coast and lasted from July 24 to September 16. The crew netted $175 per man for the 19 days fished during the 54-day trip. Trips to

Greenland and Iceland by East Coast fishermen consumed the whole season.

The *Mollie Adams* decided to go to Cape Flattery for a fresh-fish trip, but experienced such poor fishing it returned to Rose Spit for another salt fletching trip. Delayed by bad weather the fish were in poor shape on arrival in Seattle. The *Mollie Adams* was then sold to become the *E.P. Marvin*, a sealer out of Victoria, B.C. Thus terminated the contributions of the *Mollie Adams* to history, having made the first successful fletching trip as well as having discovered the famous Two Peaks Grounds in northern Hecate Strait.

The *Edward E. Webster*

The *Edward E. Webster* also owned by Captain Sol Jacobs and under the command of Captain John L. Harris, left Gloucester, Massachusetts, on October 27, 1887, in company with the *Mollie Adams*, but sprung her foremast in a storm off Argentina and had to go to Montevideo for resetting. It was further delayed en route by scurvy in the crew and arrived in the Northwest in early June, nearly three months after the *Mollie Adams*. Captain Charlie Norton then took her sealing. In 1889 the *Webster* did some sealing, and under Captain Silas Calder, it went fletching halibut in Hecate Strait. What final disposition was made of the *Edward E. Webster* is not known, but it discontinued halibut fishing in 1889.

The Steamer Era

The steamer era encompassed a long and important period in the history of the Pacific halibut fishery. The development and activity of that fleet dominated the fishery and determined its character from 1890 to about 1920, the first third of its now 90-year history.

Soon after the commercial fishery commenced it became obvious that if it was to reach any stature vessels other than sailing vessels would be required. The sailing vessels of the East Coast bank fishery could not negotiate the fjords nor the rocky harbors of the Northwest Pacific Coast with their strong tidal flows. Also, auxiliary power was required to permit winter fishing when the markets for Pacific halibut on the East Coast were the most favorable. Fast and dependable vessels were required to speed the fish from the grounds to rail points to offset the effects on quality of the time-consuming transcontinental rail haul to Atlantic Coast markets.

The first vessels of the steamer period had not been constructed for the purpose of fishing halibut. Most of them were conversions from other uses, but they were a good proving-ground for developing effective techniques and strategies for steamer halibut fishing. This led to the building of the S.S. *New England*. Both the vessel and the steamer era were launched in 1898 and the S.S. *New England* was eminently successful from the outset.

Most of the steamers were large and with big crews when judged by present standards for halibut fishing vessels. They were owned and operated by fishery companies that usually also engaged in the distribution of halibut and in many cases of other fishery products as well. The steamers usually fished with from 8 to 14 dories according to size of the vessel, although three of them used only longline gear during most of their careers. With dories fishermen were able to deploy a large amount of gear on the heavy concentrations of fish that were encountered in the early years. They had hold capacities that would carry from 150,000 to 310,000 pounds of iced halibut. During the early and middle years of the steamer period most of the vessels could each account for about 3 to 3.5 million pounds of halibut annually.

The varied fleet of Canadian and United States vessels that can be considered to be of the steamer class is listed in Table 5 along with some characteristics of each vessel.

The fleet of each flag can be divided into two categories: those units that had meaningful production records and those whose imprint upon the fishery was more historical than productive.

Canadian (or British) Vessels

The 12 Canadian or British vessels with meaningful production records are divided into two groups in the table. There were three steamers—the *Coquitlam* (Figure 13), *Capilano* (Figure 14) and *Thistle*—in the 1890s that often served as chartered tenders or transporters for small boats fishing out of shore-based camps or from barges anchored in sheltered harbors on the British Columbia coast in the vicinity of the fishing grounds. In the later years of their operations after 1895, the *Capilano* and *Coquitlam* assumed more of the typical dory fishing vessel functions. It serviced the small fishing boats or one-man dories directly from the vessel and for successive days would be independent of the land-based camps or barges. Initially the *Capilano* and *Coquitlam* had been primarily small freighters and both were chartered from the Union Steamship Co. of Vancouver. They were named after Indian tribes located in the vicinity of that city. The *Thistle* operated from Victoria, B.C. It also serviced a camp operation near the fishing grounds in Hecate Strait. It was owned by the Victoria Fishing and Trading Co. which was closely connected with the International Fish Co. in Tacoma where the *Thistle* landed some of its fares. It was under the command of Captain Wallace Langley.

TABLE 5. The Steamer Fleet

UNITED STATES REGISTRY

FISHED • • • (OT = Otter Trawl)

VESSEL	Home Port	Built Date	Built Place	Net Tons	Capacity	Owners	Fished (1893–1930)	REMARKS
New England	Vancouver, B.C.	1897	Camden, N.J.	71	220	New England Fish Co.	Cost $25,000 • 1898–1930	Iron Ship — Sold For Scrap, WWII
Manhatten		1902	Philadelphia	134	225		• 1907–1917	Lost Lituya B. Nov. 1917
Kingfisher		1904	San Francisco	141	250		• 1902–1914	Ore Carrier, 1916
San Juan	Seattle	1904	Seattle	128	180	San Juan Fishing & Packing	Value 1905: $45,000 • 1905–1917	Cannery Tender
Starr		1904	Seattle	187	N.A.		• 1914–1918; • 1912–1913	Mail Boat, W. Alaska
Grant		—	—	180	180		• 1907–1911 Had "wireless" in 1909	Former U.S. Revenue Cutter — Lost Dec. 11, Banks Is., B.C.
Edith	Tacoma	1882	San Francisco	78	175	International Fish Co.	• 1894–1905 Value $30,000	Became a Coastal Freighter — Condemned 1924
Zapora		1905	Tacoma	196	250		Cost $50,000 • 1905–1920; • 1924–1925	Lost Feb. 14, 1937 S.E. Alaska
Weiding Bros.	Seattle	1909	Seattle	125	170	Weiding Independent Fish Co.	• 1909–1914	Lost July 1913
Independent		1910	Seattle	123	230		• 1911–1916	Lost Middleton Is., Oct. 6, 1916
Rush		—	—	254	N.A.		• 1914	Former U.S. Revenue Cutter — Sold out of Fishery 1915
Chicago	Seattle	1910	Seattle	129	300	Chlopeck Fish Co.	• 1908–1920; • 1923	Became towboat in B.C.
Rainier				109	150		• 1902–1903	Wooden Ship 8 dories — Lost Icy Strait Oct. 1903

BRITISH (CANADIAN) REGISTRY

VESSEL	Home Port	Built Date	Built Place	Net Tons	Capacity	Owners	Fished (1893–1930)	REMARKS
Group I								
Capilano	Vancouver, B.C.	1891	Scotland	157	—	Union S.S. Co.	• 1893–1902	Fabricated in Scotland and assembled in Vancouver —
Coquitlam		1891	Scotland	157	—	Union S.S. Co.	• 1893–1899	Fabricated in Scotland and assembled in Vancouver — Wrecked 1905
Thistle	Victoria, B.C.	1890	Victoria	147	—	Victoria Fishing and Trading	• 1894–1897; • 1900	Lumber Transporting March 18, 1920 — Sealing in 1891
Group II								
Celestial Empire	Vancouver, B.C.	—	England	79	200	Canadian Fishing Co.	Used OT first trip • 1905–1919	When Owned in France was Named Jupiter — became tugboat — Cape Scott, in 1921
Flamingo		—	England	125	250		• 1905–1918	Bought by New England Fish Co. for debts
Columbia		—	Gothenburg		N.A.		• 1903	Formerly U.S. SAGA Bought by New England Fish Co. for debts 1904 — Wrecked 1915
Andrew Kelly	Prince Rupert	1912		96		Canadian Fish and Cold Storage Co.	• 1912–1919; • 1921–1924	Longlined 1913-1920 — Became towboat then herring barge
James Carruthers		1912	England	96			• 1912–1919 (OT 1916); • 1921–1924	
Geo. E. Foster		1912		96			• 1912–1919 (OT 1917); • 1921 OT, • 1923, • 1924 OT	
BCP	Steveston and New Westminster	1912	Ladner, B.C.	83	200	B.C. Packers	• 1910–1911; • 1913–1916	
Roman		—	England	144	300		• 1908–1909; • 1911–1915	Lost S.E. Alaska Nov. 18, 1916
Onward Ho		—	do.	130	300		• 1913–1914	Lost off Yakutat Jan. 1916
Canada	Alford Bay	1882		66	N.A.	B.C. Fishing Ltd.	• 1912; • 1916–1917; • 1919	Bankrupt. Bought by Canadian Fish Co. in 1916 — Packing in 1916
Imbricaria		1893	England	66	N.A.		OT OT OT	Bankrupt. Bought by Canadian Fish Co. in 1916
Triumph		1907		121	N.A.		• 1912	Bankrupt. Bought by Olson & Kendall, Halifax, N.S. — Sunk by U-boat Aug. 20, 1918
Kingsway	Pacofi	1906	England	85		Pac. Coast Fishing	• 1908–1909; • 1912	

The following nine vessels were the more typical halibut fishing steamers. All used dories except for some initial but unsuccessful attempts to use bottom nets or otter trawls. Three of the nine, the *Celestial Empire, Flamingo* and *Columbia,* were owned by the Canadian Fishing Co., or Canfisco, which after 1908 was the Canadian subsidiary of New England Fish Co. The *Columbia*, which had been the United States *Saga* before purchase in 1903 by Canfisco of Vancouver, had a very short career in the halibut fishery under its new name and flag. In contrast, the *Celestial Empire* (Figure 15) and *Flamingo* (Figure 16) had long and productive records in the fishery. The *Celestial Empire,* built in Hull, England, had been wrecked on her maiden voyage to Iceland, salvaged, and then lengthened 12 feet amidship. It was then sold to French owners and renamed *Jupiter* and used in the Iceland fishery out of Lorinth, France. It was sent to England for resale and bought by Canfisco in 1906 and reverted to its original name, *Celestial Empire.*

The S.S. *Flamingo,* acquired by New England Fish Co. from a defunct Vancouver fishing company, became a major producer of halibut under the command of Absalom Freeman. It landed some of the largest fares ever made on the Pacific Coast, many being caught in only one to four days of fishing.

Another three vessels in this group, the *Andrew*

FIGURE 13. S.S. *Coquitlam* entering Burrard Inlet.
E.W. Wright, *Marine History of the Pacific Northwest*

FIGURE 14. S.S. *Capilano* entering Burrard Inlet.
E.W. Wright, *Marine History of the Pacific Northwest*

Kelley (Figure 18), *James Carruthers* and *George E. Foster* (Figure 19) were owned by the Canadian Fish and Cold Storage Co. of Prince Rupert, B.C. Although initially fitted for otter trawling when built for the company in England, they fished halibut with setline gear. Most of the years a longline method of setting and hauling from the main vessel rather than dories was used. They introduced the stern chute from England for setting the gear and patterned their operations after the "liners" out of Fleetwood, England. The vessels were named after officers of the company: President Andrew Kelly of Winnipeg, Vice President James Carruthers of Montreal, and George E. Foster, each of whom had played a prominent role in organizing the company in 1907.

The remaining three vessels, the *BCP, Roman* (Figure 20) and *Onward Ho,* were owned by the British Columbia Fishing and Packing Co., the B.C. Packers, Ltd. Though primarily concerned with salmon, the company has had a long and continuing interest in halibut. The *BCP* was built in Ladner, B.C., in 1912 with a capacity to carry 300,000 pounds of iced halibut. It fished until the end of World War I before engaging in other pursuits. The *Roman* and the *Onward Ho* were purchased in England—the former from the Imperial Steam Fishing Co. in Hull. Both vessels had halibut-carrying capacities of about 300,000 pounds. The *Onward Ho* had a tragic ending, foundering in a freezing winter gale in 1916 off Cape Fairweather, with the loss of the crew of 36 men.

FIGURE 15. The S.S. *Celestial Empire*
International Pacific Halibut Commission

In addition to the foregoing 12 vessels, making up the main effective Canadian steamer fleet, there are other steamers whose contributions were more of historical interest than productive. Some features of these vessels are mentioned below.

Three large vessels, the *Canada, Imbricaria,* and *Triumph* (Figure 21), were brought to the Pacific Coast from England by an over-promoted and soon-to-be-defunct company, the British Columbia Fisheries, Ltd. The vessels arrived completely equipped to fish with otter trawl despite the fact that it had been repeatedly demonstrated during the previous 20 years that it was not an effective method of operating for halibut on the Pacific Coast. It was also known that there was no profitable market in North America for the other demersal species that the trawl would catch. The company went into receivership soon after the arrival of the three vessels. The disposition of the vessels is shown in the accompanying Table 5.

The *Kingsway* was another vessel brought from England equipped for otter trawling. In 1910 its owner, the Pacific Coast Fish Co. had built a cold storage plant at Pacofi, an isolated place on the east coast of Queen Charlotte Islands. This company also terminated activities shortly after the arrival of the vessel. The venture was apparently financed by German capital. Many years later in 1938 when the old Pacofi plant was being torn down by the new owners, the B.C. Packers Ltd., elaborate reinforced concrete installations were found under the cold storage plant. Examination by naval experts revealed that they had been designed for use as secret submarine pens.

FIGURE 16. S.S. *Flamingo*
E.W. Wright, *Marine History of the Pacific Northwest*

FIGURE 17. Part of a catch of 320,000 pounds of halibut taken by the S.S. *Flamingo* in August, 1912.
International Pacific Halibut Commission

FIGURE 18. The *Andrew Kelly*
International Pacific Halibut Commission

There are several smaller vessels in this group of Canadian steamers that contributed to the history of the fishery. The *Iona,* built at the Morse Mill in Vancouver, for the B.C. Fishing and Trading Co., formerly Fader Brothers, has the unique distinction on the Pacific Coast of having been first to have fished for halibut with a 50-foot beam (not otter) trawl. While it caught up to 30,000 pounds of halibut in a 10-day trip to the Queen Charlotte Islands, it was not a success and was sold for debts after two seasons. A 37-ton steam fur sealer, the *Eliza Edwards,* fished halibut one season but decided that treasure hunting in the tropical weather of the Cocos Islands was more attractive and possibly more profitable.

The *Dauntless,* a small steamer of 34 tons with eight dories owned by a New Westminster company fished from November 3, 1901, to the spring of 1902, when it engaged in other pursuits. It was later chartered in late December, 1904, and fished a few trips under Captain A. Dawe with a crew of steamer fishermen that were out on strike against the New England Fish Co. The vessel withdrew permanently from the halibut fishery after that episode.

The *Mischief,* formerly of United States registry out of Yaquina Bay, Oregon, operated out of New Westminster for two seasons—1892 and 1893. The *Velos,* a

FIGURE 19. The *George E. Foster*
Photograph by permission of Larssen and Andrews, courtesy of the International Pacific Halibut Commission

FIGURE 20. The *Roman* in 1915. Her dories are "on the gear" and their place on the stern is vacant.
International Pacific Halibut Commission

FIGURE 21. The S.S. *Triumph*
Canadian Fisherman

FIGURE 22. The S.S. *Kingfisher*
Canadian Fisherman

small steamer, operated out of Vancouver in the very early 1890s. Contrary to what has been sometimes reported both the *Mischief* and *Velos* were not fishing vessels. They were transporting vessels, picking up fish caught sometimes by Indians, at various coastal points. They more closely resembled the present day packer or mobile buying vessel.

United States Vessels

The 11 important steamers of United States registry are grouped in Table 5 by their ownership. The *New England, Manhatten,* and *Kingfisher* (Figure 22) were owned by the New England Fish Co. and were operated out of Vancouver. Their names tend to relate to their owner, all being reminiscent of the northeastern United States where New England Fish Co. had originated. The *New England* was the first steamer built expressly for halibut fishing, and it operated for three decades in the fishery, much longer than any other of that class of vessel.

Of the three large steamers of the San Juan Fishing and Packing Co. of Seattle, the origin of the name of one, the *San Juan* (Figure 23), is obvious. The *Starr* was for the family name of the wife of William C. Calvert Sr., president and founder of the company. The *Grant* (Figure 24) purchased by the company was formerly a United States revenue cutter named after the 18th President of the United States, Ulysses Simpson Grant.

The *Edith* (Figure 25) of Tacoma was one of the earliest of the regular fishing steamers. Once the private yacht of W.C. Ralston, president of the Northern Pacific Railway, it was named after his wife. It had lain a long time rotting in the Sacramento River and was rebuilt from the waterline in 1896 by Edward E. Johnson of Tacoma for the International Fish Co. of that city. The *Zapora* (Figure 26), built in Tacoma by the International Fish Co., took over in 1905 when the *Edith* was retired from the fishery.

The *Independent* (Figure 27) and *Weiding Bros.* (Figure 28) were owned by the Weiding and Independent Fisheries of Seattle, formed in 1911. The *Weiding Bros.* had previously fished for the Occidental Fish Co. of Seattle. Andrew, one of the Weiding

FIGURE 23. The S.S. *San Juan*
Webster and Stevens

FIGURE 24. The S.S. *Grant*
Puget Sound Maritime Historical Society

brothers, had the distinction of being first mate on the *New England* in its first season of fishing out of Vancouver in the winter of 1898-1899. In their short careers—they were lost at sea—both vessels of the company were large producers. From 1911 to 1914 the *Independent* averaged 2.5 million pounds a year. It had the distinction of landing the largest fare in the history of the fishery—362,572 pounds in Seattle in 1912.

With the loss of the *Weiding Bros.*, the *Rush*, a former United States revenue cutter, was converted for halibut fishing. It operated one season, but the death of its captain, Andrew Weiding, in August, 1914, brought on the sale of the vessel.

The *Chicago* (Figure 29), owned by Chlopeck Fish Co. of Seattle, had one of the largest carrying capacities of the United States steamers—300,000 pounds of iced halibut—and on one occasion had over 320,000 pounds, the second largest catch for any halibut vessel. It regularly carried 14 dories. Its name derives from the company's close connection with Booth Fisheries of Chicago which absorbed Chlopeck in 1909.

As in the case of the Canadian fleet of steamers, there were some United States steamers of varying size that contributed to this period of the fishery. The *Geo. H. Chance,* 87 net tons, of Portland was launched at Yaquina Bay, Oregon, February 13, 1889, and made several trips for halibut under Captains George W. Adams and Peter H. Crim. There was little profit, and the vessel entered the coastal trade in 1892. It was lengthened in 1893 and renamed the *Bendorille.* The *C.H. White* of San Francisco, 84 net tons, fletched halibut out of Seattle and made trips to Cape Flattery and Hecate Strait, near Bonilla Island. It lost two men off Cape Flattery and returned to San Francisco. The *Francis Cutting* (Figure 30) fished a couple of years out of Seattle in the 1890s. It was seized in Canadian waters in 1908. The S.S. *Saga* fished for a few seasons but was sold for debts to New England Fish Co. of Vancouver, who changed the registry to British and the name to *Columbia.*

Another vessel with a varied career was the *Queen of the Pacific.* It was built in Bellingham, Washington, as a sailing schooner for the Nome, Alaska, trade, but was converted to a steamer in 1904 with a capacity to carry 100,000 pounds of iced halibut and a crew of 18. Since, for sundry reasons, it seemed to spend a greater amount of time in harbor than fishing, it was known derisively in the halibut fleet as the "Bull of the Woods." It did not have a good production record and was later sold to Wallace Fisheries in British Columbia who planned to use it to supply halibut to its cold storage at Claxton in northern British Columbia.

FIGURE 25. The *Edith* in 1899.
International Pacific Halibut Commission

FIGURE 26. The S.S. *Zapora*
Webster and Stevens

FIGURE 27. The S.S. *Independent*
International Pacific Halibut Commission

FIGURE 28. The S.S. *Weiding Bros.*
Pacific Fisherman

FIGURE 29. The S.S. *Chicago*
Puget Sound Maritime Historical Society

The Rise and Decline of the Steamer Fleet

During the first decade of the steamer era, 1895 to 1904, the combined fleet of both countries consisted of not over four vessels in any one year. With the proven success of the *New England* and *Kingfisher* of the New England Fish Co., it was inevitable that the fleet would expand as other companies sought to emulate the profitable experience of New England Fish Co. Commencing in 1905 the fleet increased steadily in number reaching a maximum number of 16 in 1914. During World War I it declined, and by 1918 there were 10 vessels. Thereafter the rate of decline was even more rapid with only one boat fishing in 1925. That vessel, the S.S. *New England*, operated until 1930, the end of an important era in the fishery.

There were a number of developments that led to the decline and final liquidation of the steamer fleet. The initial advantage of the large steamer in being able to deploy large amounts of gear on the heavy concentrations of fish initially found on the grounds, particularly south of Cape Spencer, had practically disappeared with the decline in the stocks. Capital for the construction of competing types of vessels had become more generally available. Many former steamer fishermen oftentimes aided by capital from supply houses or ship chandlers invested their savings in vessels of varying size. These vessels had much lower manpower requirements than the steamers and were exempt in both countries from the Steamboat Inspection Regulations that specified the number and qualifications of deck and engine room personnel and other restrictions. The manpower matter generally became an acute one for the company steamers during World War I, particularly as to the availability and remuneration of nonfishing crew members. The fishermen were paid according to the amount of fish caught. It was even difficult for many of the larger independently owned vessels whose crew were all on shares. Some owners even secretly paid some of the trip costs to avoid losing crew members to other boats.

The decline in the steamer fleet was accelerated by the wreck of four of the largest of the steamers in 1916. With building costs escalating and shipyards completely preoccupied in cargo-vessel construction to offset massive Allied losses to German U boats, the vessels could not have been replaced even if it had been economical to do so. The fishing companies were also diverting more of their energies and capital to building up their interest in the usually more lucrative salmon fishery. Furthermore, it had become very evident that halibut fishing could be carried out more efficiently and economically by owner-operated vessels. Records of one company showed that the production costs of halibut landed by its three steamers rose from 3.45¢ per pound in 1915 to 8.53¢ in 1917, not including capital or maintenance costs, which had also risen sharply. It became cheaper to buy from independent vessels.

It was apparent that the changing times had brought changed conditions that finally ended an important episode in the history of the halibut fishery. World War I and some loss of vessels probably only accelerated the inevitable.

FIGURE 30. An early steamer out of Puget Sound, the *Francis Cutting*.
International Pacific Halibut Commission

FIGURE 31. The S.S. *New England* when it was a coal burner.
Canadian Fishing Co., Ltd.

The S.S. *New England*

The S.S. *New England* had a long and eventful career in the Pacific halibut fishery. It bears the distinction of having been the first steamer built expressly for the fishery, and it was also the last of that class of vessel to operate in the fishery. During 32 years from its departure on its first halibut trip from Vancouver on September 28, 1898, to the landing of its last halibut fare on October 1, 1930, at Ketchikan, it established a production record of halibut for a single vessel that will never be eclipsed. The *New England* also participated in several noteworthy events in the maritime history of the Pacific Northwest and Alaska. The exploits and operations of this vessel will be recorded in some detail as they portray an important period in the history of the Pacific halibut fishery, the era of the company-owned steamers, and their rise and decline. The writer accompanied the *New England* on several of its trips.

The design of the *New England* was based on the experience gained by the New England Fish Co. from the time they started their Pacific Coast operations in the fall of 1894. The vessel would have to be self-sufficient and capable of being independent of any land base for weeks at a time. Reasonable speed was a necessity to assure a fast return to port and thus conserve the freshness or "shelf life" of the product for the long rail haul to the East Coast. The carrying capacity of iced halibut had to be equal to what might be caught in a reasonable length of time and what could be expeditiously unloaded and transshipped at the landing port.

The vessel was built by the Crump Shipyards, Camden, New Jersey, in 1897 under the supervision of Captains Hanson B. Joyce and Absalom Freeman who had directed the New England Fish Co. operations on the Pacific Coast since their outset in 1894. The *New England* was 121.8 feet long, 24 feet wide, 12 feet deep, 230 gross and 70 net tons and was registered at Portland, Maine. It could accommodate 12 dories and a crew of 37, including 24 fishermen and was able to carry 220,000 pounds of iced halibut. At the time of construction it burned coal but later converted to oil. Oil was available at more locations than coal. It was also cleaner, required less space and labor to load and usually required two fewer firemen in the crew.

Contrary to some the term "smoke boat" was directed by steamer crews to the schooners with semi-diesel engines. However, when the schooners had gas engines and the steamers burned coal, as did the *New England* in Figure 31, the reverse designation could have applied.

It commenced its active career two days before Christmas on December 23, 1897. In a howling winter storm the *New England*, with a crew of 18 under command of Captain J.W. Small of Provinceton, Massachusetts, made a hasty departure from Boston, then the home office of the New England Fish Co. Immediate war with Spain over its treatment of insurgents in Cuba was being vigorously supported and exploited by the press of that time. The possibility of being commandeered for war duty prompted the precipitous departure. Until the raging storm abated, shelter was sought in Wood's Hole, then the vessel ran directly to St. Thomas, Virgin Islands, without taking coal en route, to avoid the continued possibility of being requisitioned in a United States port by the Navy.

The vessel made a much-needed coaling in Rio de Janeiro, Brazil, and then at Montevideo, Uruguay. They were soon at the southern extremity of their long 15,000-mile voyage, a distance of more than halfway around the world. After a daylight passage through the Strait of Magellan the vessel proceeded northward to coal again at Talcahuana, Chile. Unlike the generally good weather encountered in the Atlantic Ocean, violent storms were the rule in the Pacific, although so named by Magellan for its peaceful character. Another coal stop was made in Callao, Peru, thence on to San Francisco arriving March 18, 1898. Five days later the *New England* arrived in Vancouver on March 23, 1898, three months to the day out of Boston, 14,770 miles away, a long voyage but without mishap.

Captain Small returned to the East Coast having completed his assignment, the safe delivery of the vessel to Vancouver. Since it was the end of the winter fishing it was not a propitious time to start fishing. Instead the vessel, two days after its arrival in Vancouver, proceeded to Seattle under Captain Hanson B. Joyce.

The *New England* then successfully towed two Yukon River steamers, the *Rock Island I* and *II*, to St. Michael, Bering Sea. With the Klondike gold rush at its height, two more river steamers were towed by the *New England* that summer to St. Michael at the mouth of the Yukon River. Southbound from this second tow the vessel brought down 20 Dawson miners with their gold, *the first gold to reach Seattle from the Klondike.* Although they had been signed on as crew members because the vessel was not licensed to carry passengers, the miners paid their fares in gold dust. This historic occasion of bringing out the first Klondike gold to Seattle was also a most profitable one for the New England Fish Co. The towing and passenger business made a net profit of $14,000. This was a high return for the vessel which was then valued at about $25,000.

The *New England* left Vancouver on its first halibut trip in late September, 1898, under Captain H.B. Joyce with Captain A. Freeman as pilot and Andrew Weiding as first mate. These three men played very prominent roles in the subsequent development of the fishery. Captain Freeman's position as pilot stemmed from the fact he was a Canadian and could not hold a licensed position on a United States bottom. That first winter fishing season from September, 1898, to April, 1899, 42,000 halibut were landed amounting to about 1.6 million pounds.

To supplement the earnings from winter halibut fishing the *New England* engaged in towing operations to St. Michael, Bering Sea, for two summers. In the first year—the summer of 1898 and the winter of 1898-1899—the vessel grossed $100,000, $75,000 from fishing and $25,000 from other pursuits, with a net return of $30,000. The *New England* was a productive and financial success from the outset.

Fishing was conducted during the first season off Butterworth Rocks, Rose Spit and Banks Island in Hecate Strait. The vessel carried 10 two-man dories the first two seasons. Each dory was provided with 10 tubs or skates each of 10 lines of 50 fathoms of #32 tarred cotton ground line with gangings spaced 9 feet apart. In the first two or three years of operations the *New England,* on the average, fished fewer than 100 such units of gear or skates of about 310 hooks each per trip. Indicative of the density of the more or less primeval stocks, catching 1,000 pounds per skate was regarded as only "fair" fishing and under such conditions the vessel would usually change grounds.

The vessel fished the following winter season of 1899-1900 from September 1 to the end of April with the full complement of 12 dories. Since it started about a month earlier than in 1898, the market conditions in the East required the company to set a limit of 75,000 pounds per trip to be removed in November. The limit was reinstated in the following March. Notwithstanding the restrictions the second season's catch totaled 15 million pounds.

Year-round fishing by the *New England* commenced in 1901. To avoid summer gluts the vessel was placed on a 50,000 to 75,000 trip limit between May and September. In the fall of 1900 Captain Joyce had been replaced by Captain Gott, his brother-in-law. Captain Freeman continued as pilot and was in charge of all fishing operations and fishermen.

In calendar year 1902 Captain Freeman assumed full command of the vessel having secured United States papers. Captain Joyce took over the S.S. *Kingfisher*, a new steamer acquired by the New England Fish Co.

About this time a moderately satisfactory method of transferring the fish from the dories to the steamer was devised. Each dory had several straps on which they threaded the number of fish making it possible to transfer 500 to 700 pounds of fish at a time to the steamer. In 1902 Captain Joyce introduced a net sling into the bottom of the dory, and the whole dory load could be transferred to the steamer in a single lift, a big change from lifting three or four fish with straps around their caudal peduncles.

In 1903, the *New England* made 37 trips landing 221,676 fish, about 4.4 million pounds. The vessel fished 136 days during the year with some trips consisting of one day of fishing. During the first part of the season they fished in northern Hecate Strait. In July and August fishing was conducted on Cape Scott and Goose Islands grounds and later in the year in middle Hecate Strait.

In 1904, despite some delay due to a strike the *New England* made 32 trips landing 164,292 fish having fished 91 days. Having avoided fishing on Goose Islands grounds, the average size of the fish was above that of 1903 with a total weight of about 4 million pounds. In 1905 and 1906 the vessel respectively landed 164,548 and 211,747 fish on 28 and 29 trips.

In 1906 Captain Freeman left the *New England* and in company with other persons formed the Canadian Fishing Co. The *New England* continued a profitable and productive operation under other commands. In 1909, after 10 years of continuous operation, the vessel was given a complete overhaul at a cost of $15,000. It received a new pilot house, fo'c'sle head, double stock, and refurbishing throughout.

In 1915, a deck gurdy was installed to permit hauling the gear from the vessel, i.e., longlining, in an emergency. After the grounds west of Cape Spencer were first fished in 1913, the *New England* transferred its operations to that section of the coast. In 1917, Captain Michael B. Scott took over command and was in charge until the vessel withdrew from the fishery in 1930, the end of the steamer era. The writer made several trips on the *New England* in the mid-1920s.

In 1920, the vessel fished for the first time in Cook Inlet, and in the next decade many trips were made to that area by the *New England.* In 1921 despite fishing only half the year, the vessel landed 1.25 million pounds. The crew made $1,500 per man compared to an $800 average of the independent vessels for a longer season. In the year 1922 economic conditions further deteriorated and many vessels were tied up, including the *New England.*

After 1910, the bonanza annual catches of the previous decade were of the past. Yet the vessel averaged about 1 million pounds annually from the grounds west of Cape Spencer in the 1920s. However, by 1930 it became uneconomical to operate the *New England,* first and last of the company-owned steamers.

The S.S. *New England* leaves a record of a total career catch of about 40 million pounds of halibut, a record that is unlikely will ever be exceeded or equaled by a single setline fishing vessel.

The Schooner Period

The type of vessel characterizing the schooner period was propelled primarily by internal combustion engines using various fuels, distillate, gasoline or, in later years, exclusively, the less costly diesel oil. The vessels were of wood and smaller than the steel or iron company-owned steamers they superseded. For the most part they were owned or partly owned by one or more operators, most of them former fishermen. Some ship chandlers or supply houses frequently participated in some of the financing of the vessels.

FIGURE 32. The M.V. *Tordenskjold,* a dory vessel, in moderate to heavy weather with sail for stability.
International Pacific Halibut Commission

The designation as schooner arose from the fact that they had two masts, and initially, the vessels carried sails for stability (Figure 32) and as auxiliary power in case of engine breakdown. Also, some of the early schooners were conversions and possessed lines of the true sailing schooner. However, unlike the latter vessel, the halibut schooner had the mainsail on the foremast because the after-deck pilot house usually interfered with having a mainsail on the after or mainmast. Sails were gradually dispensed with except for the mainsail which was retained by some vessels as a stabilizer in bad weather. However, when fishing no longer continued into late fall or winter, most vessels no longer carried any sail. Nearly all by then possessed radio telephone in the event of breakdown.

There were a few vessels in the early 1900s that qualified under the foregoing definition of a schooner. The *Northland* (Figure 33) built in 1905 relied mainly on two 25 h.p. gas engines rather than sail for power. It had the typical two masts and after pilot house and might be considered the first of the schooner type.

FIGURE 33. The M.V. *Northland*
International Pacific Halibut Commission

Another distinctive vessel was built in 1909, the *La Paloma* (Figure 34). It represented a transition between the old and the new, the East Coast sailing vessel and what became the classic power schooner of the Pacific. The *La Paloma* (Figure 35) was later converted more to the power schooner type but could not hide its East Coast heritage. The vessel was 72 feet in length and 40 net tons. When launched it had a 75 h.p. Frisco Standard gas engine.

The first decade of the twentieth century was about over before the first typical schooner, such as the M.V. *Tordenskjold* (Figure 36), emerged. It was an extremely seaworthy vessel and was able to cope with the worst of the severe weather encountered in North Pacific winters. The Aleutian Low, with its almost continuous winter fronts of bad weather, seems to locate itself during most of the winter in the general region of the main halibut grounds, west of Cape Spencer.

The schooner initially used dories to concentrate as much gear as possible on the abundant stocks. As the fish stocks declined and crews became less interested

FIGURE 34. The *La Paloma* as a sailing vessel.
International Pacific Halibut Commission

FIGURE 35. The *La Paloma* in later days as a power schooner.
International Pacific Halibut Commission

in the hazards of dory fishing, a shift from such fishing commenced, first during winter operations. It subsequently extended to summer operations as well. The conversion to longlining had reached a high point by 1926. By 1932, dory fishing was restricted to two vessels. Shortly thereafter, the commission prohibited the use of such gear in Area 2 in 1934 and in all regulatory areas by 1944.

The schooner vessel type dominated the fishery west of Cape Spencer for many years, even though the last vessels were built before the end of the 1920s. The *Masonic* and *Aleutian* of Seattle were among some of the last to be built. The durability and effectiveness of the schooner are attested by the fact that many are still fishing halibut after 50 or 60 years.

While the basic type has remained the same over the past three quarters century, many improvements and adaptations have been made. Developing technology was utilized to make construction and equipment changes in the vessels to meet the needs of new fishing techniques and changes in fishing strategies.

FIGURE 36. The M.V. *Tordenskjold* on its trial run.
International Pacific Halibut Commission

FIGURE 37. The M.V. *Polaris* as of 1978.
Ray Krantz, Seattle Marine Photography

One might trace significant changes that have been made in a typical schooner, the *Polaris* (Figure 37) of Seattle, built in 1913 and still in the fishery. Figures 38 and 39 profile the vessel's record and epitomize the schooner period.

The M.V. *Polaris* of Seattle

The *Polaris* was 73.4 feet long, 66 gross and 45 net tons. Initially it fished dories with a crew of 15 including captain, cook, engineer and 12 fishermen for the six two-man dories. In 1927 it changed to longlining with a crew of 10. Pilot house controls of the engine that freed the engineer to relieve the captain in the pilot house were installed. On some schooners that converted to longlining, both the cook and engineer would bait a few skates of gear along with their other duties. For several years in the 1930s an additional man was added to the crew as permitted under the fleet's voluntary production control program that limited the catch allowed per-man-per-trip. In 1953 the total crew was reduced to 7 men. This sharply increased the man-share of the proceeds. To maintain the same level of production, changes were made in the shifts on deck at any one time and full

FIGURE 38. Production profile of a typical schooner, the M.V. *Polaris*, showing crew size and number of trips.
F.H. Bell

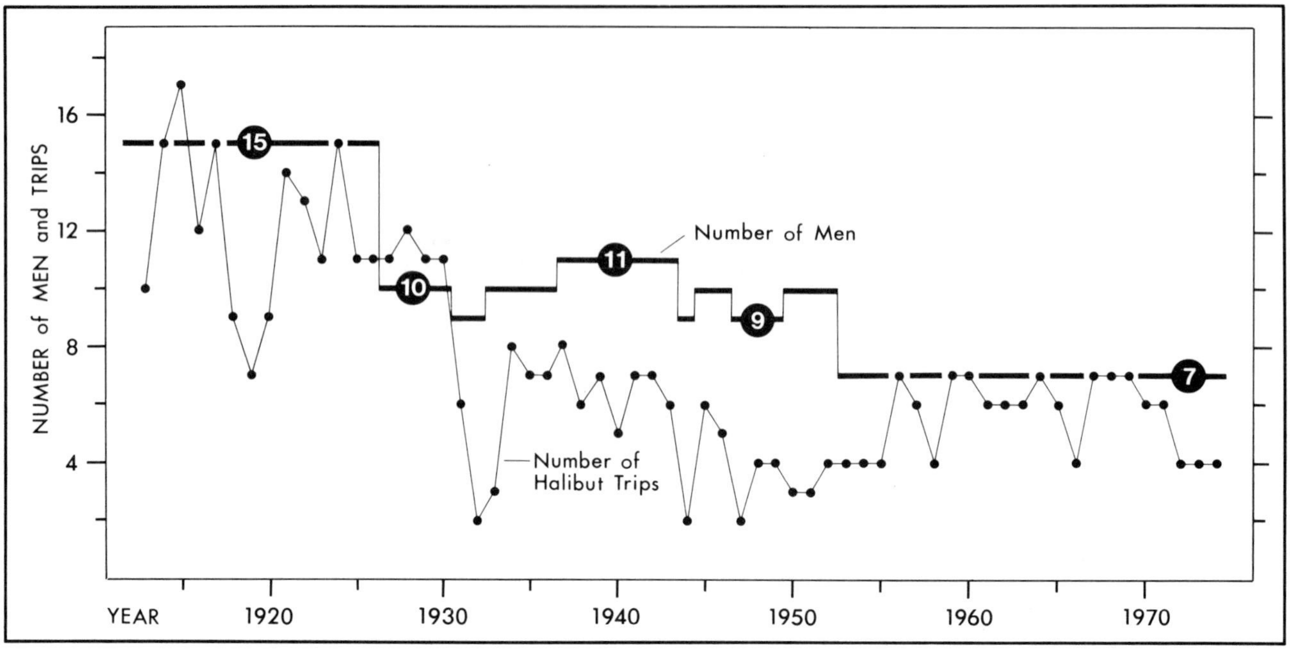

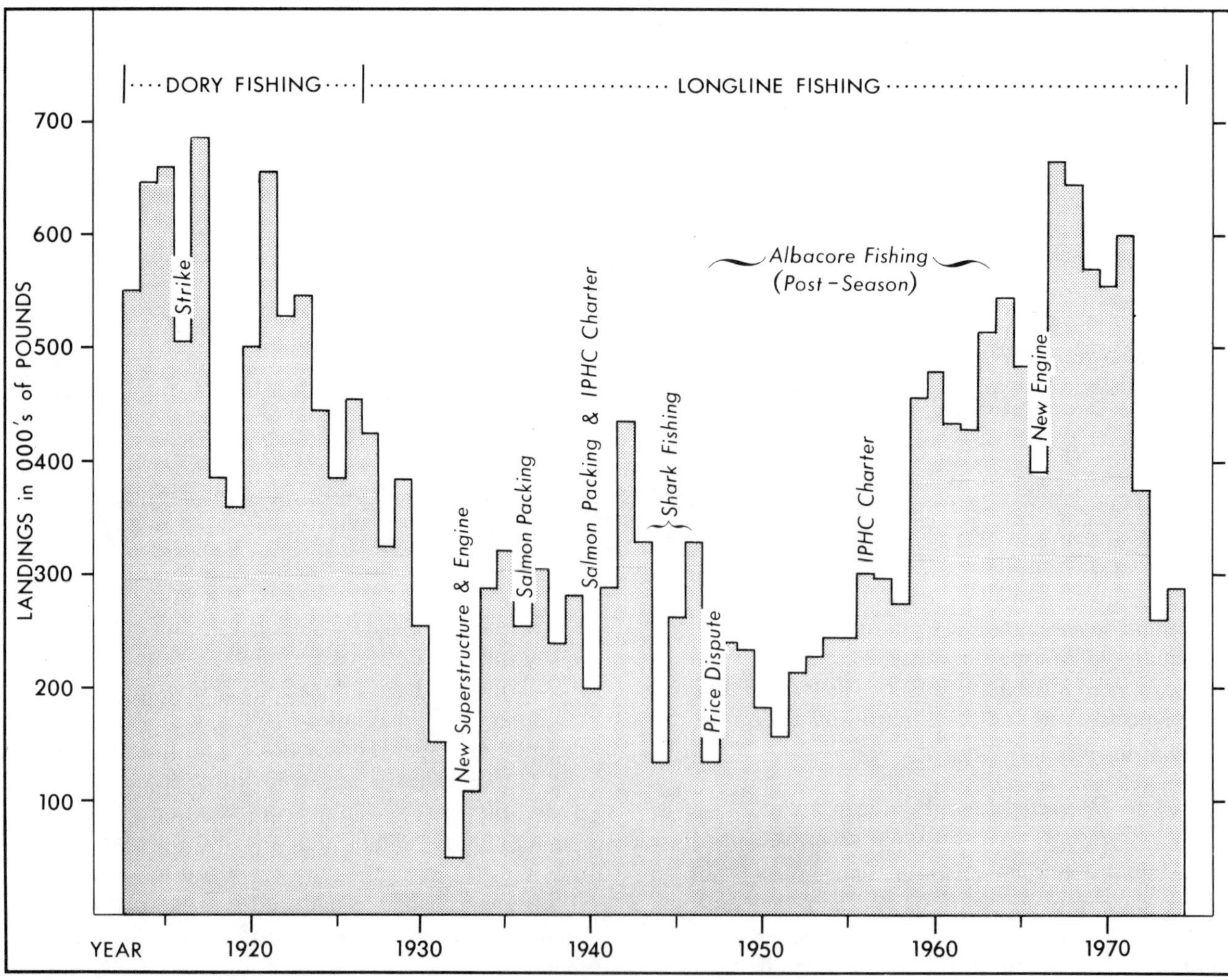

FIGURE 39. Annual production of the M.V. *Polaris* from 1913 to 1974 and some postseason activities.
F.H. Bell

advantage was made of the many advances made over the years in the vessel's operating equipment.

As with most schooners the *Polaris* made a number of replacements in motive power over the past 65 years. When built in 1913 it had a 100 h.p. Frisco Standard gas engine. In 1920 a Frisco Standard semi diesel was installed, which was replaced in 1932 with a 175 h.p. Atlas full diesel. In 1966 a 340 h.p. General Motors high rpm diesel was installed.

Significant above-water-line construction changes were made in 1932 and 1933 involving a new pilot house and deck, an enlarged fish hold for 7,000 pounds iced halibut over and above the original 80,000 pound capacity. The fo'c'sle head and floor were raised to install new and larger tanks, and the after bulkhead was moved aft. In 1956 the after cabin with its bunks was taken out and converted to a refrigerated hold space to stow an additional 18,000 pounds of iced halibut bringing the total capacity to over 100,000 pounds. It also served as a storage for frozen bait. Glass-lined water tanks were installed in 1956.

Hydraulic controls were installed in the late 1940s for steering, then for the gurdy and anchor winch. In the early 1950s the refrigeration drives were hydraulic controlled to maintain a more uniform temperature.

Navigation aids underwent a number of changes and additions. The *Polaris* had its first radiophone in 1936 and by the mid-1970s carried three sets of various ranges and sophistication. Radar was installed in 1956 and the vessel now carries both a 24-mile and 48-mile set of high resolution. Loran was first installed in 1940 with the latest models and types which were eventually commonly installed in the mid-1970s. The vessel carries three fathometers of varied capacities and types. In the 1970s it had the latest model of an autopilot. This device is particularly useful when trolling for albacore tuna.

The production profile of the *Polaris* as shown in Figure 39 reflects the general trends of the fishery. The decline in production in the 1920s resulted from a steady deterioration in the stocks. In 1932 and 1933 the *Polaris* underwent a major rebuilding of the superstructure and fished but a few trips. Though the stocks did improve during the 1930s and 1940s, the shortening of the fishing seasons reduced the annual production of the individual vessels. Most of the vessels turned to other pursuits at the end of the halibut season such as albacore tuna fishing, shark fishing in the 1940s, and fish packing in the salmon fishery. In the 1930s annual production of the individual vessel was also reduced by the fleet's voluntary production control program. With the lengthening of the season after the early 1950s and continued improvements in the stocks, the *Polaris* production increased with larger numbers and sizes of fares.

The full impact upon the stocks of foreign fishing in the late 1960s is evident in the decline in the annual catches of the *Polaris*, a decline that continued through the 1970s.

With some differences in timing and in magnitude the changes undergone by the *Polaris* are reasonably representative of what occurred in the schooner fleet as a whole. This profile of the *Polaris* is also representative of the caliber of the people involved in the ownership and operation of the schooner fleet. Arne S. Einmo, operating owner of the *Polaris* for the past 50 years is such a person. During that one-half century there were good years and bad years for the fleet, periods of fish scarcity and periods of abundance, prices sometimes good and oftentimes poor; bad weather and good weather, breakdowns and accidents, but always long hours and hard work. It was a testing ground for both men and vessels.

The schooner era was not only a productive period of the fishery, it also saw the emergence of the independent owner who soon became the dominant entrepreneural group. Company ownership of fishing vessels declined sharply, particularly in the United States. A very large proportion of the early owners had served as fishermen on the company-owned steamers. Most were of Norwegian origin with some from Nova Scotia and Newfoundland. With savings from their labor and supplemental capital from the supply houses or ship chandlers, they built their own vessels.

There were many independent owners who approximated the above general pattern. At the back of this book there is an account in some detail of one, Mattias Madsen, an outstanding example of one from that group of men.

The Saga of the *Scandia*

The M.V. *Scandia* (Figure 40) out of Seattle was one of the larger of the early schooners. While it did not have a particularly eventful career fishing halibut, it came to a memorable end one stormy winter night on a rocky reef in Western Alaska. Its last resting place possessed a foreboding name—Graveyard Reef.

The *Scandia* was 90.8 feet long with a 160 h.p. Bolinda semi-diesel engine and a capacity of 140,000 pounds of iced fish. Built in Seattle in 1914, it was owned and operated after 1922 by Captain Olaf O. Hvatum. It originally carried eight dories, but later it had six with 12 fishermen and the captain, cook and engineer. In the spring of 1926, during my research

FIGURE 40. The M.V. *Scandia* making a landing in Juneau on February 16, 1927, while under charter to the Halibut Commission.
F.H. Bell

that required many months at sea with vessels of the fishing fleet, I accompanied the *Scandia* on a dory-fishing trip to Shumagin Islands.

Subsequently in June the commission chartered the *Scandia* after competitive bidding to conduct a series of successful summer tagging trips off the coast of British Columbia. The longline method of fishing was used because the fish had to be brought on board alive for tagging. During the following fall and winter the *Scandia* was again chartered by the Halibut Commission.

With the closed winter season in effect, the vessel would be operating alone in the stormy waters of the Gulf of Alaska. It was therefore decided to place a radio receiver and transmitter on board and employ a radio operator. Learning that the 13th United States Naval District, Bremerton, Washington, had on hand some obsolete World War I spark transmitters, arrangements were made to have one such set transferred to the United States Bureau of Fisheries, a domestic agency, which then loaned it to the Halibut Commission, an international agency. The Bureau of Fisheries was charged $1 by the Navy, which for the Halibut Commission later proved to have been a fortunate investment. Commander Bouson of the United States Navy Radio Laboratories, Bremerton, personally assisted the commission with the transfer and the installation of the equipment aboard the *Scandia.* It was not until many years later when radio telephone became available that halibut vessels carried communication equipment, although the S.S. *Grant* carried a spark set in 1911.

After installation of the radio, the vessel departed for a fall tagging operation on the spawning stocks in the Gulf of Alaska. During January, 1927, hydrographic and plankton studies were conducted as far west as Kodiak Island. The *Scandia* returned to Juneau and after reoutfitting and clearing the ice, the vessel departed on February 18, equipped for fishing and tagging operations west of Kodiak Island between Trinity and Chirikof islands. This author was senior scientist accompanied by three members of the small staff of the commission and the 10-man crew. After enduring heavy weather in the Gulf of Alaska, the vessel stopped briefly at Seward for incidental supplies and water. Although the winter storm continued, the *Scandia* departed Seward early on February 23, for the Trinity Islands fishing grounds. The storm worsened and, rather than battle the mountainous seas, it was decided to seek shelter in Kodiak. Winter darkness comes early in those latitudes. The course of the *Scandia* was confirmed by the bell buoy on Williams Reef. Full watch was kept on deck and in the pilot house, with everyone straining through the snow to pick out the buoy on Hutchinson Reef. The buoy was never heard through the now raging winter gale. It had been torn away by the storm, and was later found ashore some miles away.

Suddenly *Scandia* struck hard, shuddered, rose like a cork and fell upon the jagged basaltic pinnacles of the reef. Snow and blackened sky obscured all vision beyond a few fathoms from the stricken vessel. Seas swept over the ship and water flooded the engine room through a jagged hole in the hull. Fortunately, the storage batteries, higher in the engine room, remained above water temporarily.

It was immediately evident that the ship was doomed and would soon be torn apart on the rugged rocks of the reef. The 19-year-old radio operator and deck assistant, Paul Roegner, managed to reach the spark radio transmitter in the pilot house to transmit an SOS for help and the ship's location, before the batteries flooded.

The order came to abandon ship. Three seaworthy dories were nested aft as life boats. Their launch from the listing vessel into the heavy seas was difficult and dangerous. The first dory was swept away, and the second swamped before it could reach the lee side of the vessel. The last dory nearly swamped, but a crew member bravely jumped into the dory and bailed it out as it shifted to the lee of the doomed ship.

Evacuating the 15 men with one dory was now urgent. The men decided that eight would go toward shore, in the first load, six of whom would leave the dory outside the surf, to make their way to shore with life jackets. Two would return to the stricken ship to remove the remaining 7 men.

To guide the return to the ship in the snow-thick blackness, a crew member, Olaf Eide, went down into the deeply-flooded fo'c'sle and retrieved several hundred fathoms of buoyline from the floor locker.

In the meantime, the lone radioman on duty during a normally quiet night shift at the Naval Radio Station near Kodiak had removed his earphones before our call for help and was leaving the room to make coffee in the galley. However, the signal was strong enough in the earphones on the desk and still audible to the technician. He quickly returned to the receiver to detect the weakening distress signals as the batteries of the *Scandia* started to flood.

The message was telephoned to nearby Kodiak, but the position of the *Scandia* had been misread due to the faintness of the signal. That William "Bill" Stone of *Duncan No. 1*, a cannery tender for a clam operation in Shelikof Strait was in Kodiak, was a lifesaver in the literal sense of the word. If there was one man who had the courage and the local navigational knowledge to come out that night on a rescue mission, it was Bill Stone. He was also able to deduce the correct location of the *Scandia.* A number of halibut crews from Seattle and Ketchikan which had earlier sought shelter in Kodiak, also prepared to face the winter storm, despite their lack of knowledge of the unchartered reefs outside.

Bill Stone proceeded to where he figured that the *Scandia* was impaled upon the rocks, circled and worked back between the shore and the reef. Miraculously, the snow lifted briefly and Stone saw the first dory load. Alert to the dory's danger, Bill Stone went full speed to head it off from the shore, thus preventing certain death for the six who were to jump into the surf—the shore was nothing more than a sheer wall up from the sea. Stone took them aboard the cannery tender without mishap except for the loss of one finger tip by this author.

Continuing the evacuation plan, Conrad Knutsen, deep-sea sailor and experienced doryman, would not allow his brother Berger to accompany the dory back to the *Scandia* as he felt one member of a family was enough for such a dangerous mission. Another volunteered and the remainder of the original crew of 15 were safely removed from the vessel as large slabs of the heavy wooden hull were ripped off by the relentless seas and the foremast toppled.

The first reports reaching the states had the entire research staff of the commission, including those not on the vessel, along with the 10-man crew lost at sea. On arrival in Kodiak, a prompt cable from this author corrected the first reports, but not before Henry O'Malley, Commissioner of Fisheries for United States, in Washington, D.C., called Herbert C. Hoover, Secretary of Commerce, at his home to arrange a major search for possible survivors by the United States Navy and the Coast Guard.

The 15 persons would have perished that winter night of February 24, 1927, except for a succession of fortunate occurrences and the courage of several. Of the commission's staff on board, Richard VanCleve attained the height of his profession as Dean of the College of Fisheries, of the University of Washington. Harry Hansen, after this episode found freshwater fisheries research more attractive. Olaf E. Erikson, except for an interval as a fish and liver buyer in the industry, retired from the Halibut Commission after a service of many years. Paul Roegner, the radio operator, became active in the early radio broadcasting industry of Seattle.

Of the crew, Captain Olaf O. Hvatum continued in the fishery with the M.V. *Dorothy* (Figure 41). Olaf Eide became editor of the *Washington Posten,* second-largest Norwegian language newspaper in North America. Conrad Knutsen continued in the halibut fishery and was on several vessels that the commission chartered from time to time in subsequent years. Later he was master fisherman on one of the large research vessels of the United States Fish and Wildlife Service. Other crew members continued in their chosen field, halibut fishing. Many were encountered during the subsequent 43 years this author was with the Halibut Commission. The bravery and skill of Bill Stone continued to be uppermost in our thoughts, and the one dollar investment in the second-hand Navy radio spark set never regretted.

The Seine-Boat Fleet

By the mid-1920s it was evident that the dominance of the schooner was in decline. Although the schooner possessed flexibility, it could not adapt sufficiently to meet the changing conditions and opportunities in the fisheries. The last standard schooner was built in the late 1920s.

FIGURE 41. The M.V. *Dorothy,* formerly the *Tyee.* One of the largest schooners in the halibut fleet.
International Pacific Halibut Commission

There was a need for vessels that could participate in more than a single fishery, longlining. When most of the winter halibut fishing stopped by the three-month closure by the commission, the superior capability of schooners to survive bad weather could be sacrificed. Also, schooners had higher mast and deck rigging maintenance costs than did the seine boats. Crew quarters and galley on schooners were confined to below water level fo'c'sles or to after cabins compared to the more comfortable deck-level facilities on many of the seine boats.

The development of the multi-purpose or combination vessel as it was called was also stimulated by the expansion of the seine-net fishery for the California sardine. In the early 1930s several units of the halibut fleet outfitted for this fishery. Even a few schooners, one in particular, the *Clipper*, underwent a major conversion to a seine-boat type (Figure 42).

In Canada the combination vessel reached a high point of development. The great expansion in British Columbia of the winter herring fishery during World War II resulted in many large and seaworthy vessels being built in the post-war period that fished halibut in the late spring to early fall before engaging in the profitable winter herring fishery.

The availability of such year-round and profitable occupation of the capital invested was the main reason for the growth in the Canadian share of the annual halibut catch allowed by the commission. Some of the vessels had a federal subsidy.

The combination vessel *Silver Dolphin* (Figure 43) was a larger boat and specifically designed to be multipurpose. Such vessels were well suited not only for setline halibut fishing but also were most effective in the salmon and herring seine net fisheries. Some of these boats were responsible for large halibut fares. A few such catches were that of the *Northern Dawn* with 195,000 pounds from Area 3 in 1966; that of the *Silver Bounty* of 154,000 pounds from Hecate Strait in 1955; and that of the *Northern Dawn* of 189,000 pounds from Western Alaska in 1962. The foregoing vessels carried eight to nine men.

The largest trip of the steamer era was that of 362,572 pounds landed in Seattle in 1912 from Hecate Strait by the S.S. *Independent* which had a crew of 37 men with 14 dories.

In addition to the large combination vessels, there were many smaller vessels fishing chiefly in Area 2. They carried up to six men and a number of them could and did convert to salmon seining in the summer. In Southeastern Alaska the boats had an "Alaska limit," namely under 50 feet in length to meet the legal length placed on salmon seiners in that part of the state.

FIGURE 42. The M.V. *Clipper*
International Pacific Halibut Commission

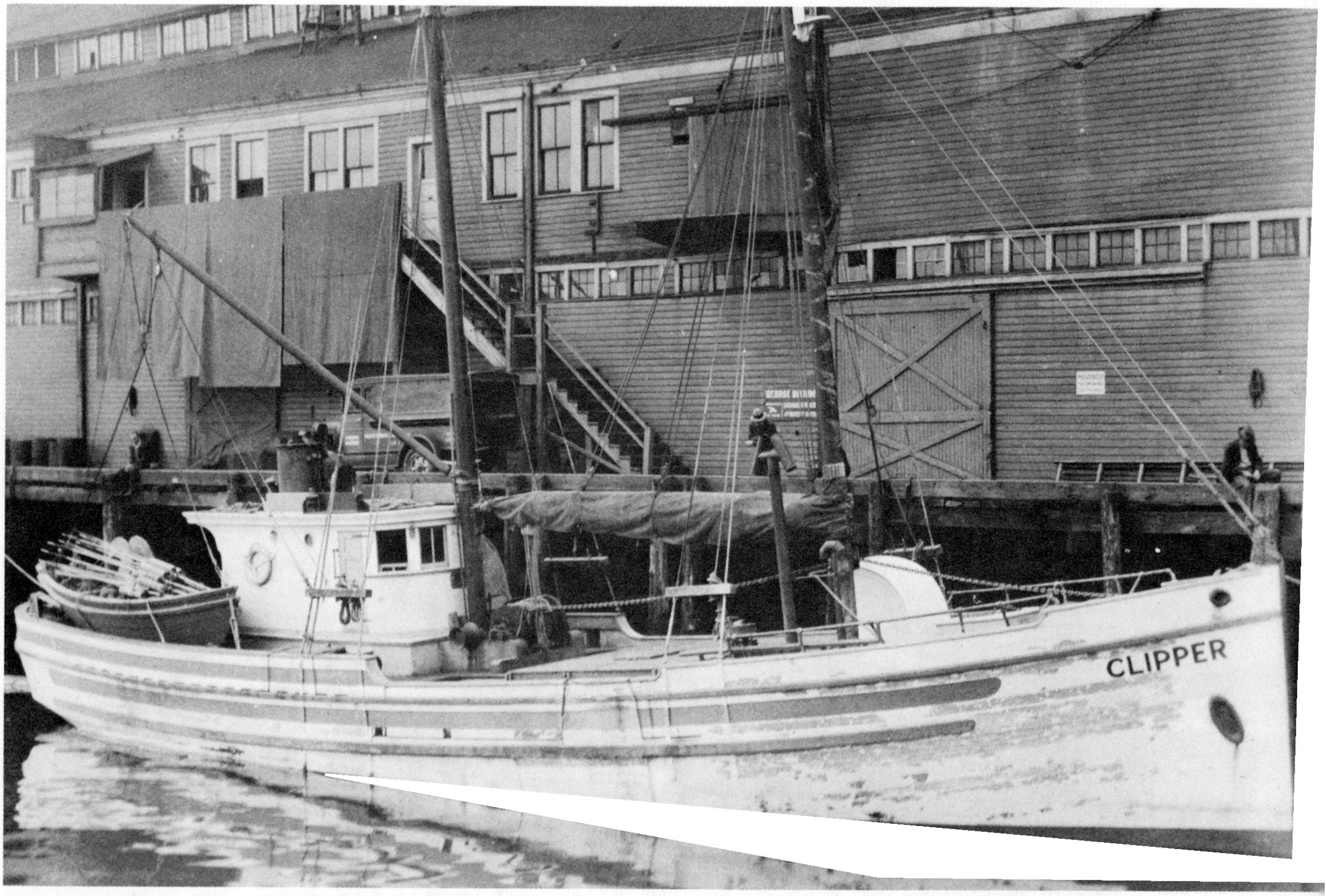

FIGURE 43. The *Silver Dolphin*, a modern steel seine-boat type with pilothouse forward and covered baiting and setting area on stern.
International Pacific Halibut Commission

FIGURE 44. The *Presho*, typical of the small seine-boat vessels that carry 5-7 men and were for many years the dominant type vessels in use south of Cape Spencer, Alaska.
International Pacific Halibut Commission

There were a number of rather nondescript vessels that were not adapted to seining that engaged in salmon trolling or setlining for blackcod after closure of the halibut fishing season.

As a curiosity, one could include in the category of nondescript boats the *Puzzle* (Figure 45) of Prince Rupert — probably the smallest functional setliner in the halibut fleet.

In the early 1930s, undaunted by the times, *Puzzle's* enterprising owner successfully converted little more than a large rowboat into a setliner, stern chute and all. How he functioned on such a craft was indeed a puzzle, but function he did catching fish on close-in grounds.

FIGURE 45. The *Puzzle* of Prince Rupert.
International Pacific Halibut Commission

The Small-Boat Fleet

In addition to the various types of larger vessels that have made up the regular halibut fleet, there are a number of small boats usually under five net tons that may fish halibut with a variety of hook and setline gear. They tend to operate during the early part of the halibut season prior to the salmon fishery. Many are from the salmon gill-net fishery. When fishing halibut they may carry from one to three men.

In addition to these small boats fishing setline gear, very large numbers of salmon trollers may catch varying amounts of halibut incidentally to trolling for salmon. The amounts caught will vary from one fish up to many thousands of pounds depending upon the area trolled and the methods of fishing. Most of such halibut are sold along with their salmon catches although many trollers will use the occasionally-caught halibut for personal use. Much of the poundage is landed at fishing camps serviced by packing or collecting vessels out of the major ports.

The small-boat fleet was of negligible proportions until the late 1930s when a successful cooperative in northern British Columbia contracted for the catches of a considerable segment of the Canadian halibut fleet. To counter the effect of the loss to the open market of this captive fish the independent dealers encouraged large numbers of salmon gill-netters to outfit for halibut prior to their salmon season.

The impact of the small-boat fleet on the fishery is best judged by the magnitude of its catches rather than by the numbers of boats involved. Recent annual reports of the Halibut Commission in the mid-1970s attempt to enumerate the size of that fleet and includes all boats even though many may sell but a few random halibut in a year. While this results in very impressive totals reaching into the thousands, it is a deceiving census figure. Prior to 1970, the small-boat fleet was canvassed in detail from time to time. By using certain objective criteria for a small boat to

TABLE 6. Landings From Area 2 in Thousands of Pounds Eviscerated, Heads-off, made by Small Boats, Including Salmon Trollers

Year or Period	United States	Canada	Total	Percentage of Area 2 Total
1931	413	469	882	3.7
1934	396	572	968	4.2
1939	765	1,231	1,996	8.6
1943	758	1,824	2,582	9.2
1950-54 (average)	931	4,183	5,114	15.5
1955-59 (average)	1,564	3,192	4,756	15.1
1960-64 (average)	1,809	3,484	5,293	19.8

qualify as a halibut boat, it was estimated that the small-boat fleet, not including those trolling for salmon, increased from about 125 in the early 1930s to 650-700 by 1952 and later years in Area 2. There was no small-boat fleet as such in Area 3 until a few started to show up in the mid-1960s. By 1976 it had attained considerable proportions, 343 being recorded for that year (IPHC Annual Report, 1976). This number probably included boats that landed an occasional halibut in the year.

During the decade of the 1950s the proportion of the total catch in Area 2 taken by the small-boat fleet including trollers fishing primarily for salmon remained quite stable, about 15% annually. There was a sharp increase in the catch of United States small boats with an offsetting decline in the Canadian in the late 1950s. Canadian companies maintaining camps for the small-boat fleet were finding the fish so produced too expensive.

The development of the small-boat fleet created problems in designing lay-in programs under the fleets self-imposed-between-trip tie up system. The small boats usually made daily landings and mostly at isolated camps. In addition there were the salmon trollers which accumulated their incidental catches of halibut over several days or even up to two weeks. The Halibut Commission at the request of the fleets made many studies so that equitable formula might be applied by the fleet in their lay-in programs each year.

With the high halibut prices of the 1970s and the availability of large numbers of small salmon boats prior to the opening of the salmon season the proportion of the Area 2 catch taken by small boats, including that taken incidentally while trolling for salmon, reached 27%. Indications were that the proportion would increase. By 1977 it resulted in the abandonment of the fleet's voluntary effort to spread the commission's catch limits over a longer season. Apparently it became impossible for the fleets to reach a consensus on a lay-in program that was agreeable to all elements involved including the growing small-boat fleet not only in Area 2 but in Area 3 as well.

Initially the regular halibut fleet attempted to discourage the development of the small-boat fleet by claiming that their catches were not reported to the commission, that the gear used was destroying small undersized fish and that all boats of any size should be licensed and be required to clear port after each landing. Repeated studies by the Halibut Commission found no grounds for such discriminatory claims, and they were rejected. On the other hand the commission fully supported the efforts of the fleet to extend the length of the season, an apparent ambivalence that was not always understood by many people.

Size of the Regular Halibut Fleet

The size of the fleet as measured by the number of vessels and manpower involved is useful from an economic or census point of view but is not necessarily a satisfactory measure of available fishing

TABLE 7. The Average Annual Number of Canadian and United States Vessels Over Five-net Tons in Halibut Fleet and the Manpower of Respective Crews

	Vessels			Crew*		
Period	United States	Canada	Total	United States	Canada	Total
1928-30	344	108	452	1,860	490	2,350
1931-35	314	105	419	1,615	427	2,042
1936-40	353	162	515	1,893	688	2,581
1941-45	374	178	552	2,004	823	2,827
1946-50	519	229	748	2,492	1,181	3,673
1951-55	439	245	684	2,138	1,295	3,433
1956-60	387	186	573	1,736	1,025	2,761
1961-65	282	207	489	1,222	1,054	2,276
1966-70	217	164	381	884	851	1,735
1971-74	189	135	324	741	648	1,389

*Including Captain

TABLE 8. Number of Vessels and Men by Sections of the Coast Where Their Home Ports Were Located

	Number of Vessels			Crew		
Locale of Home Port	1960-64	1965-69	Decline	1960-64	1965-69	Decline
Seattle	111	72	35%	753	407	46%
Southern British Columbia	85	76	11%	541	469	13%
Northern British Columbia	122	92	25%	524	410	22%
Alaska	194	147	24%	700	499	29%
Total U.S. Flag	305	219	28%	1,453	906	38%
Total Canadian Flag	207	168	19%	1,065	879	17%
Grand Total	512	387	24%	2,518	1,785	29%

power. For the latter purpose adjustments must be made for the size of the individual vessels, the many technological improvements as well as the length of time each vessel may fish each year.

After the end of World War II the United States fleet of regular halibut vessels reached a maximum of slightly over 500 vessels manned by about 2,500 fishermen. During the next 30 years that fleet declined by the first half of the 1970s to somewhat under 200 vessels manned by about 750 men, a 60% decline in vessels, a 70% decline in manpower. The Canadian fleet reached a maximum of 245 vessels in the early 1950s manned by about 1,300 men. By the mid-1970s the Canadian fleet numbered 135 vessels with 650 men. This was a 45% decline in fleet and a 50% decline in crew.

Some of the sharpest changes that have occurred in the Pacific halibut fishery took place in the decade of the 1960s. There were major reductions made in the annual catch limits allowed the Canadian and United States fleets due primarily to the increased foreign fishing and possibly to some decline in recruitment. This situation had a major impact upon the size of the fleets.

The fleet size, including crews of Canadian and United States vessels, are shown in Table 8 for the first and second halves of the decade of the 1960s by sections of the coast where the home ports of the vessels are located.

Though the Canadian flag vessels during this decade do not show the long term trend to smaller crews, it is noticeable in the United States sector. The reduction in crew size over the past 30 years has been made possible by the application of many technological improvements in the halibut fleet.

After 1974, the category of vessels heretofore classed as regular halibut vessels was amended to include all vessels of 5 net tons and over, licensed for the halibut fishery. Such a category is meaningless as a census statistic. It could be merely a measure of how diligent officers are in issuing licenses in particular years to the small 5- to 10-ton class of boats. Many small boats secure such no-cost licenses and may only catch a few hundred pounds, or even less, of halibut before engaging in other pursuits depending on conditions which vary from year to year. Equally meaningless are the numbers of unlicensed boats that include boats that may have sold only a single 10-pound halibut in the year. The numbers also exclude many hundreds of trollers who eat all the halibut they catch!

In the 1974 Annual Report of the commission the catch per trip of each class of vessel is given. Such a statistic is practically meaningless as the data of Areas 2 and 3 are not separated. It was also noted that the catch per trip was higher in 1974 than in 1973 due to "reduction in gear competition." The latter statement is one for which there is no evidence.

3 FISHING METHODS

The Pacific halibut fishery is a hook and line fishery, although any type of gear or device may be used to catch halibut except those involving nets of any description. The latter prohibition was instituted in 1944. The commercial fishery has used setlines in which long lines with baited hooks attached at intervals are set on the ocean bottom. There are other types of hook and line gear including salmon trolling gear and handlines that catch halibut with variable effectiveness.

Halibut setlining can be divided into two main types, dory fishing, in which the gear is set and retrieved from dories or small boats that are carried and set out by a "mothership," and longlining, where the gear is set and retrieved directly by the erstwhile mothership.

Since the only way to learn to fish halibut is to "ship out" on a commercial halibut vessel as an "in breaker" or an apprentice, usually on half share, no attempt will be made to include here an instruction manual on halibut fishing. It would be a waste of time, to say the least.

DORY FISHING

Dory fishing after the type used in the Northwest Atlantic was the first method to be used on the Pacific Coast. The *Oscar and Hattie* and companion vessels the *Mollie Adams* and *Edward E. Webster,* which inaugurated the fishery in 1888, carried six dories each manned by two fishermen. Some small sloops and schooners carrying two to four dories continued the fishery into the twentieth century. The building of the *New England* in 1898 ushered in the company-owned steamer era, all fairly large vessels that fished from 8 to 14 dories depending upon their size. A few such steamers in Canada, brought from England in 1913, were equipped for longlining in the Fleetwood, England, tradition and continued to do so in Canada.

By the second decade of the twentieth century, independently owned schooners propelled by internal combustion engines and not sail entered the fishery in numbers. They carried five to eight dories usually in one but sometimes in two nests on either side of the stern. The steamer era waned by the end of World War I and owner-operated internal combustion schooners carried on the dory fishing tradition until gradually shifting to longlining. In the transition years some vessels would fish dories in the summer on grounds south of Cape Spencer then longline west of the cape during other seasons of the year.

FIGURE 46. Putting the dories overboard from a steamer at half-speed. Each dory is set in a line of dory "berths" so many minutes run of the "mothership" apart, depending upon the grounds being fished.
International Pacific Halibut Commission

Dory fishing was well suited for the more prolific stocks in the early years of the fishery. Gear could be concentrated on the then more dense congregations of fish. There was also some economic advantage to the vessel owner whose proportionate share of the proceeds from dory fishing was somewhat higher than with longlining. Dory fishing was much more arduous than longlining although the number of hours on deck fishing each 24 hours was usually less. The dories were very seaworthy in the hands of the experienced, but the occupation was fraught with many dangers. Numerous lives were lost in the stormy waters of the North Pacific particularly during winter fishing prior to the closed winter season that was instituted in 1924 under the 1923 Halibut Convention.

FIGURE 47. A dory at work. The line is taken over the gurdy, the fish thrown in the landing net, and the gear coiled in front of the standing fisherman. The dory in the next berth can be seen in the distance.
International Pacific Halibut Commission

As the stocks declined in the 1920s and as the inclination to undergo the hazards and rigors of dory fishing lessened, the changeover to longlining accelerated. The approximate proportions of the total coast catch taken by dory vessels declined in the following manner: 1920, 60%; 1923, 40%; 1926, 15%; 1929, 5%; 1931, 1%.

In Area 2 from 1924 to 1929 an annual average of 14 dory vessels fished some or all trips with dories for an average annual total catch of 2,745,000 pounds. From 1930 to 1933 the respective annual averages were six vessels and 1,371,000 pounds. In 1934 three dory vessels caught 296,000 pounds in Area 2 of which one vessel the *La Paloma* accounted for 211,000 pounds.

Investigations by the Halibut Commission indicated

FIGURE 48. Preparing to set.
Puget Sound Maritime Historical Society

FIGURE 49. A dory on board the mothership. Note the rubber-covered sheave on each side of the dory for reeling in the line and the bearings attached to the rails of the bulwark.
International Pacific Halibut Commission

that dory fishing tended to be concentrated particularly on grounds south of Cape Spencer that were characterized by small fish, those below the optimum harvesting size for the species. In 1933 the commission announced that dory fishing would be prohibited in Area 2 (south of Cape Spencer) after 1934. This notice period permitted the last few remaining dory vessels in Area 2 to phase out their dory gear. In 1944 the prohibition was extended to all convention waters. The latter action for all practical purposes was academic as all regular dory fishing had terminated in 1934. Infrequently after 1934, an occasional vessel fishing in Area 3 would use a dory to set a string of

FIGURE 50. A fisherman hauls a halibut over the rail, one step in the longline system of halibut fishing. As the ground line is reeled onto the boat by a gurdy amidship, it is drawn over the roller at the rail, shown at the lower left corner of the photograph. The halibut is hauled to the surface of the water, attached by a ganging and hook to the line, and the fisherman hooks it in the jaw with a gaff to haul aboard. The ganging can be seen running from the mouth of the fish back to the ground line. A large fish would require the help of the crew member that would otherwise be busy coiling the line into skates. Note the strength visible in the musculature of the halibut.
International Pacific Halibut Commission

FIGURE 51. One of the first attempts at longlining or fishing directly from the deck. This is the *James Carruthers* off Yakutat in 1913. Note the ice on deck, and the fishermen helping the line overboard. The stern chute came into general use shortly after this picture was taken.
International Pacific Halibut Commission

their longline gear in water too shallow for the vessel to navigate.

Dory gear used lighter ground line than that in longlining as it was not subjected to the strains in being hauled aboard the larger "motherships." The hooks were attached at intervals of 9 feet which permitted the man coiling the gear to be free of handling a hook when the gurdy-man might be occupied with clearing a halibut off the next hook. The 9-foot intervals also made the coils of gear a manageable diameter and caused the beckets holding the gangings to the ground line to line-up, one above the other, in the coil. The 13-foot, 18-foot and wider spaced intervals used by longliners also resulted in the line-up of beckets in the coiled skates. Such conformity facilitated coiling when hauling and baiting.

LONGLINING

Longlining was introduced to the Pacific Coast shortly before World War I. It is probable that the S.S. *James Carruthers* of Prince Rupert, B.C., under Captain Knighthall, formerly of Grimsby, England, was one of the first large vessels to use the typical longlining method. This was in the tradition of the "liners" that fished for demersal species including halibut out of Fleetwood, England. The *James Carruthers* made its first such trip in September, 1913. Small vessels had used the method several years earlier.

The following quotation from Report No. 5 (1930)

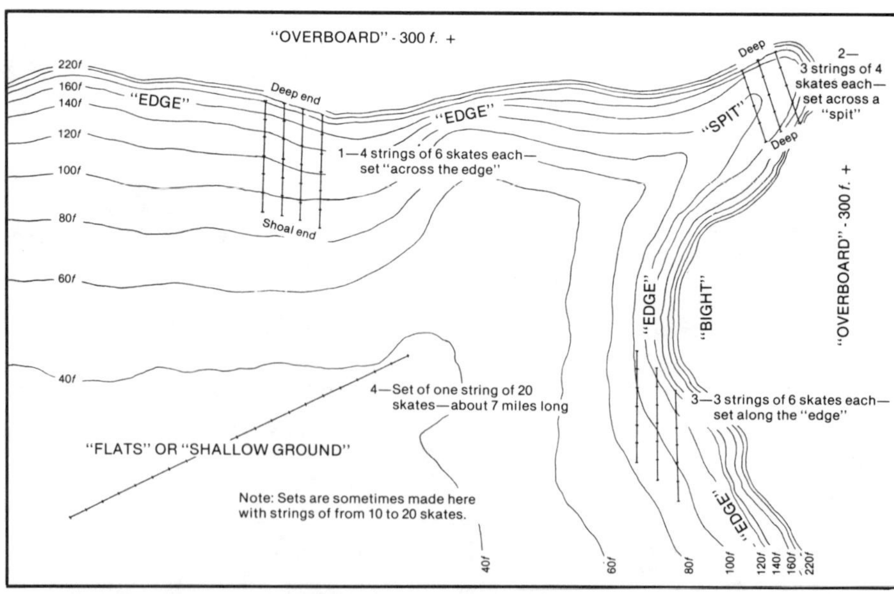

FIGURE 52. Diagram of manner in which strings of halibut gear are set on different types of grounds, depths shown by contour lines at 20 fathom intervals out to 220 fathoms.
International Pacific Halibut Commission

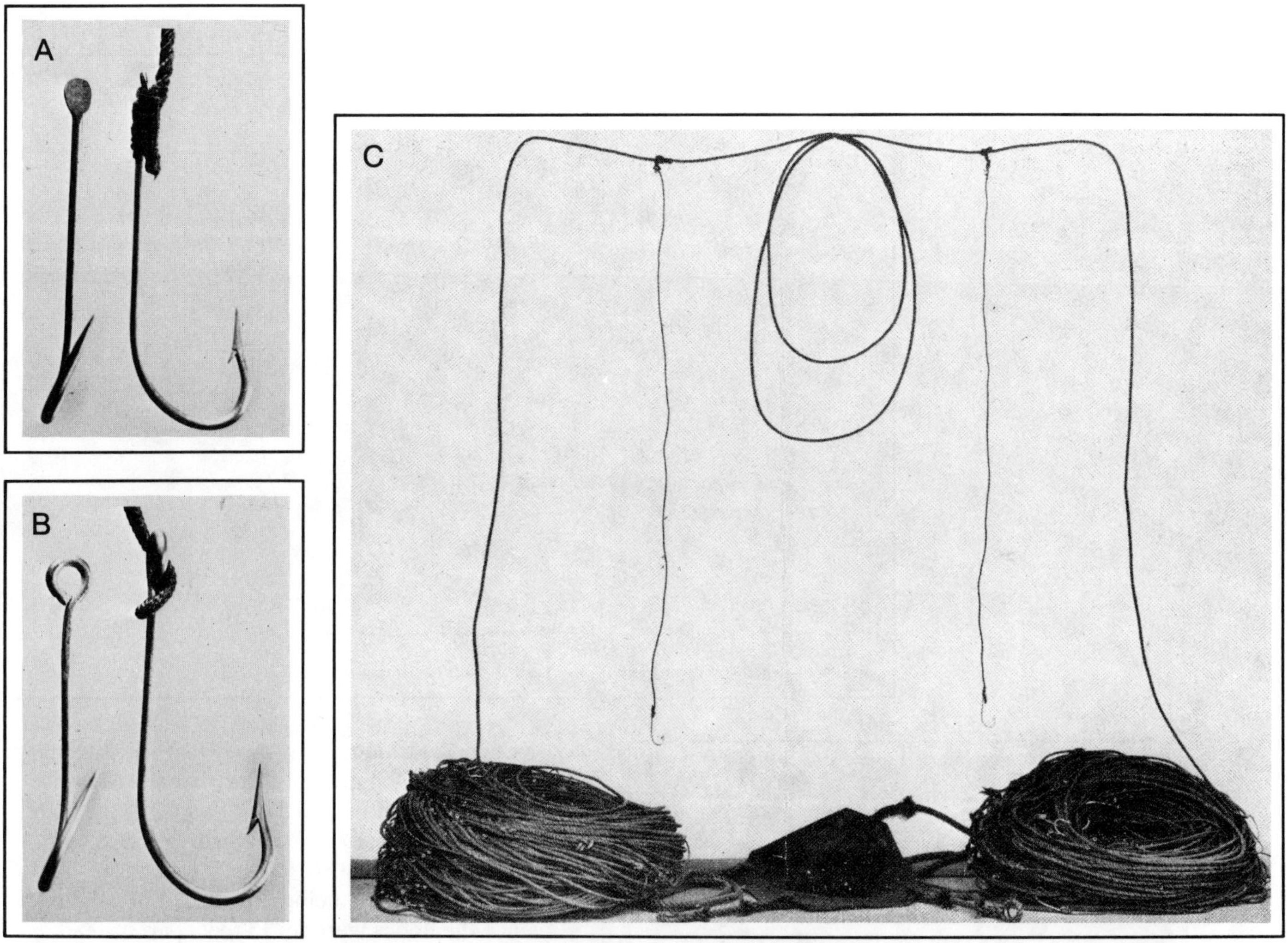

FIGURE 53. Various features of setline gear and setline fishing. (A) Offset hook, ganged or bound to hemp ganging with pine and tar-treated thread, used prior to 1965; (B) Offset eyed-hook, used since 1965, held to loop of nylon ganging; (C) Skate of gear, partly coiled, extended to show ganging and hook; (D) Coiling a skate; (E) Hauling gear over rail roller with power gurdy or winch.
International Pacific Halibut Commission

FIGURE 54. Diagram of halibut gear as set.
International Pacific Halibut Commission

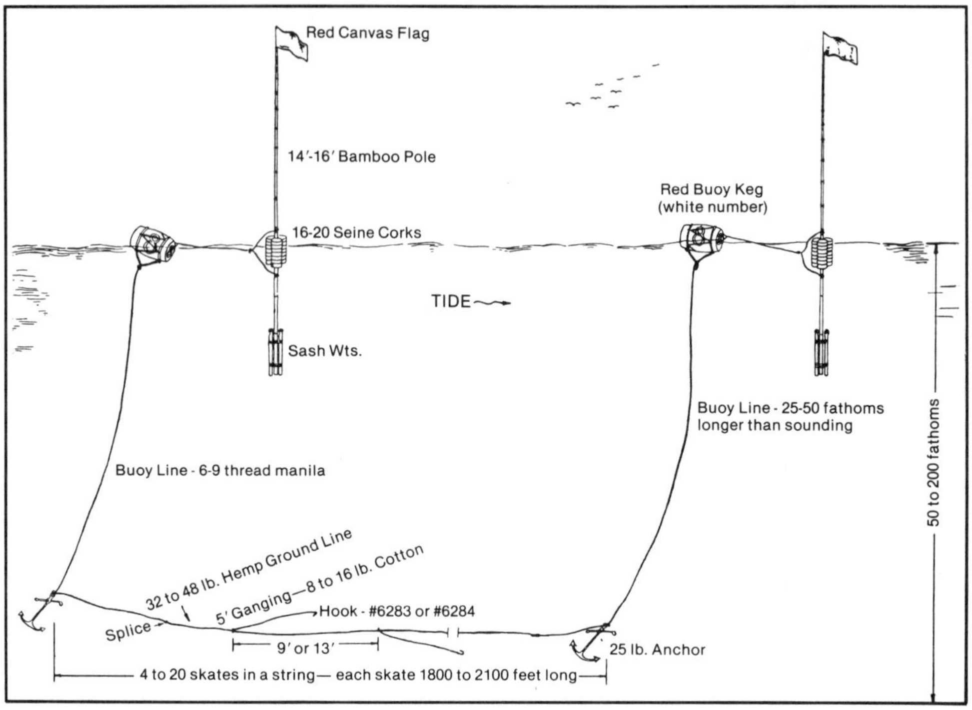

FIGURE 55. A gaff used to hook the halibut in its jaw in order to haul the fish out of the water and onto the vessel.
Christine Olsen

of the Halibut Commission contrasts the two methods of fishing, dory fishing and longlining.

> This method [longlining] made many changes in the established dory fishing, and it was rapidly improved. In the first place, it required a different method of laying out the gear. Whereas before the dories were set out in a line, and parallel strings of two or three skates each set, now this need no longer be the case, and setting the gear was done as convenient, not all at once, nor in a block. The number of men was reduced. Fishing in worse weather became possible, although many dory men will rather doubt this even yet. To avoid danger to the man setting the gear at the stern, a chute was invented over which the line was raised and paid out without being touched. For convenience the hooks were set 13 feet apart instead of 9, as was best in the dory. The change in number of hooks was not a distinct loss, as the catch when fish are somewhat scarce is, after all, determined by the extent of ground covered than by the number of hooks, and the long-line method of handling the gear gave very nearly as high an average catch per skate as when the 9 foot dory gear was used. In this case, too, it is hard to convince the dory man that his gear does not in the long run catch more than the long-line gear, and only the averages of many catches of both types of boats can be relied upon. In large schooners the long-line method has not completely displaced the dory, as our records show, but the change has been complete in the smaller vessels and threatens to become so in the larger. In the end a more efficient type of vessel has been developed, one that has been able to fish 24 hours per day when needed, even in the darkness of the long winter nights of Alaska, and it has become possible to handle the gear much more rapidly.

The basics of the regular setline halibut gear and methods of setting as they were in the 1920s are shown in Figures 51 to 54. During the succeeding 50 years many changes were made in the materials used and in the design of some of the components, but the basics remain unaltered, namely, ground line, beckets, gangings, offset hooks, buoy line, buoy floats and flag poles.

In "snap-on" gear, which came into wider use in the 1970s, the gangings are clamped on the ground line as it comes off a drum when setting. Small boats and medium-sized vessels tend to use this gear more than the larger vessels. It is a development from the stainless steel gear introduced in the 1950s, but was used by only a few of the smaller vessels, usually where the crew were all part-owners. The gear possessed handling problems as well as cost-sharing complexities. The initial cost was relatively high which under the union lay is borne solely by the owner. Depreciation was low, a cost which is borne by the crew.

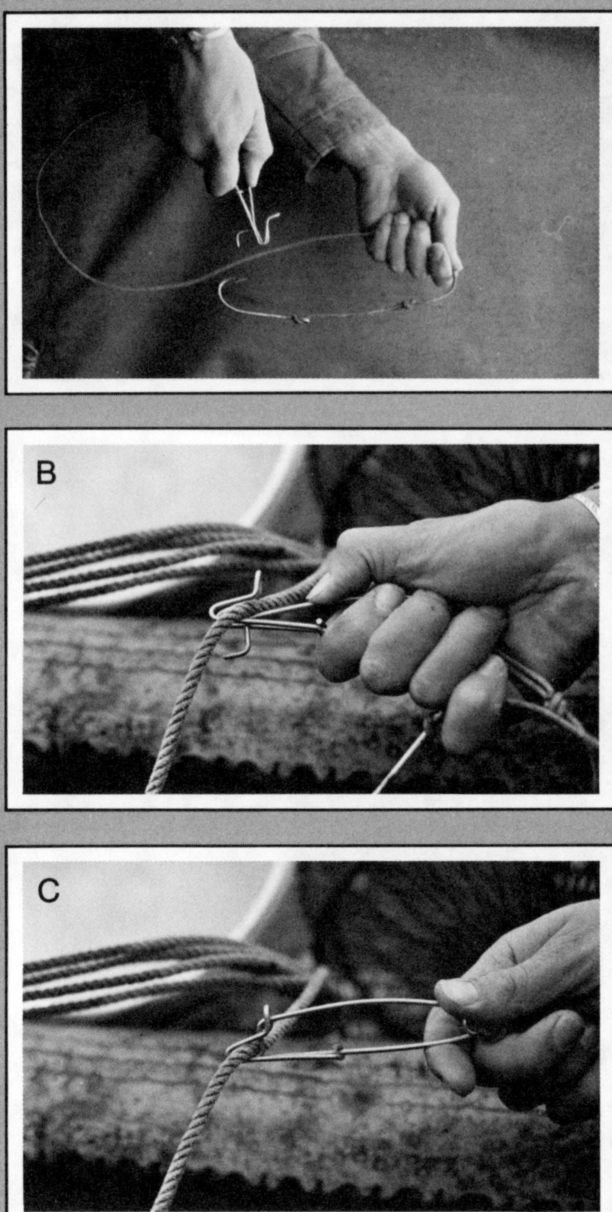

FIGURE 56. Snap-on gear. (A) Snap, ganging and hook; (B) Snap ready for attachment; (C) Snap attached to the ground line.
International Pacific Halibut Commission

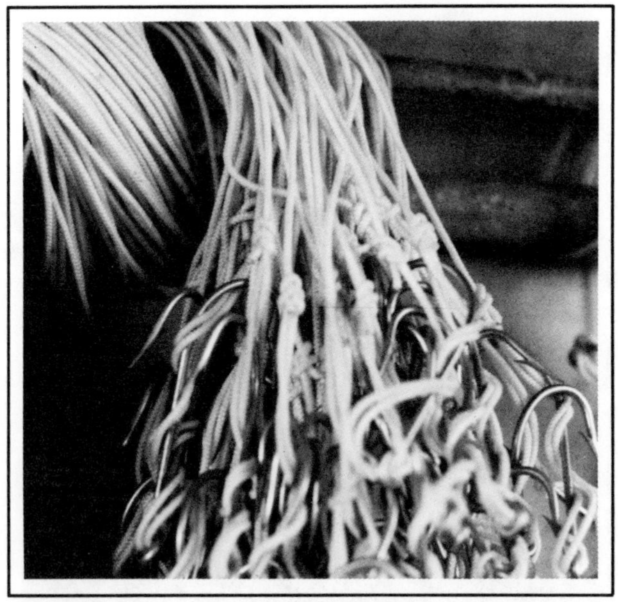

FIGURE 57. A "bunch" of nylon ganging and steel hooks such as this is placed near the incoming ground line so that frayed ganging and bent hooks can be readily replaced.
Christine Olsen

At the present time the ground line is usually nylon instead of a succession of natural fibers as in the past depending either upon their availability or the choice of the individual vessels. The ground line now comes in continuous coils of any required length eliminating the former splicing of 50-fathom lines to make up a skate of any desired number of lines. Beckets are spliced into the ground line at the desired intervals as heretofore and gangings are looped into the becket. Offset hooks are still used but are now eyed which dispenses with having to seize the shank of the hook to the end of the ganging with linen thread soaked in pine tar. Ganging hooks was a tedious and time-consuming task, and some fishermen were more adept than others at the task.

Buoys for the buoy line are now inflated bags of plastic materials rather than wooden kegs and the flagpoles are now of light metal alloys rather than bamboo. Plastic floats on the pole have replaced the

FIGURE 58. Over the last half-century elements of halibut longline gear have remained basically the same yet with some modifications. Inflatable plastic bags are one form of buoy that have replaced wooden kegs, and the bamboo poles shown, although not as modern as metal poles, do show the use of styrofoam rather than cork as floats.
Northwest & Alaska Fisheries Center

FIGURE 59. Now commonly used hard buoys.
Christine Olsen

FIGURE 60. An element of gear that is becoming more common in the Pacific halibut fishery is "tub" gear as shown in this photograph. Tub gear is used widely in halibut fishing off the East Coast of North America, where the ground line is not coiled into skates. That storing line and hooks in the tubs is easier than coiling them may be one reason it is seeing growing use in the Pacific halibut fishery.
International Pacific Halibut Commission

FIGURE 61. Christmas tree gear also replaces the need for coiling skates and probably reduces labor costs by increasing the number of hooks that can be set per man per day. It is not as widespread as drum gear in regards to new types of gear found in the Pacific halibut fishery.
International Pacific Halibut Commission

FIGURE 62. The stern chute, used in setting ground line.
International Pacific Halibut Commission

FIGURE 63. Drum gear is used on many small boats fishing for halibut and is coming into more general use on the medium-sized vessels. (A) Storage of hooks; (B) Common deck layout showing ground line drum and hooks; and (C) the ground line being set.
International Pacific Halibut Commission

natural cork. In addition to using lights on the buoy poles for night fishing, some radar reflectors are used for fishing at night as well as in heavy fog. The foregoing changes in the gear are a reflection of the advancing technoloy of which the fleet has taken full advantage. They are only a few of the many advances that deny the widely held myth of some economists and others that the setline halibut fleet was a decadent, obsolete and economically expendable entity. The improvements have benefited the vessels' owners and the crews who are joint shareholders in the enterprise.

4 PRESERVATION METHODS

NATIVE METHODS

Preservation, chiefly for subsistence purposes, by the native peoples consisted of air drying supplemented with smoking in wet weather to reduce the water content of the flesh to the point where bacterial growth and other spoilage factors would be inhibited. The large fish were first sliced into very thin fletches and piled under heavy stones which pressed out much of the "free" water before hanging out to dry. Such "water horsing" is still done in the commercial dried fish business. The smoke itself may have also conferred some chemical bactericidal factors sometimes collectively described as pyroligneous acid. Smoked halibut is still an article of commerce, but today it is typically frozen or refrigerated to prolong its shelf life.

SALTING

The earliest commercial halibut fishing on the Pacific Coast consisted of fletching trips that involved the salting of the large lengthwise fillets or fletches (flitches) of halibut in the hold of the ship. This method was brought from the Northwest Atlantic where it was used by sailing schooners out of New England ports in the distant water halibut fisheries of the Grand Banks, Davis Strait and Iceland.

The first fletching trip on the Pacific Coast was by the *Mollie Adams* in 1888. That vessel along with the *Oscar and Hattie* and *Edward E. Webster* in 1888-89 all made protracted salt fletching trips as far north as Dixon Entrance and undertook some fresh halibut trips, usually to Cape Flattery grounds.

A limited amount of salt fletching was continued through the decade of the 1890s and several companies engaged in the business as late as the early 1900s. On March 7, 1901, the Pacific Fish Co. of Nanaimo, B.C., dispatched the *Nellie Thurston* on a fletching trip. Concern was felt for the vessel as she was not heard from until found ashore four months later on the 1st of July in Skidegate on the east coast of the Queen Charlotte Islands. She had 160,000 pounds of salted fletches on board, taken off Cape Scott. In 1904 the Nanaimo Halibut Co. established a shore-based halibut saltery at Skidegate and shipped the product south by the *Nellie Thurston* and other freight steamers. The Western Canadian Fish Co. of Barnett, B.C., in 1903 fitted out the brigantine *Blakeley* to cure (salt) cod and halibut. All these projects were fairly short-lived.

With artificially produced ice and freezing facilities becoming generally available after the turn of the twentieth century, halibut salting soon became a process of the past. Yet as late as the mid-1920s old-time fishermen on halibut vessels could be observed salting down a keg of halibut for personal use.

SMOKING

While the volume of halibut that is now preserved commercially by smoking or "barbecuing" is not large, it continues to have some market acceptance in North America. Since World War II the amount produced in United States has reached as much as 500,000 pounds of product weight annually. It is usually frozen to extend its keeping qualities. The smoked walls of the body cavity or "bellies" are an acceptable product and more reasonably priced than the main body flesh.

CHILLING

The use of some refrigerant such as water ice to keep the fish fresh while not freezing it has been engaged in since the fishery first began. In the very early years the high cost of ice or its scarcity greatly impeded the development of the fishery. Vessels out of Puget Sound sometimes went far afield to secure glacier or lake ice. In 1888 the *Oscar and Hattie* had to sail 800 miles to Glacier Bay, Alaska, for a load of ice in order to fish on grounds far to the south.

Later, around the turn of the century the availability of glacier ice along with abundant halibut stocks in protected inside waters led to a substantial winter fishery in Southeastern Alaska by Puget Sound vessels. The fishery out of Vancouver, Victoria and Tacoma in the 1890s was also conducted in winter months. Not only was the market in the East stronger due to winter curtailment of fishing on the offshore banks of the Atlantic, but the halibut from the Pacific Coast arrived in better condition with the colder transcontinental weather en route. Also, lake ice and snow were available to the vessels for stowing their catches and to the railroads for filling the car bunkers. Artificially produced ice soon became generally available, although one company in Prince Rupert, the Atlin Fish Co., cut and stored lake ice well into the 1920s. Such use continued in Wrangell until 1923.

The halibut fleet has directed a great deal of attention to the quality of ice used. "Green" ice that was recently frozen and not yet "cured" to a low temperature of at least 20 °F. was to be avoided at all costs. Flake or drum-frozen ice developed after the Second World War alleviated the green ice problem. Such ice, when "peeled off" the rotating drum, is about 20 °F. and requires little or no curing through further reduction in temperature. Drum ice was uniformly fine and light and reduced the bruising of the fish flesh. The production unit was very compact compared to that required using ice that had been frozen in blocks then crushed.

The use of refrigerated sea water (brine) rather than ice to chill but not freeze halibut was the object of extensive study in Canada during the 1950s. While there were many technical advantages for its use, it was not generally accepted by the halibut industry because the process gave a leached-out appearance to the skin as well as off flavors. By the late 1970s some of the objections were alleviated and a number of halibut vessels found the method effective and labor saving.

The installation of mechanical refrigeration in the holds to lower temperature and conserve the ice without freezing the halibut has been an important development. The first installation in a halibut vessel occurred many years ago. Today most vessels are so equipped. The equipment had been in general use in the European distant waters fisheries. Large scale experiments in England had shown, however, that good hold insulation was even more important, although refrigeration did lower the temperature of the fish a further 0.8 °C over using ice alone. Trawlers out of New England and vessels fishing the warm waters out of California have also long used such refrigeration with a great saving in ice coupled with reduced spoilage. In the Pacific halibut fishery it permitted the fleet during the low stock levels of the 1970s to fish longer in distant waters and thus assure a trip of profitable size and acceptable quality. Such refrigeration would have helped the halibut fleet during the low stock levels of the late 1920s and early 1930s.

All the fisheries, including those for Pacific halibut, have experimented with additives as spoilage inhibitors. To date it seems that no additive has been particularly effective at sea. They are no substitutes for timeliness, cleanliness and properly iced fish.

Except for a brief period when considerable amounts of halibut were salted on the fletching trips of the nineteenth century, chilled or fresh halibut dominated the North American market until shortly before World War II. The successful continental distribution and merchandising of such a perishable product required keen management in the fishing industry. This management seems diverted to other corporate functions such as meeting the regulations and demands of federal, state and municipal governments.

To attain the continental distribution of the halibut catch and other perishables, many advances were made in refrigerated rail cars. The Fisheries Research Board of Canada contributed much to improving the rail transport of halibut and of fish in general. Refrigerated trucks with their greater flexibility for direct pickup and delivery soon began to displace the rail car and direct shipments were made to nearly all states of the union and all provinces of Canada. Air shipment is a growing trend for some products.

Many factors led to the declining proportion of the halibut catch being marketed as fresh fish. The enormous growth of the chain store and of frozen foods was the main reason for the shift away from the fresh to the frozen product. While the shortening of the halibut fishing season after 1924 contributed to the shift, it has been one of the minor reasons. The frozen fish trade was greatly stimulated by technological developments in freezing since the 1920s, leading to a greatly improved product.

FREEZING

Frozen halibut is the chief way in which the catch is now distributed. In many instances, it is previously thawed before sale to the consumer, unless it is a prepackaged product.

Frozen halibut was eviscerated and its head cut off, but otherwise left whole. It was frozen in "sharp freezers," glazed to reduce dehydration (freezer burn) and oxidation (rusting) and stored in large piles. The piles were sprayed with water to replenish the glaze on the exposed outside fish. They were shipped in wood boxes and distributed as whole fish to the retail trade, where they were butchered, usually into steaks.

In the course of time many improvements have been made in the procedures to freeze whole halibut by the sharp freezer method. The loading of sharp freezers, a very labor-intensive procedure, has been facilitated by the use of conveyors. The recent use of vacuumized plastic film bags instead of glazing was a major advance. Labor-saving palletized wire tins or cages are now used to store the fish. Shipping in fiberboard cartons instead of wooden boxes saves freight costs. Along with many other labor-saving changes there have been technological advances in the freezing process and in the storage of frozen fish, including the jacketed cold storage with lower and more constant temperatures. These have resulted in a greatly improved product for the consumer.

The major change in the merchandising of frozen halibut has been in the butchering and packaging before distribution. A large proportion of frozen whole halibut is now pre-sawed into steaks which are individually glazed and packed in small cartons for the retail trade. Large fish are frequently fletched before freezing into boneless strips that are then cut into pieces of uniform size and weight and frozen for what is known as the "portion pack." Halibut is a versatile species that can be presented to the consumer in many forms, as steaks, fillets, uniform portions, as chunks for roasting, or minced in fish sticks, uncooked or cooked and breaded in convenient frozen fish dinners.

CANNING

The preservation of halibut by canning in heat-treated vacuum-sealed containers has been attempted many times in the past 100 years, but in no instance has it been a commercial success.

One of the earliest attempts in North America was at Klawock in Southeastern Alaska where the first salmon cannery in Alaska was built in 1878. Along with the salmon pack that same year a 200-case lot of halibut was canned. It is rather coincidental that 66 years later in 1944 and 1945 Libby McNeil and Libby put up about 10,000 cases of canned halibut in the same general area. In 1944 they used the Borg process, a patented method involving royalties, to reduce the water content of the flesh with a centrifuge before processing. In the second year a steam process was used. A recovery rate of one case of one-half-pound flats for each 100 pounds of dressed, heads off halibut was attained. It was projected that the Libby's Taku, George Inlet and Yakutat plants would be used for halibut after or before the salmon season.

FIGURE 64. Fletching halibut.
International Pacific Halibut Commission

FIGURE 65. Frozen halibut sawed into steaks.
National Fisherman

However, this did not materialize, for like all attempts in the past, this one was not a commercial success.

In 1878 it was reported in Victoria that an Indian had landed 12 tons of halibut in two hours fishing off Cape Flattery, and that it was to be canned.

About 1881 in Stavanger, Norway, Mr. Richard Blaauw canned halibut in brown gravy with a bay leaf. He also experimented in the United States with frying the halibut before packing it in one-pound oval cans. A few orders were filled including one for 400 cases. However, the opportunities seemed greater in salmon canning and the halibut project faded. In 1945, a modest 200-case pack of presmoked halibut was canned at Tenakee in Southeastern Alaska. A later attempt of some magnitude was that by the Halibut Producers Cooperative, which packed a few hundred cases under the *White Wave* label in the 1960s out of Seward, Alaska. Another ill-starred venture was that of the Aleutian Canning Co. of Seattle in post-World War II years.

One of the problems to be met in canning halibut is reducing the water content of the flesh and replacing it with oil. Most products to date have too short a shelf life, developing undesirable organoleptic qualities such as gray or brown colors, dryness and strange flavors. Its white meat lacked consumer appeal when compared to the more robust and uniform flavor of the albacore tuna.

5 HALIBUT PORTS AND THEIR LANDINGS

Although halibut has been landed at about 150 or more places on the Pacific Coast, most of those locations have consisted of no more than buying barge or camp located in many of the hundreds of small-boat anchorages on the fjord-like coast. At such camps or barges most of the poundage was taken incidentally while trolling for salmon. The remainder was taken mainly by small one- or two-man boats setlining for halibut before engaging in the gill-net salmon fishery. The places or ports where vessels of the regular halibut fleet landed were relatively few in number. Until recent times the more important of such places were the railhead ports of Seattle, Vancouver and Prince Rupert. At times, Ketchikan, which had a cold storage since 1910, also played an important role. After World War II, locations in Central and Western Alaska became increasingly important as landing places for the fleets fishing west of Cape Spencer. In later years, some of those locations have far eclipsed the railhead ports in volume of their halibut receipts.

Much of the history and development of the Pacific halibut fishery is observable in the rise and decline in importance of the various centers where halibut has been landed. The very early development of each landing center can be described only in general terms. However, for the last 45 to 50 years, back to about 1930, and in some instances much earlier, there are statistics available for the more important landing ports. Landings on each broad section of the coast, namely California, Oregon, Washington State, British Columbia and Alaska, can be extended at least back to 1911, and earlier in some instances (Table 9).

As might be judged from the early development of the landing ports, various sections of the coast have dominated the fishery from time to time. The shift from one to another has been occasioned by such basic considerations as the completion of transcontinental railroads, location of productive fishing grounds, their depletion, their subsequent rebuilding and later decline, changes in the structure and motive power of the fishing vessels, the length of the fishing seasons, wartime conditions and many other economic as well as biologic factors.

Since fishing began on grounds west of Cape Spencer, Alaska, in 1913 (55,000 pounds were taken in 1912), the relative magnitude of the landings on any section of the coast was, to a large extent, determined by the amount landed from west of Cape Spencer by the larger vessels fishing that region. Their

TABLE 9. Average Annual Landings of Halibut by Sections of the Coast in Millions of Eviscerated, Heads-off Pounds

5-Year Period	CALIFORNIA and OREGON	WASHINGTON	BRITISH COLUMBIA	ALASKA	COAST TOTALS
1911-15	.3	31.6	22.5	9.9	64.3
1916-20	.3	13.2	22.2	9.2	44.9
1921-25	.6	9.5	29.3	10.7	50.1
1926-30	.8	12.3	27.6	12.9	53.6
1931-35	1.0	20.5	17.6	7.0	46.1
1936-40	.8	21.3	20.5	8.6	51.2
1941-45	.9	14.6	22.2	15.3	53.0
1946-50	.7	10.1	23.0	23.5	57.3
1951-55	.7	14.8	26.1	20.4	62.0
1956-60	.6	18.5	25.3	23.2	67.7
1961-65	.3	13.9	26.2	27.5	67.9
1966-70	.1	9.6	23.2	23.1	56.0
1971-75	.1	2.5	11.8	19.6	34.0
1976-78	.1	1.5	7.0	15.4	24.0

cruising range and mobility permitted them to land at ports and places that were economically most advantageous at the time. The smaller vessels tended to land at their respective home ports or in those ports in some proximity to the grounds they had fished.

Combined annual landings of halibut in the states of California and Oregon during the past 50 years have generally been under 1 million pounds. Prior to 1915 a very limited fishery was conducted from Oregon and northern California ports, but no landing totals are available for those early years. After 1923, annual landings fluctuated between 600,000 and 1.4 million pounds. Decline in the stocks and to a lesser degree the shortening of the fishing season resulted in the landings in the two states declining to levels consistently below 1 million pounds annually. In more recent years landings in the two states have been at negligible levels despite a long fishing season.

The catches, for the most part originated from the grounds off the coast of Oregon and northern California. In some years a portion of the Oregon landings originated on grounds off Washington State and British Columbia and very occasionally off Alaska. In turn, quantities of halibut caught off the Oregon coast have from time to time been landed in Washington State and are recorded in the totals for that state. Also in 1916 some halibut originating from grounds off Oregon were landed in Vancouver, British Columbia.

CALIFORNIA

With the waters off California lying at the southern extremity of the commercial range of the species, it was not unexpected to find that some of the Pacific halibut landed by fishing vessels in California had been caught on grounds to the north, and occasionally as distant as the coast of British Columbia.

For the later years between 1930 and 1955 the above totals included documented poundage landed illegally chiefly by trawlers fishing for other demersal species. Such poundage was not included in Technical Report No. 14 (1977) of the Halibut Commission.

San Francisco and Eureka are the only two places where any consequential amounts of Pacific halibut have been landed. However, from time to time small amounts of halibut have been landed at Point Reyes, Tomales Bay, Bodega Bay, Point Arena, Albion, Shelter Cove, Fields Landing, Trinidad and Crescent City. In addition, random individual Pacific halibut have been landed at practically any place where bottomfish or troll salmon may be landed.

TABLE 10. Average Annual Landings of Halibut in California in Thousands of Eviscerated, Heads-off Pounds

Period	Pounds	Period	Pounds
1926-30	464	1951-55	114
1931-35	723	1956-60	134
1936-40	354	1961-65	34
1941-45	243	1966-70	35
1946-50	236	1971-75	20

San Francisco

This large city was founded in 1835 and named after the mission San Francisco de Asis, established by the Spaniards in 1776. With the completion of the Southern Pacific Railroad in 1869 it soon became the most important fishing center on the Pacific Coast. It had four salmon canneries, 16 to 20 large codfish schooners fishing the Bering and Okhotsk seas, 1,500 Chinese persons fishing shrimp in San Francisco Bay and sturgeon in the tributary rivers, a large ring-net crab fishery and a paranzella net bottom fishery. The latter two fisheries were dominated by people of Italian origin. San Francisco was also the marketing center for fish landed elsewhere in the state.

In this background one finds many references to the landing of Pacific halibut. As early as 1855 a few halibut were reportedly taken from the vicinity of the Farallon Islands and marketed in San Francisco. Halibut of 40 to 50 pounds were observed for sale in the San Francisco markets at 50¢ per pound (Ayres, 1855). A schooner-load of halibut from Vancouver Island delivered in San Francisco on June 12, 1879, flooded the market (Lockington, 1879). In a later report, Lockington (1880) stated that while halibut were taken at the Farallon Islands, he had seen only one such specimen in the San Francisco markets and it was selling at 50¢ per pound. Lack of an active demand prevented any systematic effort to supply San Francisco with halibut from nearby grounds except for the occasional fish brought in by trawlers (Collins, 1892). However, some shipments continued from the north from Puget Sound, requiring several days' travel. The quality of the fish was good, and the large sizes sold for 10 to 15¢ per pound. For this early period there is no indication of an active halibut fishery indigenous to northern California ports (Lockington, 1881).

It was reported that a vessel originating from some unspecified region sold 9,000 pounds of halibut at San Francisco for 30¢ per pound in 1899, but usually the small amounts brought in by the two-vessel paranzella net trawlers continued to be adequate to supply the local demand (Wilcox, 1902).

The importance of San Francisco as a primary fishery center soon faded. Except for a temporary resurgence of sardine canning and reduction in the "Bay" area in the mid-1930s the region became largely a marketing area for fishery products.

Annual landings of Pacific halibut in San Francisco probably never exceeded one quarter million pounds brought in chiefly by the paranzella trawlers and later by their replacement, the otter trawlers. The landing statistics for this area have to some degree been affected by failure to always distinguish between California halibut *(Paralichthys)* and Pacific halibut *(Hippoglossus)*. Also, considerable quantities of illegally caught Pacific halibut were shipped into San Francisco from Washington and Oregon locations, particularly in the 1930s and 1940s.

Eureka

Eureka, a city of about 30,000 persons (1960), on Humboldt Bay and located about 240 miles north of San Francisco was founded in 1850. Though initially a mining center, logging and lumber soon took over and is still the dominant industry. In 1857 Chinese fishermen started a shark oil and dried fish operation in the area, and from the early days Humboldt Bay crab (Dungeness) has been the most important fishery product of the region. After rail connection with San Francisco was achieved in October, 1914, Eureka became a more important fishery center.

By 1923 a modest fishery for Pacific halibut had become established in Eureka. Prior to that time catches of halibut by salmon trollers during the off-season for salmon had been sufficient to supply the San Francisco market (Downing, 1923; *Pacific Fisherman*, 1923). Halibut yields from this northern California area attracted the interest of some setline vessels from Puget Sound and by 1927 a few such vessels were fishing out of Eureka.

Operations were considered somewhat hazardous due to the prevalence of fog, the necessity of fishing in the coastal steamer lane, and the difficult harbor-entry across a bar. Notwithstanding, the fishery continued to attract halibut vessels from Oregon and Washington ports. The increased fishing effort was reflected in the rising trend of halibut landings as shown in the reports of the California marine fish catch. In 1934 over 750,000 pounds were delivered in California, mostly into Eureka.

However, this increasing trend of landings was short-lived and by 1940 the total landings of halibut in California had fallen below 200,000 pounds. The decline occurred in the halibut landings of all classes of vessels as shown in Table 11. Although the shortening of the halibut fishing season contributed to the decline there was also a reduced interest in the taking of halibut as the various accumulations of the fish on the banks proved to be incapable of sustained yields.

With the lengthening of the halibut season after 1945 there was a brief period of increased landings. This temporary increase was chiefly at San Francisco where some variable criteria had been used in separating the Pacific from the "California" halibut during that period. However, there is insufficient information available at this late date to evaluate the effect. Illegal halibut from Puget Sound was brought into San Francisco at this time, which probably also confused the statistics.

The sharp decline in poundage landed by each segment of the California fishing fleet between 1934 and 1940 are shown in Table 11.

By the late 1940s, about 90% of the landed total in California was made by less than 20 small boats fishing with halibut setline gear. The remaining 10% was landed by nearly 200 smaller boats, such as salmon trollers and boats using other than regular setline gear and primarily interested in the capture of other species. There was also a sharp decrease in the number of vessels fishing for sablefish *(Anoplopoma fimbria)* in the Eureka area, declining from 55 in 1946 to 9 in 1952.

By the late 1960s there were no hook-and-line vessels directing their efforts primarily to the taking of Pacific halibut off California despite the region's long open halibut fishing season. Landings were at a low ebb.

The trend of landings of Pacific halibut in Eureka,

TABLE 11. Halibut Catches by Segments of the California Fishing Fleets

	1934			1940		
	Number of Boats	Pounds	% of California Catch	Number of Boats	Pounds	% of California Catch
Paranzella Trawlers (Two-Boat Sets)	12	160,098	21	7	40,385	22
Setline Vessels Targeting Halibut	28	586,072	76	17	142,867	77
Smaller Hook and Line Boats, Including Troller, Not Primarily Seeking Halibut	205	20,053	3	40	2,924	1
Total	245	766,223		64	186,176	

which accounted for most of the halibut receipts in California, is similar to that shown for the state.

OREGON

While the landing of halibut in Oregon has a long history, no one location would have qualified as a halibut landing port. On the other hand Portland, 90 miles from the mouth of the Columbia River, Astoria and Warrenton in its estuary, and Newport and Coos Bay on the coast have, over the years, received significant quantities of halibut. From time to time smaller amounts have been landed at such coastal communities as North Bend, Charleston, Reedsport, Port Orford, Florence, Tillamook, Hammond and Depoe Bay. In the small towns, most of the catches were taken incidentally by fishermen trolling for salmon. Occasional vessels, bottom fishing with trawl gear, would land and illegally retain halibut at both the small and large ports.

After World War I and through the 1920s there were several small three- to five-man setliners that included such names as the *Clarion, Kiwanda, Crescent, Enterprise* and *Jennie F. Decker* fishing quite consistently off the Oregon coast. They tended to move from one small concentration of fish to another, and they landed their fares at points all the way between Coos Bay to Portland.

TABLE 12. Average Annual Landings of Pacific Halibut in Oregon in Thousands of Eviscerated, Heads-off Pounds

Period	Pounds	Period	Pounds
1926-30	306	1951-55	536
1931-35	304	1956-60	428
1936-40	547	1961-65	222
1941-45	740	1966-70	98
1946-50	519	1971-75	80

Portland

The local Portland market was sufficiently attractive at times for a few halibut fishing vessels to negotiate the long run up the Columbia River.

As early as 1885, the *Carrie B Lake* brought to Portland halibut and other bottom fish that had been taken with bottom trawls off the coast of Washington Territory (State). In 1889, the *George H Chase* operated out of Portland and fished off Cape Mendocina. This vessel was built in San Francisco expressly for this purpose at the cost of $23,000 by a recently formed company in Portland, the Deep Sea Fishing Co.

This company, managed by Luther Maddocks, had ambitious plans to make Portland a center for the halibut trade, to vie with that already developing in Vancouver, B.C., and Tacoma, Washington. The firm was also part-owner of a cold storage in Portland which at least assured a supply of ice when ice was often in short supply. However, the project was doomed to failure almost from the outset. Before the Portland company got under way, the Tacoma-based enterprise had experienced logistic and other difficulties in shipping fresh halibut across the continent. The eastern market was difficult to penetrate, being closely controlled by a group with a Pacific Coast base in Vancouver. The long run upriver to Portland plus not-too-plentiful supplies of halibut off the Oregon coast, did not help the situation.

The manager and prime mover of the Portland company, Luther Maddocks, had to terminate his connection with the company and return to the East leaving another in charge. Early in 1890, Peter Nelson who had been mate of the *Oscar and Hattie* from the time she left New England, was mate on the *George H Chase* when it foundered off the mouth of the Columbia River. This ended the attempts in Portland to compete with Tacoma and Vancouver in the transcontinental halibut trade.

Astoria

Astoria was founded by John Jacob Astor of the Northwest Fur Co. in 1811 in competition with Fort Vancouver of the Hudson's Bay Co., located farther up the Columbia River on its north side, in what is now Washington State. Astoria is the oldest white man's settlement in Oregon.

While occasional small landings of halibut were made at Astoria in the early years, salmon were the primary interest of the early settlement, both for trade as well as sustenance. There was a modest halibut fishery during the early twentieth century out of Astoria, but it was not until the mid-1940s that it could be regarded as a significant activity. At that time Astoria became the landing point for about a dozen four- to six-man halibut vessels. Most of their fish were brought from the coast of British Columbia with a few from the Area 3 grounds in the Gulf of Alaska. By the 1940s the banks off Oregon contributed a relatively small share of the total receipts of halibut in that port.

Receipts of halibut in Astoria accounted for about 60 to 70% of the total landed in Oregon. Portland did receive significant amounts in the early years, but after 1925 little was landed there. However, it continued as a distribution center for halibut landed elsewhere.

Warrenton

Warrenton, a few miles from Astoria, was the Oregon home of the San Juan Fishing and Packing Co. of Seattle. From time to time some Astoria-based vessels and a few Seattle vessels landed halibut there.

Newport

Newport, on Yaquina Bay on the central Oregon coast gained importance as a halibut center because of its protected harbor and proximity to one of the more productive halibut grounds found south of Cape Flattery, namely Heceta Bank. Prior to the discovery of that bank in 1914-1915, Newport was of minor importance as a port for landing halibut.

Coos Bay

Coos Bay, known earlier as Marshfield, farther south than Newport on the outside coast, was a major lumber shipping point with a protected harbor. It was an active fishing port for salmon, halibut and, in later years, bottom fish caught by trawlers. Though it played a relatively modest role in the halibut fishery, it was the scene of an enterprise which developed from small beginnings into an activity of consequence. It could be called "The Charles Fellers Story," a story in the true American tradition.

Having heard, as a young man in England, about Rogue River apples imported into Europe from Oregon, young Fellers was impelled to visit that distant land. Arriving in Marshfield with his last $2 in his pocket, he chose to invest it in a large salmon, pieces of which he peddled from door to door. In a few days he had sufficient capital to make payment on a pushcart. This increased his volume and enlarged his marketing territory. In not too many years such modest beginnings evolved into a fully integrated and major fishery operation in southern Oregon, Chas. Fellers Inc.

While a number of factors joined to cause the decline and almost extinction of halibut landings in Oregon, the buying strategies of some of the larger Pacific Northwest fish companies had the most impact. Some firms no longer were as interested in buying halibut and others shifted their buying of halibut back to Seattle. Furthermore, the diversion of the few Oregon-based halibut vessels to other more profitable pursuits such as salmon or tuna trolling or to trawling for other bottom fish was also an important factor in the decline in halibut landings in Oregon.

In the later years, a number of places in Oregon had gained the unenviable reputation of receiving quantities of unreported halibut. On some occasions, the amount unreported exceeded the amount reported. Much of that unreported consisted of halibut taken incidentally to setline fishing for other species in the halibut closed season in excess of the proportion permitted by the halibut regulations. In some places, bottom trawlers also landed not inconsiderable amounts of halibut caught incidentally, and illegally retained. In addition, in the 1930s a few vessels of the Puget Sound halibut fleet chose to unload in Oregon, where catches in excess of the amounts allowed under the fleet's voluntary program to control the rate of landing were more difficult for that group to detect.

WASHINGTON

Washington, first as a territory then as a state after 1888, has been an important center for the Pacific halibut industry.

Except for a short period prior to about 1900 when Vancouver dominated the young industry, Seattle has been the main center of the halibut industry. Despite being eclipsed from time to time in ex-vessel landings, Seattle has been the leading continental distributing point for halibut since early in the twentieth century. Most of the Pacific Coast dealers in halibut have their head office in Seattle, and most of the halibut landed in Alaska by the United States fleet is transshipped to Seattle. Seattle has been the headquarters of the International Pacific Halibut Commission since 1953, and of its predecessor, the International Fisheries Commission, since its formation in 1924 under the 1923 convention.

TABLE 13. Average Annual Landings of Halibut in Washington State by United States and Canadian Vessels in Thousands of Eviscerated, Heads-off Pounds

Period	United States	Canada	Total
1916-20	13,150	16	13,166
1921-25	9,449	1	9,450
1926-30	12,318	4	12,322
1931-35	20,511	7	20,518
1936-40	21,236	53	21,289
1941-45	14,547	31	14,578
1946-50	10,070	62	10,132
1951-55	14,661	105	14,766
1956-60	16,428	2,057	18,485
1961-65	10,942	2,952	13,894
1966-70	7,067	2,539	9,606
1971-75	1,449	1,010	2,459
1976-78	583	918	1,501

For a brief period Tacoma was the leading landing point for halibut in Puget Sound, but was soon superseded by Seattle about 1905. Port Townsend in early years and Bellingham in later years played modest roles in the Puget Sound industry. Places of lesser significance such as Kalama and Longview on the Columbia River, Grays Harbor and Willapa Bay on the coast and Everett and Anacortes in Puget Sound also contributed to the development of the industry, particularly Everett.

As indicated in the early development of the individual landing ports, significant landings of halibut in Washington State had their beginnings in 1888 at Tacoma. Development proceeded slowly, annual landings being less than 10 million pounds until the turn of the century. Large company-owned steamers and many independent auxiliary-powered sloops were introduced, and Washington annual landings passed the 30 million mark in 1911. In 1916 a sharp fall from this level occurred because of the many landings by United States vessels going to Prince Rupert and the restrictive effects of World War I. Decline continued until 1924, when the total was slightly over 7 million pounds.

From 1924 to 1931 a steady increase in Washington landings occurred. A primary factor was the development of vessels with motor power permitting the economical exploitation directly from Washington State ports of banks west of Trinity Islands, Alaska. In addition, the closure of the winter season under the Halibut Convention in 1923 brought the last trips of the season of Washington State vessels to their home ports.

From 1932 to 1936 voluntary tie-ups between trips and generally low prices resulted in a considerable increase in the amount of fish landed in Seattle, the home port of most of the United States fleet that fished west of Cape Spencer. After 1936, in spite of the continuation of the voluntary tie-up program, landings in Seattle declined due to improvement in the relative prices paid in northern ports and a steady increase in the number of vessels attempting to augment their number of trips per year by landing closer to the fishing grounds. As more vessels pursued this strategy, it was self-defeating because of a fixed catch limit for the season set by the Halibut Commission. Conditions arising chiefly from World War II caused a further decline in Seattle landings after 1941.

Through the 1940s landings in the state remained at a level of about 10 million to 14 million pounds annually. With the general increase in the 1950s of the permitted catch limits, Washington State receipts showed some increase, but it could not be sustained into the 1960s because of the increasing share of the still-large coast totals being taken by the growing Canadian fleet. Also the Goose Island Grounds, long an important producing area for vessels from Seattle, were no longer as productive. Furthermore, the voluntary between-trip lay-in plan reinstated in 1957 did not attract the vessels to land in their home ports in the south as was the case with the lay-in program in the 1930s. Instead, many crews could now fly home from distant unloading ports in Alaska in a matter of a few hours for their voluntary between-trip "rest and recreation" period of eight days. Fishermen were more affluent and air schedules from Alaskan ports more frequent than in the 1930s.

By the late 1960s ex-vessel landings of halibut in the state reached levels as low as at any time this century and the decline continued into the 1970s. Only the landings by a few Canadian vessels in the state prevented the totals from falling below the 4 million-pound mark. The decline has been accentuated by the sharp reductions made by the Halibut Commission in the permitted overall catches during the past decade, the deterioration in the stocks caused chiefly by foreign trawling and setlining and an increasing proportion of the catch being landed in Alaska by Washington State vessels.

Canadian vessels landed inconsequential amounts of halibut in Washington ports up to 1955. The increasing proportion of the larger total annual catches taken by Canadian vessels after the mid-1950s resulted in a number of the larger Canadian vessels selling in Vancouver for delivery in Washington State, particularly in Bellingham. This sometimes occurred to relieve port congestion in Vancouver particularly during the summer salmon season. Also, some Seattle buyers found it expedient to purchase catches by radiotelephone to the southbound vessel for Bellingham delivery and storage. These developments were facilitated by the building of a modern cold storage plant in Bellingham in 1946, and its adaptation in 1960 to handling halibut. By 1971, landings in Bellingham by Canadian vessels exceeded United States landings in Seattle.

In the 1970s, vessel landings in Puget Sound ports including Seattle had reached a very low ebb. Nevertheless, Seattle still remained the most important halibut distributing center as most of landings in Alaska were transshipped frozen to Seattle for distribution throughout the United States.

Tacoma

Tacoma was the first major landing point in Puget Sound for halibut vessels powered other than by sail. Auxiliary power was a necessity as it sometimes took three to four days to sail from Port Townsend to Tacoma, and towing costs were prohibitively high.

Tacoma has the historic distinction of being the place where on September 20, 1888, Northern Pacific Railroad car Number 14044 was loaded with boxes of iced halibut caught off Cape Flattery by the *Oscar and Hattie*. This was the first carload lot of fresh halibut to be shipped East from the Pacific Coast. Although it arrived in Boston in poor condition, it can be regarded as the beginning of the commercial halibut fishery on the Pacific coast.

Tacoma also received halibut by cargo steamer from Victoria for shipment East. In the winter season 1895-1896 the S.S. *Edith* commenced fishing out of Tacoma. Also, the same winter, the S.S. *Thistle* of Victoria landed three of its halibut catches in Tacoma and transshipped three others to the city by the freighter S.S. *City of Kingston* for a total of about 300,000 pounds.

The city became a major halibut center with the building and launching in 1905 of the S.S. *Zapora* by the International Fisheries Co., which had been organized on October 22, 1902, as a branch of the American Fish Co. of Boston. This was aided by the Northern Pacific Railway upgrading its service in an attempt to equal that made available in Vancouver, B.C., by the Canadian Pacific Railway. The S.S. *Edith* joined the S.S. *Zapora,* which had a capacity of 175,000 pounds of iced halibut, and produced a steady supply of halibut for the port. The International Fisheries Co., which in 1907 was taken over by Booth Fisheries of Chicago, brought halibut from Alaska using the S.S. *Katahoon* and S.S. *Jupiter* as packers. The S.S. *Edith* was later withdrawn from the fishery and not replaced. The S.S. *Zapora* continued fishing out of Tacoma for a number of years. With the growth of Seattle's halibut fleet in the early 1900s, Tacoma was soon eclipsed by Seattle as the major Puget Sound halibut port.

Except for an occasional small fare of halibut landed in Tacoma during the succeeding 50 years, Tacoma as a halibut port passed into oblivion.

Seattle

Seattle had long been an active local market and distribution point for halibut, both fresh and salted. It was destined soon to become the center of the halibut trade on the Pacific Coast either as a port of landing by fishing vessels or as a distribution point for halibut landed elsewhere or both.

Years before the Seattle fire of 1888, a few small boats would fish for the Seattle market off the south end of Whidbey Island and off Point-No-Point, but not on the scale that prevailed out of Port Townsend. The early schooners from New England, the *Mollie Adams* and the *Edward E. Webster* made Seattle their headquarters during their brief and rather unprofitable sojourn in the halibut fishery. They landed both fresh and salted fletches in 1889. They were primarily interested in sealing. The *Mollie Adams* had arrived from New England early enough in 1888 to land four fresh halibut trips in Seattle that year totaling about 150,000 pounds worth $4,500 and 180,000 pounds of salted fletches worth $6,300. The latter was the first fletching trip made on the Pacific Coast. In the summer of 1889, the *C.H. White* out of San Francisco, fished halibut out of Seattle under charter.

During the 1890s Ainsworth and Dunn of Seattle were pioneer fish dealers on the Pacific Coast. At the time there were a number of small five-man, two-dory sloops fishing out of Seattle. Later in the decade of the 1890s some of the vessels commenced going to Southeastern Alaska in the fall where they fished the inside channels from about October 1st to the end of March. The fish were shipped iced in boxes to Seattle by coastal steamer, except for the last trip in the spring which was usually brought down by the fishing vessel itself.

The difficulties of securing ice, the relatively high transportation costs, and the long distance to the large volume markets in the East slowed down the development of the fishery in Puget Sound. Also, some engine power was an urgent necessity on the fjord-like coasts of the Northwest. It was not until the turn of the twentieth century that Seattle began to assume significance as a halibut center.

The San Juan Fishing and Packing Co. was organized in 1899 by the Calvert brothers, William and James, and Edwin Ripley. They bought the Seattle Fish Co., which was a subsidiary of the once large and then defunct Pacific Packing and Navigation Co. of Bellingham. In turn, the Seattle Fish Co. was successor to Ainsworth and Dunn. Chlopeck Fish Co., later acquired by Booth Fisheries, was active in the Seattle halibut industry early in the century and had

TABLE 14. Average Annual Landings of Halibut in Seattle by United States and Canadian Vessels in Thousands of Eviscerated, Heads-off Pounds

Period	United States	Canada	Total
1931-35	20,402	7	20,409
1936-40	20,699	53	20,752
1941-45	14,015	31	14,046
1946-50	8,849	57	8,906
1951-55	12,727	105	12,832
1956-60	15,334	1,694	17,028
1961-65	9,429	809	10,238
1966-70	6,097	318	6,415
1971-75	814	41	855
1976-78	283	14	297

the S.S. *Rainier*, a wooden vessel, fishing with eight dories. It was lost in 1903 near Hoonah, Alaska, while halibut fishing in Icy Strait under Captain Churchill. When early in the 1900s, the San Juan Fishing and Packing Co. acquired the large steamers, *San Juan* and *Grant,* and the Chlopeck Fish Co. owned the S.S. *Edrie* and S.S. *Zapora,* the combined production of these vessels and the still-growing fleet of independently-owned small schooners and sloops, made Seattle the "Halibut Capital of the World" by about 1906. Tacoma and Vancouver were dethroned. Fame being ephemeral, the title would not be held long. Ten years hence came Prince Rupert and it was realized that, "Uneasy lies the head that wears a crown."

The landings shown for Washington State as a whole in Table 13 for all practical purpose also represent the trend of receipts in Seattle prior to 1931. Landings in Bellingham or elsewhere in Washington had not been of any statistical consequence prior to 1931.

Bellingham

Bellingham, which in later years was to become a significant landing port for halibut, had an early though modest background in the fishery. It was, however, the center of an active salmon fishery. In the 1890s, the Bellingham Ice Co. was a supplementary source of ice for both Canadian and United States shippers of halibut and supplied herring bait to Puget Sound halibut vessels at nearby Lummi Island. Shipments of halibut to the East in the 1890s originated in Fairhaven, now a suburb of modern Bellingham. The Hackett Cold Storage Co. of Boston built a cold storage plant in Bellingham in 1904. The company had grandiose plans for a large halibut operation, including the building by Bellingham interests of a $60,000 fishing steamer. Little came of the stock promotion.

For the next 50 years there was practically no halibut activity in Bellingham except for some landings by an occasional Seattle halibut vessel and, from time to time, by otter trawlers, who after 1945 were in contravention of the Pacific Halibut Regulations. With the installation of facilities in 1960 for handling halibut at the Bellingham Cold Storage, the port became a significant factor in the legitimate halibut trade in Puget Sound.

TABLE 15. Average Annual Landings of Halibut in Bellingham by United States and Canadian Vessels in Thousands of Eviscerated, Heads-off Pounds

Period	Canada	United States	Total
1946-50	—	123	123
1951-55	—	154	154
1956-60	347	226	573
1961-65	2,142	1,251	3,393
1966-70	2,221	858	3,079
1971-75	968	383	1,351
1976-78	569	413	982

Everett

Everett, about 25 miles north of Seattle, was the terminus of the Great Northern Railroad. Almost from the outset the lumber industry dominated the Everett economy and to a large degree has continued to do so. Fishing was a secondary activity, although the first salmon cannery on Puget Sound was built in 1877 in Mukilteo, almost a suburb of Everett.

There was no early development of a halibut fishery in Everett as was the case of Tacoma and Seattle, although in 1910 the 83-ton vessel *Mary D. Hume* was reported fishing halibut out of Everett. Two small wholesale fish companies operated out of Everett for many years handling some halibut. In later years they consolidated and continued to handle about 100,000 pounds of halibut annually.

Toward the end of World War II the New England Fish Co. secured a ten-year lease on a Great Northern dock in Everett and installed a three-line salmon cannery along with freezing and cold storage facilities. Commencing about 1948 they handled about 400,000 pounds of halibut annually increasing to about 1.5 million pounds by 1952. It remained about that level to the mid-1950s when the lease ran out and New England Fish Co. shifted their halibut purchases to Seattle.

Port Townsend

The United States Customs port of entry for Washington Territory in the 1880s supported a moderately-sized fleet of small boats manned by a total of about 40 men that fished for halibut for local consumption. Fishing was conducted in close proximity to the town, in Washington Sound, off Whidbey Island, off Port Angeles and sometimes as far south as Point-No-Point. Later, Port Townsend was the home port for the first vessel fishery for Pacific halibut with the arrival of the *Oscar and Hattie* in 1888. However, most of the fresh fish caught by that vessel were landed in Tacoma, the terminus of the transcontinental Northern Pacific Railroad. Outside of being the port of entry to Puget Sound in the early days and the home port of the historic *Oscar and Hattie*, Port Townsend played a relatively small role in the early or subsequent development of the vessel fishery for halibut. A small number of the smaller halibut setliners out of Seattle in the 1930s fished occasionally on the few early-defined halibut banks in Washington Sound off Port Townsend.

Other population centers in Washington also have been receiving points for moderate amounts of halibut at various times during the past 75 years. These include such places as Kalama and Longview on the Columbia, Grays Harbor and Willapa Harbor on the coast, and Anacortes in Puget Sound. Neah Bay close to Cape Flattery has a long history as the center for halibut fishing by the Makah Indian tribe. Further references to some of these lesser ports will be made in reviewing other aspects of the fishery including those of Indians.

BRITISH COLUMBIA

Although Pacific halibut are found in commercial quantities along about 2,000 miles of the North American coasts, British Columbia has a greater importance in the fishery than its 500-mile section of the Pacific Coast might suggest. No portion of the Pacific Coast possessed such large primeval concentrations of halibut—concentrations whose productivity persisted over a long period. It is also the terminus of two railroads, one in the north and the other in the south part of the province. The northern railhead, Prince Rupert, being closer to the fishing grounds off Alaska became a major landing point by 1915 for both Canadian and United States vessels from the Gulf of Alaska. Also, from time to time halibut landed in Alaskan ports was transshipped via Prince Rupert by rail to eastern United States points.

TABLE 16. Average Annual Landings of Halibut in British Columbia by United States and Canadian Vessels in Thousands of Eviscerated, Heads-off Pounds

Period	United States	Canada	Total
1916-20	13,356	8,816	22,172
1921-25	20,184	9,090	29,274
1926-30	18,982	8,629	27,611
1931-35	9,105	8,473	17,578
1936-40	8,234	12,256	20,490
1941-45	9,157	13,051	22,208
1946-50	3,466	19,561	23,027
1951-55	3,284	22,797	26,081
1956-60	1,366	23,972	25,338
1961-65	852	25,385	26,237
1966-70	662	22,433	23,095
1971-75	645	10,937	11,582
1976-78	40	6,957	6,997

Note: The steady decline in landings by United States vessels in Canadian ports has been largely offset by increased landings by the Canadian fleet. The decline in the latter in the 1970s is a reflection of the overall decline in total coast catch.

Table 16 shows the landings of halibut in British Columbia since 1916 when it was first possible to make an accurate separation between those made by Canadian and United States vessels.

New Westminster

New Westminster, the Queen City, which earlier had been capital of the Mainland Crown Colony was the first major halibut center in British Columbia. By the early 1890s Port and Winch, operating the former United States vessel *Mischief* of Yaquina Bay, Oregon, changed to British registry, made a number of carload shipments to the eastern markets. In 1892 halibut landings in British Columbia amounted to 1,357,500 pounds worth $67,875 having increased from an average of 246,000 pounds annually during the five-year period since 1888, the first year of record. While it is not possible now to separate the totals by landing ports, much of the poundage was probably landed in New Westminster. Early in 1893 Port and Winch chartered the S.S. *Coquitlam* from the Union Steamship Co. and packed halibut from shore camps established in Hecate Strait. After 10 trips to New Westminster in the spring, Port and Winch concentrated thereafter on salmon but must be regarded as the pioneer firm in the British Columbia halibut trade.

In the succeeding winter season of 1893-94 the *Capilano* was chartered by the New Westminster Fish Co., whose Haney office was managed by Mr. Grier Starratt. The vessel was under command of Captain Absalom Freeman, a former mate on the S.S. *Comox,* a coastal freight and passenger vessel owned by the Union Steamship Co. The fish company shipped 16 carloads of halibut to New York and three to Oregon up to March 30, 1894, when the fish camp was disbanded for the season. The following winter season of 1894-95 the gear and fishing boats of the *Capilano* were taken over by the American Fish Co. of Vancouver, which had chartered the *Coquitlam.* The *Capilano* was chartered by the New England Fish Co. of Vancouver in the fall of 1894 and New Westminster was temporarily out of the vessel fishery.

Although New Westminster never regained its early dominance in the British Columbia halibut fishery, it did achieve considerable prominence immediately before and during World War I. It was then the homeport for the S.S. *Roman* and S.S. *Onward Ho,* two large halibut steamers owned by the B.C. Fishing and Packing Co. Each had a capacity of 300,000 pounds of iced halibut. They were both wrecked, the *Onward Ho* in January, 1916, and the *Roman* in November the same year. The company also owned the *B.C.P.*, also with a capacity of 300,000 pounds. It was built in 1911 and fished halibut out of New Westminster until the end of World War I.

Except for the occasional landing of small fares during the 1920s New Westminster did not gain any importance as a landing point for halibut until the early 1930s. At that time the Edmunds and Walker Co. established itself in the large public cold storage with unloading docks on the Fraser River. This firm had taken over the fresh and frozen fish business of B.C. Packers, the large British Columbia salmon canning company that had experienced severe financial difficulties. Until Edmunds and Walker were reabsorbed into the reorganized B.C. Packers in 1939 and relocated in Vancouver, their New Westminster operation resulted in considerable quantities of halibut being handled in that port during the 1930s.

After the relocation of Edmunds and Walker in Vancouver, New Westminster was no longer of any importance in the halibut fishery. Steveston, close to the mouth of the Fraser River and the location of a large diversified fishery operation of the B.C. Packers, has since the mid-1940s been used extensively for the freezing of halibut, whether unloaded ex-vessel at the company's Steveston plant or trucked from their Vancouver Edmunds and Walker plant.

Vancouver

Vancouver is another city destined to become one of the great metropolises of the world. The first Europeans to set foot on its soil were Don Jose Maria Marvaez and Don Jose Vardin, mates of the 75-foot Spanish ship *San Carlos.* Landing on Point Grey on July 5 and 6, 1791, they thought it was an island, Isla de Langara.* A year later Captain George Vancouver also landed at Point Grey at about noon July 12, 1792, and proceeded to explore English Bay and Burrard Inlet.

Vancouver was of some importance in the halibut fishery even prior to its change of name from Gastown and its incorporation as a city, both occurring in the year 1886. This was also the year of a great fire that wiped out the city in under 30 minutes on a Sunday afternoon, June 13, 1886.

In the early 1880s, small boats were catching halibut in English Bay and occasionally even in Burrard Inlet for local consumption. The completion of the Canadian Pacific Railway to Port Moody in 1885 and later its spur to False Creek in Gastown, now named Vancouver, made that port a more logical distribution point to eastern markets than New Westminster, the Queen City. The latter required difficult up-river navigation although the sojourn in fresh river water to remove the marine growth off the vessels' hulls was probably a partial compensation.

The *Iona*, fishing with beam trawl for halibut in the winter of 1891-92 under Captain Arthur Edgett with Captain Mitchell as fishing captain was the first fishing vessel to operate out of Vancouver to produce halibut for the eastern market. Previously the *Iona* and the steamer *Velos* had been engaged in coastal freight and passenger service and picked up Indian-caught halibut and salmon for the Fader Brothers of Vancouver for local consumption. They were not fishing vessels. In the fall of 1891 both vessels were chartered to the B.C. Fishing and Trading Co. but due to lack of fish (or markets) the charter of the *Velos* was dropped. Despite a moderately successful winter operation by the *Iona* there were financial difficulties and near the end of that winter season, it was held by the customs for liens.

Another early vessel to produce halibut for the eastern markets out of Vancouver was the *Eliza Edwards* in the winter of 1891-92. Captain MacKenzie of the vessel went east to establish market connections and nearly all the winter's production was shipped on the Canadian Pacific Railway's fast train, the Atlantic Express. The fish arrived in good condition and was disposed of successfully.

The initiative and cooperation of the Canadian Pacific Railway was an important factor in Vancouver becoming the leading landing point for halibut in the early industry. The Canadian Pacific Railway developed six refrigerator cars and was willing to attach them to their fastest transcontinental trains. The Northern Pacific out of Tacoma and Great Northern out of Seattle attempted to move their cars of iced halibut by fast freight and often met with disastrous results in quality on arrival on the East Coast. It is quite probable that their rolling stock being used as "reefers" did not meet standards required by law for passenger trains.

The *Eliza Edwards* itself did not actually fish, but the crew bought fish from Indians using canoes in the vicinity of Queen Charlotte Islands. It operated for but one season. Later in 1893 the *Eliza Edwards* participated in one of the ubiquitous "Cocos Island Treasure Hunts" and was sold in Mexico, probably for debts, as has been the case for many other vessels that have since made similar expeditions to Cocos Islands.

The *Velos*, whose charter had been dropped in 1891, was later successfully operated by Captain James L. Anderson. He was a master mariner born in

*Downtown Vancouver is in fact an island. Early maps show a waterway between Burrard Inlet and False Creek where Carrol Street is now located.

FIGURE 66. Henry Darling, manager of the Union Steamship Co., Vancouver, British Columbia.
E.W. Wright, *Marine History of the Pacific Northwest*

Nova Scotia in 1841 and had extensive deep sea experience. Though several times wrecked, he had never lost a man on vessels under his command. As late as September, 1894, he landed a load of 80,000 pounds of halibut in Vancouver, counting out 2,066 fish.

It was not until the fall of 1894 when a syndicate of Boston fish dealers, the New England Fish Co., established a warehouse on the Union Steamship Dock, with Mr. Grier Starratt as manager, and chartered the S.S. *Coquitlam* on September 22, 1894, that Vancouver's future as a major factor in the Pacific halibut fishery became a certainty. It was also destined soon to eclipse New Westminster as the major halibut port.

The early development of the fishery was greatly furthered by the Union Steamship Co. and by Henry Darling, its manager in Vancouver. It was a New Zealand company and Darling's father was an owner. Henry Darling, born in 1863, was a marine engineer who had worked up "through the ranks" and possessed considerable managerial talent. He formed the Vancouver company in 1890, joining with the Burrard Inlet Towing Co. which owned three vessels. Others were soon added, including the two freighters, the S.S. *Coquitlam* and S.S. *Capilano*. Both were 127 feet long, 22 feet beam. The *Coquitlam* had a nine-foot, six-inch hold, and the *Capilano* was one foot deeper. Each made about nine miles per hour.

The two steel vessels had been built in Scotland, disassembled, shipped to Vancouver and reassembled. The *Capilano* was launched in December, 1891, and the *Coquitlam* about the same time. The latter vessel was tender to the Victoria sealing fleet, was seized by a United States cutter in 1892 with $75,000 worth of skins on board but was released. Prior to the charter of the two vessels to the New England Fish Co., each had been tested by other companies as noted earlier in discussions about New Westminster. They were also used in the coastal freight and passenger trade in the summer months.

The Union Steamship Co. in the fall of 1894 provided the dock and warehouse space to the New England Fish Co. for unloading and transferring the halibut to railcars. One of the steamship company's valued personnel, Absalom Freeman, a mate on their 200-passenger vessel *Comox* became captain of the *Coquitlam* when first chartered by the New England Fish Co. on September 22, 1894.

The two vessels, *Capilano* and *Coquitlam,* were well adapted for packing iced fish and servicing the camps. As fish became less abundant near the camps the steamship company permitted their vessels to carry dories nested on deck.

This greatly increased its perimeter of the operation. It eliminated the need to return to the shore station each night as there were sleeping accommodations for 25 persons or more on the vessels. With this development their roles had changed from fish packer to fishing vessel.

While commercial halibut fishing had some very early roots at several locations on the Pacific Coast such as Tacoma with the landings of the *Oscar and Hattie,* early developments in Vancouver played a major role in setting the course for the future.

No Pacific Northwest railhead port had the facilities the Union Steamship Co. possessed. Its coastal freighters that Henry Darling first chartered as fish packers and later permitted to become dory fishing vessels and its dock and warehouse that were used for 10 years by the New England Fish Co. for unloading and transshipping their halibut to railcars were vital elements in the early success of the halibut fishery out of Vancouver. Also, many of the steamship company's personnel became mates and captains of dory fishing steamers that entered the fishery later.

The Canadian Pacific Railway made rapid transcontinental shipment of iced halibut feasible by attaching refrigerator cars to its passenger trains out of Vancouver.

Grier Starratt of Vancouver, who laid the foundations of New England Fish Co. in that port, took his experience north. He became the key figure in the building and early development of the Canadian Fish and Cold Storage plant in Prince Rupert whose cold storage, then the largest in the world, became the "heart of the halibut fishery" in that port.

Vancouver and vicinity was also headquarters for the British Columbia Packers Ltd. that, despite its early concentration on salmon, later played a major role in the Pacific halibut fishery commencing in 1902. From Vancouver the company expanded its halibut activities and in 1945 acquired the large Prince Rupert plant of the Canadian Fish and Cold Storage Co.

A Vancouver firm built the large cold storage in Ketchikan, Alaska, and helped the area "come of age" in the halibut fishery. Vancouver was the home port of the S.S. *New England*, the first of the Steamer Era built specifically for fishing halibut. For many decades Vancouver was the headquarters on the Pacific Coast of the New England Fish Co. of Boston which had the largest fleet of steamers and out-produced in some years all other companies on the Pacific Coast combined.

The foregoing early contributions of Vancouver do not detract from the later and continuing dominance of Seattle in the halibut industry. For many years Seattle has had the largest fleet of independently owned vessels, the largest annual landings and the corporate headquarters of companies that are responsible for marketing much of the North American halibut production.

Landings of halibut on the lower mainland of British Columbia include those at Vancouver, Steveston, New Westminster and a few other places of minor importance. Those in Vancouver probably have accounted for over 98% of the landings in the region over the past 90 years.

TABLE 17. Average Annual Landings of Halibut by United States and Canadian Vessels at Lower Mainland Ports of British Columbia in Thousands of Eviscerated, Heads-off Pounds

Period	Canadian Vessels	United States Vessels	Total
1886-90	222*	—	222*
1891-95	903*	—	903*
1896-00	**	**	1,887
1901-05	**	**	6,999
1906-10	**	**	13,673
1911-15	**	**	11,226
1916-20	2,194	1,383	3,577
1921-25	719	815	1,534
1926-30	778	48	826
1931-35	1,493	6	1,499
1936-40	2,446	4	2,450
1941-45	2,955	4	2,959
1946-50	1,831	—	1,831
1951-55	4,087	16	4,103
1956-60	6,577	—	6,577
1961-65	5,356	—	5,356
1966-70	5,725	—	5,725
1971-75	2,562	6	2,568
1976-78	1,867	—	1,867

*Prior to 1895 most of the poundage was landed in New Westminster.

**U.S. Flag vessels fishing out of Vancouver probably accounted for 45% of the total production landed in Vancouver from 1901 to 1915.

Sources: 1886 to 1950, Bell et al. 1952;
1951 to 1975, Tech. Report IPHC No. 14.

Victoria

Victoria was established June 4, 1843, as Fort Albert (after the Prince Consort). It was later named Fort Victoria after Victoria Regina. In 1852 it was called Victoria and was incorporated in 1862. It was the capital of the Crown Colony of Vancouver Island and later, in 1868, was the capital of the United Crown Colonies of Vancouver Island and British Columbia. Since 1871 it has been the capital of the Province of British Columbia. It had an early prominence in the halibut fishery.

In the 1880s and 1890s there was a group of small boats manned by people of Mediterranean origin, probably Italians, that supplied the local market with halibut taken near Race Rocks and other close-in grounds, much as the case with Port Townsend. With the successful vessel fishery developing in New Westminster and Vancouver, a large Victoria general store, Simon Lozer Co., outfitted the S.S. *Thistle* for halibut operations under Captain Langley, and the vessel left Victoria on its first trip on November 11, 1894. It landed loads in Vancouver and Tacoma, Washington, or transshipped them to the latter port. The *Thistle* fished the next winter season but suffered some mishap and withdrew from the fishery. In 1900 it went into the lumber trade. This ended Victoria's participation in the halibut fishery on any major scale. The Victoria company purchased the United States vessel, S.S. *Edith*, and it operated out of Tacoma under another company name.

Victoria continued to receive some halibut for local consumption and later, in 1941, the B.C. Packers of Vancouver bought the local cold storage to freeze and store halibut bought in Vancouver for Victoria delivery. Most of the halibut for local consumption was unloaded by packers from the west coast of Vancouver Island. It had been taken by trollers fishing primarily for salmon and by a number of small two- or three-man halibut boats that fished out of several small settlements on the West Coast including, Quatsino, Kyuquot, Tofino, Ucluelet and Bamfield. Much of the troll-caught salmon was taken to Vancouver.

Nanaimo

Nanaimo, the "coal city" on the east coast of Vancouver Island where coal had been discovered in 1850, was the scene of a rather short-lived operation by the Nanaimo Halibut Co. and its presumed successor, the Pacific Fish Co. It owned the sailing vessel, *Nellie Thurston,* which in 1897 engaged in fletching and salting halibut out of Skidegate, fishing off Cumshewa Rocks on the east coast of Queen Charlotte Islands. It had five small fishing boats and supplemented their catches with Indian-caught fish. In the fall the salted fish were freighted south because the load was too small to warrant return of the fishing vessel to port.

The vessel operated in the spring and summer of 1898, but in attempting passage of Skidegate Channel between Moresby and Graham islands it ran aground. After a month it was refloated and proceeded to fish off Marble Island and Moore Channel and was probably the first vessel to fish halibut on the west coast of the Queen Charlotte Islands. It finally delivered 90,000 pounds of salt fletches to the Nanaimo plant in the fall. Its return to port was a relief to many as there had been fears that the long-overdue vessel had foundered.

The *Nellie Thurston* continued to engage in the fletching trade out of Nanaimo until 1902. While this was the city's first and last major venture into the halibut fishery, Nanaimo and vicinity for the next 50 years was only to be an important source of bait herring for the halibut fleets of Puget Sound and southern British Columbia. Many halibut steamers also "coaled up" at Nanaimo and nearby places before converting to oil.

In the later years no halibut fishery established itself in this port.

Barnett

Barnett, now an industrial suburb of Vancouver, had a somewhat similar but shorter relationship to the halibut fishery than that enjoyed by Nanaimo. However, it did not become a major supply center for bait and coal. In 1903 the Western Canadian Fish Co. employed the barkentine, *Blakely*, to fletch one load of halibut for their cod-curing plant at Barnett. While the fish-curing operation seemed to be successful to observers at the time it was shortlived and Barnett soon passed into oblivion as far the halibut fishery is concerned.

Prince Rupert

Prince Rupert, though later on the scene than Tacoma, Seattle, New Westminster, and Vancouver, has over the years exceeded all those centers combined in the extent and relative importance of its participation in the Pacific halibut fishery. Since the founding of the city in 1908 to the present, the halibut industry has been an important factor in the local economy. Prince Rupert has the title, "The Halibut Capital of the World."

Prince Rupert, situated on Kaien Island, came into being on May 1, 1906, and was named after a cousin of Charles II of England who was the first governor of the Hudson's Bay Co. The name Prince Rupert was decided by public contest with 12,000 entrants. Eleanor M. MacDonald of Winnipeg won a prize of $250. Since the name exceeded the 10 letters prescribed by the contest rules, two others proposing the name Port Rupert were also given $250.

In 1907 before the incorporation of the city on March 10, 1910, the Canadian Fish and Cold Storage Co. was incorporated for $15 million. The confidence in the halibut potential of the Pacific and foresight of the organizers of the company, Andrew Kelly of Winnipeg, James Carruthers of Montreal and others, was evidenced by their planning and construction of the largest fish cold storage in the world at Prince Rupert, to cost $350,000. The company also purchased three large North Sea-type trawlers the *Andrew Kelly, James Carruthers,* and *George E. Foster.* The cold storage was granted a federal construction subsidy of $150,000 for which it was required to maintain public services such as providing ice and freezing and storage privileges to other companies. In 1907 the company employed as manager Mr. Grier Starratt who since 1894 had made the New England Fish Co. of Vancouver the most, if not the only, successful

TABLE 18. Average Annual Landings by Canadian and United States Vessels in Prince Rupert in Thousands of Eviscerated, Heads-off Pounds

Period	Canada	United States	Total
1906-10	1,133	—	1,133
1911-15	7,594	9,894*	**
1916-20	6,132	11,974	18,106
1921-25	8,087	19,368	27,455
1926-30	7,438	19,089	26,527
1931-35	5,680	9,098	14,778
1936-40	7,402	8,235	15,637
1941-45	7,075	9,150	16,225
1946-50	13,605	3,466	17,071
1951-55	14,419	3,284	17,703
1956-60	13,989	1,366	15,355
1961-65	17,518	852	18,370
1966-70	14,501	662	15,163
1971-75	7,523	645	8,168
1976-78	4,221	40	4,261

*1915 only
**No 5-year average

producer and distributor of halibut in British Columbia.

The Prince Rupert plant with its large cold storage served as the backbone for the halibut fishery in Prince Rupert. This plant was nicknamed the "Hull," after Hull, England, the original home of many executives and capital for the Pacific fishery. September 30, 1914, was another landmark date for the port when the first refrigerated railcars of halibut left Prince Rupert for the East. The development of Prince Rupert as a major halibut center is observable in the historical statistics of landings in Table 18.

Landings of halibut in Prince Rupert reached large proportions with the arrival of the three steamers of the Canadian Fish and Cold Storage Co. in 1913. Prior to then there were some small vessels fishing out of the recently established city. Landings escalated with the opening of the port on March 9, 1915, to United States halibut vessels to store and to ship the catches through Canada in bond. The landing in Prince Rupert of the United States vessel *Success* of Ketchikan on March 9, 1915, assured a new dimension for the port as the "Halibut Capital of the World." Seattle companies established branches in the port. A fish exchange was incorporated on October 15, 1915, which in the next half century was destined to sell many hundred millions of pounds of halibut.

The business that the halibut landings by United States vessels generated in Prince Rupert and the freight revenues that accrued to the government-owned transcontinental railroad were important considerations for Canada's undertaking one-half the cost of investigating and managing the resource under the halibut conventions. It offset the fact that Canada's share of the total Pacific Coast catch was only about 20% in the early 1920s at the time of the first convention.

Nearby Ketchikan concerns made many attempts to alter or eliminate the bonding privileges of the United States vessels in Canada. They wished fish to be transshipped to Prince Rupert in the expectation it would increase Ketchikan's share of the business generated by the halibut fleet. While some business elements in Ketchikan believed the latter to be true, it is almost certain that if the options of the individual halibut vessel had been curtailed, many vessels would have proceeded to Puget Sound ports to land and reoutfit, and both Ketchikan as a reoutfitting port and Prince Rupert as a landing port would have been the losers.

The decline in United States landings in Prince Rupert in the 1930s resulted from the Seattle fishermen spending their voluntary between-trip lay-ins at home. The decline continued after World War II due to some aggressive buyers in Ketchikan and a steady increase in vessels landing in Central Alaska ports.

In the mid-1930s several cooperative marketing groups were developed in British Columbia. Through consolidations the Fishermen's Cooperative Association came into being in the early 1940s. Under good management it soon became a major factor in the marketing of Prince Rupert halibut and had much to do in changing the character of the industry.

A large proportion of the Prince Rupert fleet joined the "coop" with their production captive to the coop. To compensate for this diversion from the open market, the independent companies encouraged a large number of salmon gill-netters to fish halibut before the salmon season. Most of the small boats operated out of buying stations near the fishing grounds. Their catch was brought into Prince Rupert and other ports in consolidated packer loads. The foregoing development resembled that used in the earliest years of the fishery out of Vancouver, New Westminster and Victoria.

Maintaining the camps and packers proved to be costly and some companies joined together in servicing some of their respective buying stations, in an attempt to hold down costs. With the continued improvement in the halibut stocks, companies with salmon and herring operations outfitted more and more of their seiners to fish halibut between the winter herring season and the summer salmon season. Unlike the company steamer era, these "company" vessels in most instances sold their catches on the open market.

The net effect of this was an increase in the Canadian fishing effort and a sharp rise in the Canadian share of the Pacific Coast catch. Another consequence was a further shortening of the fishing seasons.

While the independent companies have maintained an active halibut operation in Prince Rupert the Fishermen's Cooperative Association has tended to dominate the halibut scene in that port. Many of the company-owned or associated halibut vessels have tended to land in Western Alaska ports or in Vancouver or in nearby Bellingham, Washington.

Despite all the changes that have occurred and the large reductions made in the permitted catch limits for the fishery, Prince Rupert has managed to retain the title "The Halibut Capital of the World." Recent developments in Seward and Kodiak, Alaska, in the 1970s have been threatening that 60-year-old dynasty.

There were other locations on the British Columbia coast that became of secondary consequence as landing points for units of the regular halibut fleet. They were primarily salmon canning centers that possessed cold storages and bait supplies that could service halibut vessels. Many had but a transitory importance in the halibut fishery.

Claxton, located on the Skeena River was one of the earliest of the smaller centers for halibut operations. Its cold storage built in 1901 burned in 1965. In 1905, the Wallace Brothers, owners of Claxton, bought the *Queen of the Pacific* from Norwegians in Seattle. The vessel was known disparagingly as the "Bull of the Woods" because it spent long periods in harbor. It fished for the Barnett Codfish and Halibut Curing Plant briefly then went north to fish out of Claxton where the halibut were shipped frozen to Vancouver for disposal in the East.

Butedale was one of the more prominent secondary points as it was used both by Canadian and United States vessels fishing Hecate Strait as a frozen and fresh herring bait station as well as a landing point for broken trips by the United States vessels. Also many Canadian fares were bought in Prince Rupert by the Atlin Fish Co., a subsidiary of Canadian Fishing Co., which also owned Butedale, for delivery to the latter plant for freezing and storage. *Namu*, owned by another large fishery company, the British Columbia Packers, later offered Butedale considerable competition for the halibut trade. Klemtu, another salmon cannery and cold storage operation on the British Columbia coast was moderately active in the same services to the halibut fleet for a few years after World War II. Between 1947 and 1952 Klemtu landings ranged from one-half to one million pounds annually.

Among the innumerable places where halibut were purchased almost exclusively from small halibut boats or from trollers fishing primarily for salmon, Bull Harbour requires separate mention. Located on Malcolm Island on Queen Charlotte Sound, it concentrated on the early season purchase of halibut for many years and at times accounted for a considerable production that was packed to distributors in Vancouver, Victoria or Seattle. Many landings were made by boats from Sointula, a nearby Finnish community.

At one time Kildonan, primarily a salmon canning and fish reduction plant, in Barclay Sound on the west coast of Vancouver Island had facilities for the purchase and freezing of halibut and the sale of frozen herring bait to both Canadian and United States halibut vessels.

ALASKA

Waters contiguous to Alaska have produced over 75% of the Pacific Coast halibut catch since 1888 and in recent years over 75% of the total coast catch has been landed in Alaskan ports. The name Alaska, from the Aleut root meaning "great country or continent" was vaguely defined until Captain Cook, 1778, correctly recognized that Alaschka (Alaska) was used by both the Russians and the native people as applying to all of Alaska and not only to certain islands of the Aleutian Chain.

Though salmon have held the dominant position in the fisheries of Alaska since primeval times and still do, halibut was the next most important species to North Americans. Productive stocks of halibut are found along the 2,000-mile trend of the Alaskan coast from Dixon Entrance to Adak Island and into Bering Sea. It is interesting to note that the word halibut occurs, though less frequently than salmon, in many Alaskan place names. Their geographical distribution shown below by date of first use does suggest that halibut has long been a fairly visible species in many parts of Alaska.

1848 Halibut Point—4 miles northwest of Sitka
1880 Halibut Cove—East shore of Kachemak Bay
1883 Halibut Rock—1.6 miles northwest of Hoonah
1897 Halibut Point—West bank of Portland Canal
1901 Halibut Island—Port Frederick, Chichagof Island
1904 Halibut Harbor—Southwest coast of Koseivska Island
1905 Halibut Nose—North end of Tlevak Strait
1943 Halibut Bay—West coast of Kodiak Island
1951 Halibut Creek—North end of Chichagof Island
1961 Halibut Creek—Kachemak Bay
1961 Halibut Cove Lagoon—East shore of Kachemak Bay
1962 Halibut Cove—Northwest of Juneau

What is now called Sanak Island was named Halibut Island by Captain James Cook in 1778 because there his crew caught more than 100 halibut weighing 20 to 100 pounds.

Equally important as the large stocks of adult halibut in Alaska are the major spawning grounds on the contiguous continental shelf, particularly in the northern arc of the Gulf of Alaska. They are the source of eggs and developing larvae that drift far westward to settle in the bays along the Alaska Peninsula and particularly on the shallows of the southeastern Bering Sea. As a result a great concentration of young halibut existed in the Bering Sea and were also widely dispersed on the 65,000-square-mile continental shelf from Unimak Pass to Cape Spencer. These young fish provide recruits to fishable stocks including those off Southeastern Alaska, British Columbia and probably farther south. The effects of modern trawling upon the supplies of these small fish has been disastrous.

The growing importance of Alaska to the halibut fishery after World War II is reflected in the 1953 Halibut Convention's agreement that there would be six instead of four commission members. Primarily this was to permit the United States representation from Alaska. Secondly, it permitted both northern and southern British Columbia representation.

The increased importance of Alaska since World War II in the halibut fishery is reflected, too, in the increased landings by both United States and Canadian vessels there as indicated in Table 19.

TABLE 19. Average Annual Landings in Alaska by Canadian and United States Vessels in Thousands of Eviscerated, Heads-off Pounds

Period	United States	Canada	Total
1916-20	9,152	67	9,219
1921-25	10,690	6	10,696
1926-30	12,904	15	12,919
1931-35	6,955	2	6,957
1936-40	8,592	11	8,603
1941-45	15,265	75	15,340
1946-50	23,200	276	23,476
1951-55	19,059	1,368	20,427
1956-60	20,497	2,671	23,168
1961-65	22,284	5,233	27,517
1966-70	18,195	4,925	23,120
1971-75	15,907	4,292	20,199
1976-78	13,193	2,225	15,418

Ketchikan

Ketchikan is at the southern end of the Alexander Archipelago, which is often called the Alaskan "Panhandle." It was the United States Customs port of entry by ship into Alaska. For many years all halibut vessels from Puget Sound or British Columbia, both United States or British-Canadian en route to fishing grounds off Alaska made entry at Ketchikan, and many exited at that port for southern railhead ports. This and the opening of a large cold storage in Ketchikan by the New England Fish Co. in 1910 made the port a major factor in the halibut fishery of Alaska.

Ketchikan is named after Ketchikan Creek. The word derives from the Tlingit *Kitschk-him*—meaning "wings of a prostrate eagle," and came into being with a small settlement in 1885. The first salmon cannery was built in 1887. Gold was discovered the next year. The gold mining industry soon passed into oblivion, but salmon canning prospered for many years. Ketchikan came to be known as the "Salmon Capital of the World," but that title faded in later years as the stocks of salmon declined.

After the turn of the century the Puget Sound halibut boats fished out of Ketchikan in the winter, but not as many as did from Petersburg. Small boats did fish the year-round for the local market and personal subsistence, but salmon dominated the scene.

By 1907 it had become evident to Grier Starratt, then manager of the New England Fish Co. at Vancouver, that to develop the halibut fishery off Alaska would require a major base of operations in Southeastern Alaska. The New England Fish Co. selected Ketchikan as the site of a large cold storage plant and fish handling plant despite the misgivings of a few corporate directors. Ketchikan was the nearest United States port to Prince Rupert, which was to be the Pacific terminus of the Grand Trunk and Pacific Railroad then under construction. In the years following, while the proximity of the two ports was in fact mutually advantageous to one another, many unsuccessful attempts were made by Ketchikan commercial interests through political and legal means to maximize their advantage—usually to the detriment of Prince Rupert interests.

The landing on January 1, 1910, of a small halibut catch at the completed New England Fish Co. plant signaled the beginning of Ketchikan as an important Pacific Coast halibut center. In addition to independently owned vessels, several large fishing vessels owned by the New England Fish Co. began to use Ketchikan as their base of operations, including the large auxiliary-powered *Prospector* and some steamers.

TABLE 20. Average Annual Landings of Halibut by United States Vessels in Ketchikan in Thousands of Eviscerated, Heads-off Pounds

Period	Pounds
1927-30	6,143
1931-35	3,826
1936-40	3,244
1941-45	6,465
1946-50	7,925
1951-55	5,408
1956-60	7,133
1961-65	8,808
1966-70	5,097
1971-75	986
1976-78	258

Note: Small amounts of halibut were landed in Ketchikan from time to time by Canadian vessels.

Ketchikan soon became the northern springboard in United States territory for the opening of the halibut resources of the entire North Pacific and Bering Sea. It became the major outfitting and departure point for the large United States fleet of independently owned vessels, chiefly from Seattle, as they went to the newly opened grounds west of Cape Spencer in 1913. It continued to be the center of operations for a local fleet that fished both the inside channels and outside grounds of the Alexander Archipelago.

Near Ketchikan are the Indian centers of Metlakatla, Craig, and Hydaburg whose peoples fished halibut prior to engaging their vessels in salmon fishing. The extent of Ketchikan's direct contribution to the halibut fishery since 1927 is evident in the landing statistics.

At times small but inconsequential amounts of halibut have been landed by Canadian vessels in Ketchikan. The sharp fall in total landings since the late 1960s is due to the smaller coastwise catch and diversion of landings to Central Alaska ports.

Klawock, on the west coast of Prince of Wales Island, was the location of one of the first of two salmon canneries built in Alaska in 1878. The other was in old Sitka. Klawock (Klawak) is listed with those places of some importance in the halibut fishery only for the historical fact that in its inaugural year, 1878, the Klawock cannery also processed from 200 to 300 cases of two-pound cans of halibut. Thus, at least on a token basis halibut shares with salmon the distinction of being the object of the initial canning of fishery products in Alaska. Canning of halibut was abandoned after a few seasons due to poor market reception while salmon canning in Alaska was destined to become a multimillion dollar enterprise. It is, however, interesting that 66 years later in 1944 the Libby, McNeil and Libby Company in the same general region made an experimental pack of about 10,000 cases of halibut. Market reception was again indifferent.

Petersburg

Petersburg's economy has been centered upon halibut almost from its founding in 1897 by Peter Buschmann after whom the town was named. It became known as "Little Norway" because of the predominance of persons of Norwegian descent in the town. Halibut fishing has been relatively more important to the local economy than is the case in other fishery centers in Alaska.

Sporadic, but initially unsuccessful attempts had been made to develop a halibut fishery in the region. In 1889 the Icy Strait Packing Co. built a salmon cannery and a wharf in Petersburg. It was then possible to have regular steamer service to Puget Sound ports without which fresh halibut operations would not be practicable. Many Puget Sound vessels, mostly small five- or six-man auxiliary-powered sloops or schooners that fished during the months of more clement weather off Cape Flattery, soon began to proceed to Southeastern Alaska to fish the more protected inside channels of the Alexander Archipelago in the winter. The colder weather controlled spoilage of the southbound fish which were packed in 500-pound boxes with ice taken from glaciers. Prices were firmer in the south as stormy winter weather reduced the landings in the railhead ports by the large company-owned steamers that fished dories mostly in Hecate Strait.

TABLE 21. Average Annual Landings in Petersburg by United States Vessels in Thousands of Eviscerated, Heads-off Pounds

Period	Pounds
1927-30	1,459
1931-35	742
1936-40	1,700
1941-45	1,870
1946-50	2,684
1951-55	3,181
1956-60	3,441
1961-65	4,863
1966-70	3,777
1971-75	2,810
1976-78	1,774

Note: Small amounts of halibut were landed in Petersburg from time to time by Canadian vessels.

By the winter of 1906, a typical year, there were 23 Puget Sound vessels in Petersburg, 6 gasoline powered, 14 schooners, and 3 sloops, between 15-22 tons and usually with five- or six-man crews. In addition there were 8 vessels indigenous to Southeastern Alaska, including 4 gas screw boats, 5 schooners, and 9 sloops. The Puget Sound and Alaska sloops and schooners usually had small auxiliary engines.

The winter operation was a successful activity for the first 10 years of the twentieth century but the Southeastern Alaska ports gradually developed their own local fleets and Puget Sound vessels ceased their winter operations. Petersburg soon became the home port for a closely knit fleet of uniformly successful medium-sized halibut boats and continues to this date to be the mainstay of the Southeastern Alaska halibut fleet.

The halibut industry in Petersburg as a whole has consistently supported the conservation of the fish resource, and from the fleet has come two vessel owners to be United States members of the International Pacific Halibut Commission: L. Adolph Mathisen, 1965-1970, and Neils M. Evans, in office since 1970. The Petersburg halibut fleet maintained a successful marketing cooperative for many years. A number of commission buyers for Seattle and eastern halibut dealers operated in Petersburg. In recent years Icicle Seafoods Inc. owned largely by Petersburg vessel owners, has played the dominant role in the port with branches in Seward and Sitka. The president of the firm, Robert M. Thorstenson, was a member of the International North Pacific Fisheries Commission of United States, Canada and Japan. The firm with executive and sales headquarters in Seattle has become a broadly based fishery enterprise, although halibut remains an important part of its product line.

The continuity of the Petersburg interest in the halibut fishery is evident in the following summary of the landings during the past 50 years. Overall landings have been reasonably well maintained despite the decline in the total coast catch since 1960 and the shift of a few of the larger vessels to Central Alaska ports to land.

In addition to the landings cited in Table 21, Canadian vessels have, since the 1950s, landed fares in Petersburg but not totaling over 100,000 pounds in any one year.

TABLE 22. Average Annual Landings in Wrangell by United States Vessels in Thousands of Eviscerated, Heads-off Pounds

Period	Pounds
1927-30	246
1931-35	198
1936-40	437
1941-45	616
1946-50	609
1951-55	385
1956-60	645
1961-65	497
1966-70	438
1971-75	406
1976-78	400

Note: No Canadian vessels landed in Wrangell until 1978.

Wrangell

Wrangell, though one of the lesser ports for landing halibut in Alaska over the past 50 years, has had the most consistent record of halibut receipts of any place in Alaska.

Wrangell was established by the Russians in 1834 and originally called Redoubt Saint Dionysius. It was designed to block the Hudson's Bay Co. from trading up the Stikine River. In 1839, the English company leased the region from the Russians and changed the name to Fort Stikine. After the purchase of Alaska, the United States named the military post Fort Wrangell after the island which commemorated Baron von Wrangel, a governor of Russian America.

Wrangell like Petersburg, but to a lesser degree, attracted a number of auxiliary-powered sloops from Seattle in the early years of this century whose halibut catches were shipped boxed in glacier ice to Puget Sound in the winter months. Later vessels fished the year-round out of Wrangell. Fishermen used stored glacier ice in Wrangell until 1923. There was also a cold storage ice plant.

Although close to Petersburg, Wrangell was the home port for several medium-sized halibut vessels. Annual landings tended to average about 500,000 pounds of halibut a year. It has not been affected by the diversion of the larger members of the halibut fleet to Central Alaska.

Sitka

Sitka had an early but passing encounter with the halibut fishery when the *Oscar and Hattie* in 1888 spent a long and convivial time there while replacing a mast and rigging damaged in a winter storm.

Sitka had been an early Tlingit village that by force of Russian arms under Alexander Baranov in 1804 became a center of Russian colonization and fur hunting. In 1806 the capital of Russian America, which then extended from California to Bering Sea, was transferred from Kodiak to Sitka.

At 3:30 p.m. on October 18, 1867, an event occurred in Sitka of unimaginable geopolitical significance for the United States, if not for the world. The title to Alaska of 586,400 square miles was acquired by the United States from Russia on payment of $7,200,000 or 1.4¢ per acre. This real estate bargain was eclipsed in cost only by the Louisiana Purchase of 1803 when 855,000 square miles of what is now United States was acquired from France for $15 million or 2.7¢ per acre. Sitka thereupon became the seat of the United States Territorial Government for Alaska until the capital was transferred to Juneau in 1906.

Because Sitka is on the outside coast and not on the direct steamer run to the Puget Sound, a winter halibut fishery did not develop there as it had out of Petersburg and other ports in Southeastern Alaska. Later, with the building of a cold storage in the port in 1913, a small fleet of halibut vessels made Sitka their home port. Sitka eventually became a landing and outfitting point for several primarily salmon fishing vessels engaged in halibut fishing out of the nearby Tlingit community of Angoon.

TABLE 23. Average Annual Landings in Sitka by United States and Canadian Vessels in Thousands of Eviscerated, Heads-off Pounds

Period	United States	Canada	Total
1927-30	1,222	—	1,222
1931-35	588	—	588
1936-40	1,053	6	1,059
1941-45	2,005	63	2,068
1946-50	2,587	185	2,772
1951-55	1,658	176	1,834
1956-60	1,128	132	1,260
1961-65	982	133	1,115
1966-70	714	72	786
1971-75	842	41	883
1976-78	717	8	725

Abundant supplies of fresh herring in the vicinity of Sitka made that port a favorite baiting point for many of the large halibut schooners out of Seattle en route to the offshore fishing grounds west of Cape Spencer. During the prohibition era of the 1920s it seemed that a few such vessels made the port a place of "rest and relaxation" to kill off any remaining bottles of Canadian whiskey bought while selling their previous catches in Prince Rupert. Drinking, once crews left port or entered the fishing grounds, was rare in the halibut fleet.

Due to the proximity of the port to fishing grounds in the Gulf of Alaska, much of the Sitka landings by both United States and Canadian vessels originated from west of Cape Spencer. The landings by the smaller vessels came from fishing grounds off the outside coasts of Southeastern Alaska rather than from the inside channels, as in the case of other ports in the Alexander Archipelago.

The decline in the volume of landings since World War II is a reflection of the reduced number of vessels making Sitka their home port. The more recent decline is due to the overall drop in total Pacific Coast production and the diversion of some vessels to Western Alaska ports.

Juneau

Juneau, the capital city of Alaska since 1906, was known by various names until 1881 when the placer miners named it after Joseph Juneau who owned the Discovery Claim on Gold Creek. Juneau had staked out the townsite. It experienced the typical boisterous beginning of a frontier placer mining town. Later as the capital of the Territory of Alaska, government business and the famed Alaska-Juneau Consolidated hard rock gold mine dominated the economy of the town. The mine closed in 1944. Fisheries were of secondary importance, although, being on the main steamer run to Seattle, Juneau was a major distribution point for many salmon canneries in the vicinity.

The earliest halibut fishing out of Juneau was done in 1897 by Martin Holst for local consumption. He fished in Lynn Canal, and then in Chatham Strait in a sail boat of the open Columbia River type. In 1899 he acquired a 15-ton auxiliary-powered sail boat and with Jack Rowe's boat the *Prospector* the two vessels shipped their catches to Seattle for two cents per pound, which often failed to cover costs.* In 1902 Chlopeck and the San Juan fish companies of Seattle began to buy halibut from Juneau. In 1904 Holst built one of the first gas boats in Juneau and called it *Belle.*

There was an abundance of glacier ice—Mendenhall Glacier, although in a dwindling stage, was practically inside the city limits. With the construction of a cold storage in 1913, Juneau soon developed a small indigenous fleet of three- to five-man halibut boats that were well adapted to fishing the waters of Icy Strait and Cross Sound, with their heavy tidal flows. After World War II a number of salmon seiners from the nearby Indian village of Hoonah fished halibut in the spring and landed in Juneau or Pelican City.

TABLE 24. Average Annual Landings in Juneau by United States and Canadian Vessels in Thousands of Eviscerated, Heads-off Pounds

Period	United States	Canada	Total
1927-30	2,019	—	2,019
1931-35	1,309	—	1,309
1936-40	1,895	2	1,897
1941-45	1,851	—	1,851
1946-50	2,613	12	2,625
1951-55	2,325	243	2,568
1956-60	2,063	299	2,362
1961-65	1,676	81	1,757
1966-70	1,651	28	1,679
1971-75	969	2	971
1976-78	465	—	465

Yakutat

Yakutat, long a winter village of Tlingit, had some prominence in the halibut fishery. Monti Bay, inside the entrance of Yakutat Bay, was a haven of shelter for the halibut fleet fishing during the bitter winter weather in the years before the closed season took effect in 1924. Monti Bay was the only reasonably accessible anchorage along the desolate 320-mile coast from Cape Spencer to Cape Hinchinbrook.

*Holst reported catching a halibut at the head of Lynn Canal in 1899, four pieces of which when cleaned with head and fins removed weighed a total of 400 pounds. The live weight of such a fish would probably have been from 515 to 540 pounds, the largest Pacific halibut ever reported.

The economy of Yakutat was based on the local salmon cannery which at one time transported its salmon from the nearby Situk River. Stormbound halibut vessels, often in large numbers, moored at the cannery dock where crews could replenish their water supply. While in later years small amounts of incidentally caught halibut were landed in Yakutat by salmon trollers, it did not qualify as a halibut port.

Seward

Seward was founded in 1903 by the Alaska Railroad Survey team. It was to be the ocean terminus of the Alaska Railroad running to Anchorage and Fairbanks. It was named for William H. Seward, secretary of state for the United States, who negotiated the purchase of Alaska from Russia.

TABLE 25. Annual Landings in Seward by Canadian and United States Vessels in Thousands of Eviscerated, Heads-off Pounds

Year	United States	Canada	Total
1970	2,490	1,556	4,046
1971	2,442	1,169	3,611
1972	3,499	1,557	5,056
1973	2,850	1,122	3,972
1974	1,686	244	1,930
1975	3,111	825	3,936
1976	2,845	573	3,418
1977	2,562	587	3,149
1978	2,841	555	3,396

Seward became an important halibut fishery soon after the San Juan Fishing and Packing Co. of Seattle built a cold storage plant there in 1917. It provided the halibut fleet a convenient port to dispose of "broken trips," those catches too small or with fish too old to be run east to the distant railhead ports. Reoutfitting supplies were also available.

Seward satisfactorily served its primary purpose for the halibut fleet until 1931. Thereafter very little halibut was landed in Seward until in the 1960s, except for some landings from 1949 to 1951. The establishment of several fishery ventures with the associated dock facilities in the port in the 1960s made Seward a major halibut port in the 1970s. Landings in recent years are shown in Table 25.

Kodiak

Long a minor factor in the halibut fishery Kodiak underwent an abrupt change in 1962 and by 1970 gained the distinction of being the largest receiver of United States-caught halibut on the Pacific Coast, second only to Prince Rupert in total annual volume of United States and Canadian landings combined.

Kodiak, located on the northeastern coast of Kodiak Island, the largest island in the Northeastern Pacific except for Vancouver Island, was named by Captain James Cook in 1778. The name derived from the Innuit word *Kikhtak* meaning island. Until the early 1900s it was usually referred to as Kadiak.

Despite the fact that the most productive halibut banks west of Cape Spencer such as Portlock, Albatross, Trinity and Shelikof are found on the wide continental shelf contiguous to Kodiak Island, there were several reasons for the late emergence of Kodiak as a landing center. In the 1930s vessels returned to their home ports to fulfill their self-imposed between-trip lay-ins. In the 1940s an already existing salmon cannery and a cold storage facility at nearby Port William were able to handle up to 3 million pounds of halibut annually. With the acquisition by private parties after World War II of military cold storage at Sand Point in the Shumagin Islands that thereafter sometimes handled up to 2 million pounds of halibut annually, Kodiak was further deprived of potential landings of halibut from the far western grounds.

TABLE 26. Average Annual Landings of Halibut in Kodiak by United States and Canadian Vessels in Thousands of Eviscerated, Heads-off Pounds

Period	United States	Canada	Total
1961-65	1,041	1,462	2,503
1966-70	2,671	2,387	5,058
1971-75	4,092	2,348	6,440
1976-78	3,810	647	4,457

Prior to 1961 landings of halibut in Kodiak were negligible.

It was not until the 1960s with the phenomenal growth of the crab and shrimp fisheries that it was economically attractive to build facilities in Kodiak to handle halibut in volume. It then became a major fishery center attracting several buying companies. Affluent fishermen and frequent air flights made it possible for crews of halibut vessels to fly home for their self-imposed between-trip lay-ins.

In the 1970s declining stocks required more days of fishing per trip. To offset the effects on quality many vessels chose to land at ports closer to the grounds, particularly now that there was competition between buyers for the fish. Both Kodiak and Seward were close at hand. The termination of the fleet's voluntary lay-in program in 1977 and the continued low stock levels will probably keep Kodiak and Seward important halibut ports for some time to come.

In addition to the main population centers there are a number of places in Alaska with a salmon cannery and a cold storage facility where halibut may be bought. In the early years of World War II a cold storage plant was established at the western end of Icy Strait and named Pelican City. It drew upon the salmon trolling fleet fishing off Cape Spencer and west to Cape St. Elias. Starting with halibut receipts of under 500,000 pounds in 1943, it soon became an important landing port for halibut both from inside grounds and those outside Cape Spencer. The proportion of Canadian landings is relatively high as is evident in Table 27. It is probable that Pelican City diverted some landings away from Juneau.

TABLE 27. Average Annual Landings of Halibut in Pelican City by United States and Canadian Vessels in Thousands of Eviscerated, Heads-off pounds

Period	United States	Canada	Total
1941-45	788	20	808
1946-50	2,519	—	2,519
1951-55	2,097	655	2,752
1956-60	1,702	426	2,128
1961-65	1,115	397	1,512
1966-70	1,089	316	1,405
1971-75	784	433	1,217
1976-78	798	549	1,347

The decline reflects the shift of landings to Western Alaska ports and the reduction in the total catch.

A similar situation is Port William which has a salmon cannery cold storage owned by a Seattle halibut dealer. It is located on Shuyak Island close to Kodiak. Not being a customs port, Canadian vessels could not unload except in emergencies and with special United States Customs' permission.

The decline in landings shown in Table 28 after the mid-1960s reflects the shift of landings to nearby Kodiak and to the not-too-distant Seward, both of which were population centers and customs ports.

In 1948 a former military supply base was converted to private use as a cold storage plant in the Shumagin Islands at Sand Point. The development attracted vessels to fish farther west. It also served as a base for operations in the eastern Bering Sea and was used as a control point for enforcement purposes of vessels fishing in Bering Sea.

TABLE 28. Average Annual Landings of Halibut in Port William by United States Vessels in Thousands of Eviscerated, Heads-off Pounds

Period	Pounds
1940-45	1,406
1946-50	1,417
1951-55	1,156
1956-60	1,758
1961-65	1,464
1966-70	308
1971-75	174
1976-78	110

With the reduction in the total catch, particularly in the Bering Sea, along with the diversion of landings to Kodiak, landings of halibut at Sand Point reached a low level by the 1970s and were negligible after 1975.

In the late 1950s landings by Canadian vessels usually exceeded those of the United States as shown in Table 29.

TABLE 29. Average Annual Landings of Halibut at Sand Point by Canadian and United States Vessels in Thousands of Eviscerated, Heads-off Pounds

Period	United States	Canada	Total
1948-50	1,452	—	1,452
1951-55	1,136	164	1,300
1956-60	1,323	1,410	2,733
1961-65	1,373	2,410	3,783
1966-70	861	1,341	2,202
1971-75	259	146	305

In addition to the foregoing places, many other Alaskan locations have been the site of modest landings of halibut from time to time, including Seldovia, Homer, Valdez and Cordova in Central Alaska. In Southeastern Alaska many buying stations for halibut and troll caught salmon have operated for varying periods of time. These receiving points are among the several hundreds of places on the Pacific Coast where halibut in relatively small amounts have been landed down through the years. Many consisted of only a barge anchored in some cove.

There have been a few but sporadic attempts to use floating freezing plants. The Alaska Packers Association in 1911 at Kodiak was unsuccessful with the *Metha Nelson.* In 1913 an old sailing vessel, *Glory of the Seas,* equipped with freezing facilities and storage capacity, was located in Idaho Inlet in Icy Strait. The short period of operation suggests that it was not too successful. In 1929, 1930, and 1931 the *Donna Lane* of Utopian Fisheries Inc., another floating freezer, operated at Unalaska and froze about 500,000 pounds of halibut annually. It also put up a pack of salted herring, which from that section of the coast were of large size. The onset of the Great Depression probably terminated the whole project.

6 ORIGINS OF HALIBUT CATCHES

CATCHES BY SECTIONS OF THE COAST FROM 1888 TO 1978

The catches by Canadian and United States fleets by sections of the coast since 1921 are grouped by various periods in Table 30 and in Figure 68 to show the basic trends. Such grouping avoids the obscurity of detailed year-to-year tables.

Table 31 continues the data for years prior to 1921. The total catch could only be accurately segregated into that taken south (Area 2), and that taken west (Area 3), of Cape Spencer, Alaska.

By 1960 most of the major stock components had been rebuilt and their yields were close to the sustainable maximum under the then prevailing environmental conditions.

Under the pressures of foreign fishing and lowered recruitment, the yields to the Canadian and United States fleets underwent a decline in all areas in the second half of the 1960s, and it continued through the 1970s. The reductions were greatest (66%) off British Columbia and Kodiak Island where trawling had been intense and prolonged. On other sections of the coast the reduction was about 50%.

TABLE 30. Average Annual Catches of Halibut by Sections of the Coast by Canadian and United States Vessels in Millions of Eviscerated, Heads-off Pounds

60-mile sections of coast		1921-1929	1930-1939	1940-1949	1950-1959	1960-1964	1965-1969	1970-1974	1975-1977	1978
4-8		3.1	2.1	1.9	1.9	1.6	1.2	.8	.7	.3
9-13	Area 2	14.9	13.5	15.6	18.6	14.2	10.2	7.7	5.9	4.2
14-18		8.8	8.2	9.3	10.9	11.3	9.9	6.7	5.2	4.5
Total		26.8	23.8	26.8	31.4	27.1	21.3	15.2	11.8	9.0
18-23		8.1	6.3	5.0	7.3	5.5	5.8	4.1	3.7	3.1
24-28	Area 3	12.3	12.0	15.5	16.9	17.1	14.8	11.0	6.4	7.2
29-36		4.0	5.3	7.2	6.2	14.3	14.1	8.4	3.1	2.0
Total		24.4	23.6	27.7	30.4	36.9	34.7	23.5	13.2	12.3
Bering Sea		—	—	—	1.0	5.4	1.5	.5	.5	.7
Total Pacific Coast		51.2	47.4	54.5	62.8	69.4	57.5	39.2	25.5	22.0

4-8 Willapa Bay to Cape Scott
9-13 Cape Scott to Dixon Entrance
14-18 Dixon Entrance to Cape Spencer
18-23 Cape Spencer to Cape St. Elias
24-28 Cape St. Elias to Trinity Islands
29-36 Trinity Islands and west to Bering Sea

TABLE 31. Average Annual Catch From South (Area 2) and West (Area 3) of Cape Spencer in Millions of Eviscerated, Heads-off Pounds

Region	1888-1892	1893-1896	1897-1900	1901-1904	1905-1908	1909-1912	1913-1916	1917-1920
South of Cape Spencer (Area 2)	2	4	9	25	40	54	44	29
West of Cape Spencer (Area 3)	—	—	—	—	—	—	19	14
Total Pacific Coast	2	4	9	25	40	54	63	43

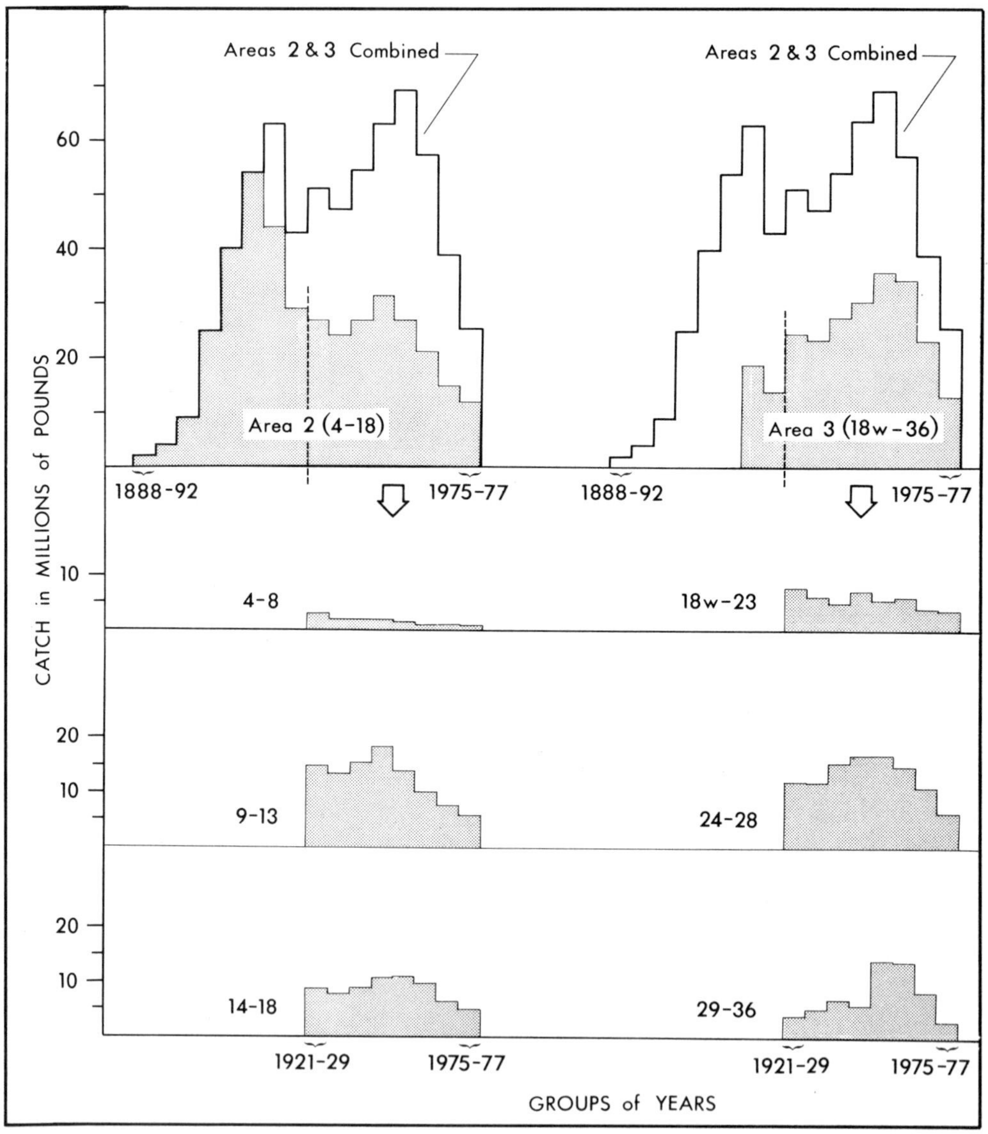

FIGURE 67. Catches by regulatory areas, Areas 2 and 3, since 1888 and by 60-mile sections of the coast 4-8, 9-13, 14-18, 18W-23, 24-28, 29-30 since 1921 for groups of years.
F.H. Bell

CATCHES BY EACH FLEET FROM WATERS CONTIGUOUS TO THE OTHER COUNTRY

The recent enactment of the 200-mile fishery zone laws in Canada and the United States brought into focus the long-standing question of the two countries fishing in waters contiguous to each other's shores.

Except in the case of certain special situations, a three-mile territorial limit from earliest times prevailed between Canada and United States with respect to fishing off each other's coast for Pacific halibut. In the very early years there were formal claims by Canada with regard to Hecate Strait that those waters were its territorial seas. In more recent years some areas in Dixon Entrance were also subject to some similar dispute. The latter developed from the fact that the boundary line separating the two countries extended offshore to a point of land on United States territory, namely Cape Muzon. In the very recent past Cook Inlet was another area that was subjected to some question as to territoriality. While a few Canadian halibut vessels had fished in that body of water from time to time, it was claimed to be exclusively United States territorial waters.

Despite such matters being the delight of demagogues, none of the foregoing claims reached any abrasive level as the waters possessed only minimum amounts of halibut, except in Hecate Strait. However, an important consideration in establishing limits of territoriality was how the baseline should be determined. During discussions after World War II to extend the limits of territorial seas from 3 to 12 miles, it was suggested by Canadian interests that the straight base-line principle be followed throughout, rather than the traditional "mean low waterline along the coast" with the "wide bay" exception. The straight base-line principle would have closed off

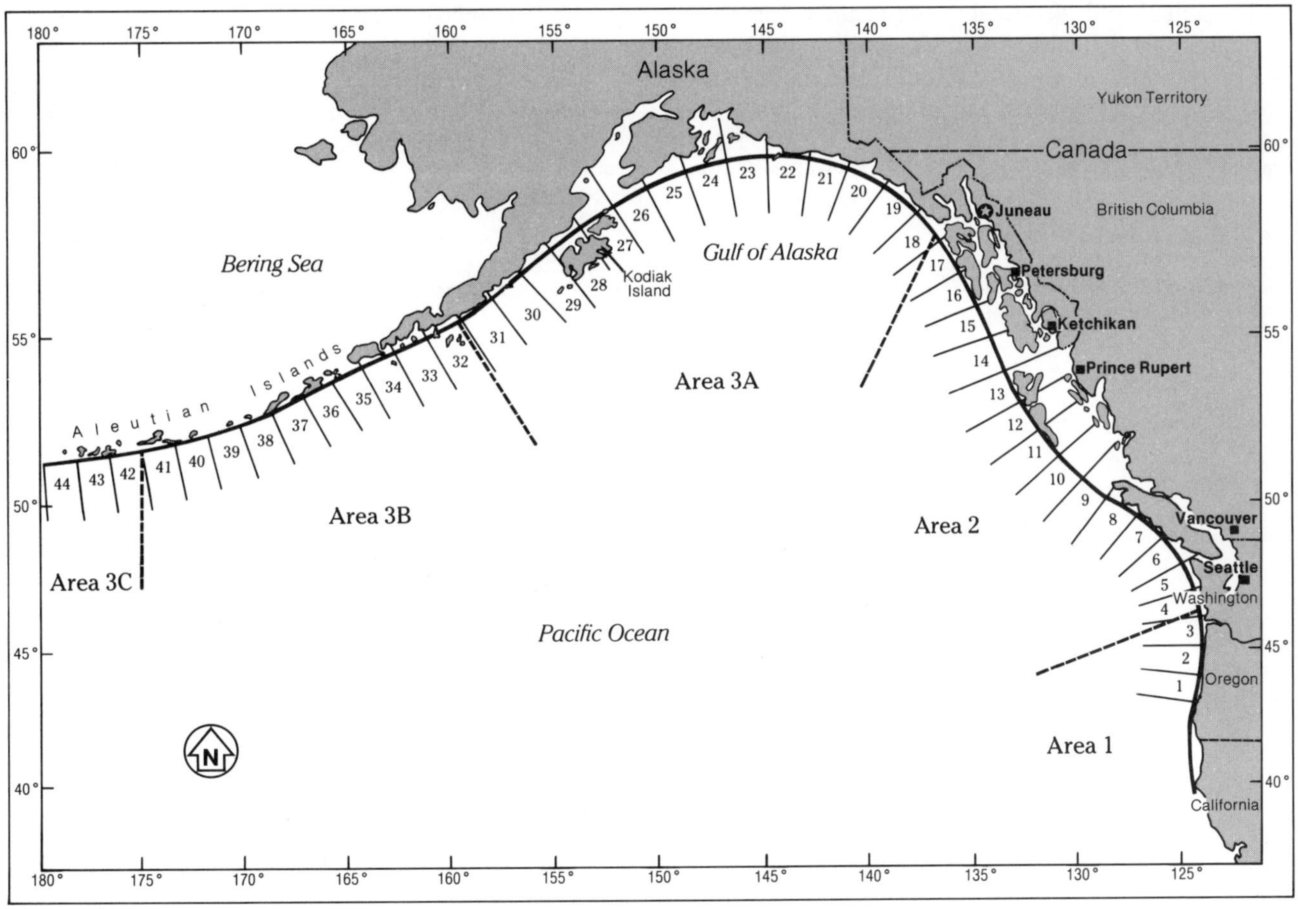

FIGURE 68. The Pacific Coast showing 60-mile statistical sections and larger regulatory areas devised by the Halibut Commission for the management of the halibut resource and fishery.
International Pacific Halibut Commission

large halibut fishing areas off British Columbia to foreign vessels including those of United States. With the strong likelihood that United States would reciprocate (or retaliate) and institute the same principle off Alaska and close large areas to Canadian halibut vessels fishing off Alaska, the proposal was not vigorously pursued by the Canadian government at that time.

Although now academic due to the unilateral implementation of 200-mile limits by the two countries, the impact of a 12-mile limit on the part of the two countries on each other's halibut fleets is indicated in Tables 32 and 33 for the conditions that prevailed in the late 1950s and early 1960s.

Since the beginning of the fishery in 1888 the amounts and proportions of the total catches taken by the Canadian and United States fleets off each other's coasts have changed radically.

United States Catches off British Columbia

The initial and historically significant commercial exploitation off British Columbia was by the United States vessel *Oscar and Hattie* that fished off Rose Spit and Tow Hill at the northern end of Graham Island in June, 1888. Between 1890 to 1898, United States vessels accounted for about 50% of the halibut taken off the British Columbia coast, which then amounted to about 2 to 4 million pounds annually.

Just before the turn of the century, fishing off the British Columbia coast by United States vessels expanded as a result of the introduction of company-owned steamers of United States registry fishing chiefly out of Vancouver. The crews on United States vessels fishing out of British Columbia ports were mostly Canadian. Such crew members were shipped under an Order-in-Council of November 8, 1897.

Until 1910, 75 to 80% of the catch off British Columbia, which had reached 50 million pounds annually by 1906, was taken by United States vessels out of Puget Sound ports and Vancouver. Between 1910 to 1920 with some of the United States catch then being taken off Southeastern Alaska and west of Cape Spencer, the amount and proportions of the total taken by that fleet off British Columbia declined although most of the large schooners and steamers still fished Goose Islands grounds and Hecate Strait in the summer. About 50% of the 400 million pounds taken off British Columbia in that decade was by United States vessels.

The extremely large removals by American flag vessels from waters contiguous to British Columbia was not favorably received in Canada. It was aggravated by the fact that much of the poundage was

TABLE 32. Catches of United States Vessels in Thousands of Eviscerated, Heads-off Pounds *In* and *Out* of the 12-Mile Limit Off British Columbia

	1962					Average 1953-1962				
	Catch			Percentage		Catch			Percentage	
Area	In	Out	Total	In	Out	In	Out	Total	In	Out
West Coast of Vancouver Island	152	174	326	47	53	241	218	459	52	48
Cape Scott and Queen Charlotte Sound	20	1,368	1,388	1	99	209	3,315	3,524	6	94
Hecate Strait	151	146	297	51	49	104	487	591	17	83
Total British Columbia coast	323	1,688	2,011	16	84	554	4,020	4,574	12	88

TABLE 33. Catches of Canadian Vessels in Thousands of Eviscerated, Heads-off Pounds *In* and *Out* of the 12-Mile Limit Off Alaska

	1962					Average 1953-1962				
	Catch			Percentage		Catch			Percentage	
Area	In	Out	Total	In	Out	In	Out	Total	In	Out
Cape Spencer — Cape St. Elias	840	2,487	3,327	25.0	75.0	625	2,842	3,467	18.0	82.0
Cape St. Elias — Trinity Islands	527	5,233	5,760	9.0	91.0	376	4,948	5,324	7.0	93.0
Trinity Islands — Shumagin Islands	1,183	4,181	5,364	22.0	78.0	383	1,832	2,215	17.0	83.0
Shumagin Islands — and West	1,280	4,554	5,834	22.0	78.0	620	1,726	2,346	27.0	73.0
Total West of Cape Spencer	3,830	16,455	20,285	18.9	81.1	2,004	11,348	13,352	15.0	85.0

The relatively small catches by Canadian vessels off Southeastern Alaska do not provide meaningful percentages.

taken out of Hecate Strait where territoriality was in question, and by the United States protectionist tariff that on an *ad valorem* basis was extremely high. Also the stocks were being severely depleted. These circumstances did not endear Canadian fishermen to companies operating American bottoms out of Vancouver and shipping the fish free of duty to the United States. It was also a period when there was a considerable amount of poaching by United States vessels inside the three-mile limit around Cape Scott, Ramsay and Banks islands.

During the next decade to 1929 the two fleets shared equally in the catch taken off British Columbia. After 1929 the total catch by the combined fleets off British Columbia increased until the 1950s, but the share taken by the United States fleet declined from about 50% to about 25% by the early 1960s. Since that time catches off British Columbia by United States vessels have reached a low level due to the reduced stocks and the decline in the size of the Seattle Area 2 fleet. In the 1970s United States vessels accounted for only about 6% of the total catch taken off British Columbia.

From once dominating the fishery off British Columbia for most of 80 years and playing the major role in the decline in the stocks up to 1930, United States vessels played a minor role in the 1970s.

Canadian Catches off Alaska

Fishing by Canadian vessels off the Pacific Coast of the United States has been largely limited to the waters contiguous to the coasts of Alaska. There has never been more than a negligible amount of halibut fishing by Canadian vessels off Washington, Oregon and California.

The initial significant commercial exploitation west of Cape Spencer by vessels of Canadian (British) registry occurred in 1913 when three large steamers out of Prince Rupert, British Columbia—the *Andrew Kelly, James Carruthers* and *George E. Foster*—fished in the region soon after the first commercial exploitation west of Cape Spencer by a United States vessel in September, 1912. No fishing was conducted off Southeastern Alaska by Canadian vessels in the early years.

In Figure 69 the extent of Canadian catches off Alaska are contrasted with the United States removals off British Columbia by decades from 1890 to 1976.

In the 1920s Canadian catch off Alaska did not exceed one half million pounds annually and was not

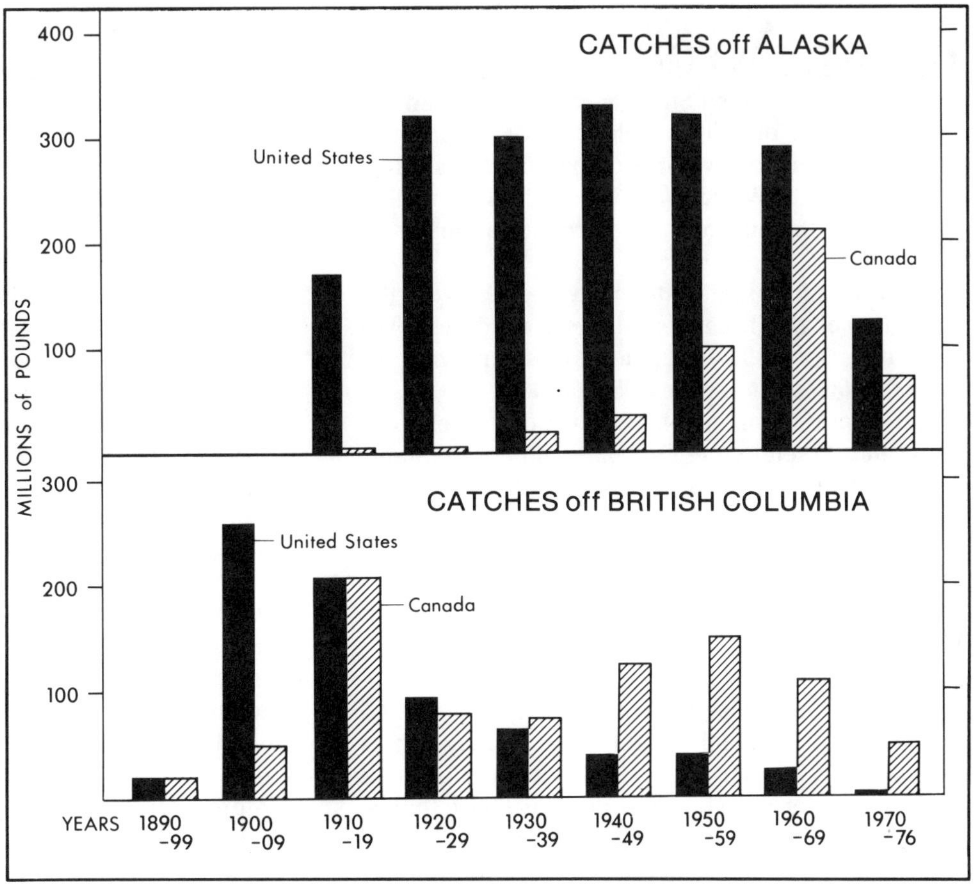

FIGURE 69. Catches of Canada and the United States off the other country's coast. F.H. Bell

over 3% of the total catch. While there was a modest increase in the total and the proportion taken by Canadian vessels after the mid-1930s, it was the mid-1940s before the total was over 4 million pounds annually, representing about 10% of the total catch by both fleets. Thereafter, the total and the proportion increased steadily and by the late 1960s nearly half of the catch off Alaska was taken by Canadian vessels. Since that time the Canadian share as well as their total catch west of Cape Spencer have declined drastically.

Canadian catches off Alaska were negligible during the period prior to 1930 when the stocks were declining because of overfishing. They became of some consequence in the 1950s and reached their maximum in the 1960s when the stocks were again at their maximum.

United States catches off British Columbia were large and dominated the removals during the period of overfishing that prevailed prior to 1930. During the subsequent rebuilding of the stocks off British Columbia, United States catches declined to a negligible level by the 1970s. Canadian removals increased until their catches also declined in the 1970s.

CANADIAN AND UNITED STATES DIVISION OF TOTAL CATCH

Not only had United States vessels dominated the removals taken off British Columbia for the most of 80 years, but the total volume of their catch was high. In years prior to World War I as much as 50 million pounds were taken from those grounds annually by the two fleets.

Similarly off Alaska, United States vessels from the beginning of the fishery of those grounds to 1975 accounted for 80% of the total removals of 2,240,000,000 pounds over a 65-year period.

The proportion of the total Pacific Coast catch from all grounds taken by the Canadian and United States fleets for the 60 years from 1916 to 1976 are shown in Figure 70, by five-year periods.

ORIGINS OF CATCHES BY DEPTH OF WATER

Setline halibut fishing is conducted on the continental shelf and slope in depths usually between 45 to 150 fathoms. Not infrequently vessels may fish as shallow as 15 fathoms, for example in parts of Cook Inlet, Alaska, or sometimes as deep as 300 fathoms. Winter fishing on the spawning grounds in the Gulf of Alaska was usually centered about 175 fathoms. There have been occasional instances where concentrations of halibut were encountered as deep as 600 to 700 fathoms. This occurred in some deep fjord-like portions of Prince William Sound, Alaska. In the Northwest Atlantic in Davis Strait, successful and sustained halibut fishing was conducted as deep as 750 fathoms during mothership operations in the 1920s.

Table 34 gives the depth by fathoms (f) or bathymetric distribution of the United States and Canadian halibut catch for a typical year (1961) by sections of the Gulf of Alaska in thousands of pounds and by percentage of each section total.

The 1961 fishing season covered the period from May to August inclusive. In other years with very early spring or, late fall or winter fishing, a somewhat larger proportion of the catch would have been taken at the 100+ f. depth. Due to the configuration of the shelf and coastline the same tabulation is not feasible for the fishery south of Cape Spencer.

SEAMOUNT FISHING

In the earlier years of the halibut fishery, government agencies were petitioned from time to time to conduct exploratory fishing to find those offshore banks where halibut were thought to migrate seasonally when not available on the accustomed grounds. These were not unreasonable requests as the detection of seamounts or shoal areas outside the continental shelf was beyond the sounding capability of the fleet at that time.

On August 1, 1950, the United States Fish and Wildlife Service vessel, the *John N. Cobb*, while on a research mission discovered a seamount about 270 miles off Willapa Bay, Washington. They returned on September 15 and 16, the same year, and made seven sets of setline gear totaling 1,470 hooks at a depth of 40 to 110 fathoms. Four halibut ranging in weight between 23 and 28 pounds were caught along with several species of rockfish.

Another known seamount lying about 120 miles westward of Graham Island, British Columbia, charted as the Bowie Seamount, was fished by a Canadian halibut vessel in the early 1960s. A moderate quantity of halibut were taken. Subsequently the Halibut Commission tagged 23 halibut on Bowie Seamount. Two recoveries were made on the same ground. The age composition which ranged from 15 to 30 years suggested that the halibut on the "mount" consisted of an accumulation of old fish that did not freely intermix with those on the continental shelf.

More sophisticated sounding equipment has resulted in the charting of many seamounts in the North Pacific ocean. Several of them showing depths

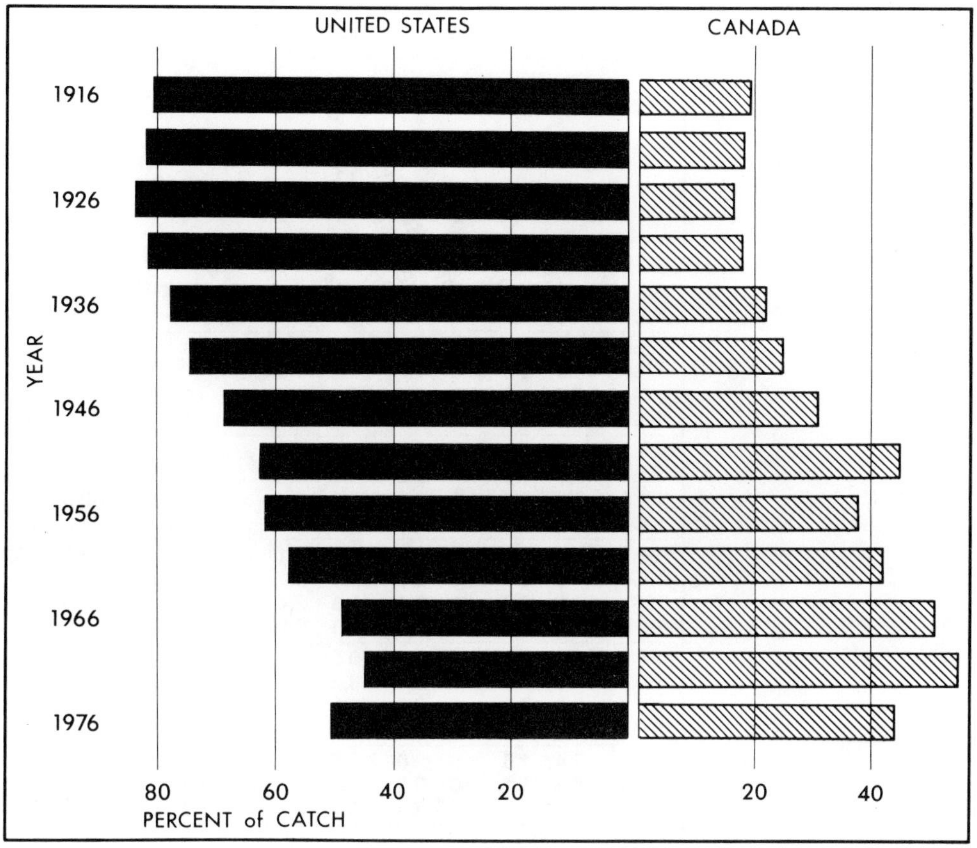

FIGURE 70. Proportion of total catch taken by each fleet by five-year periods from 1916 to 1976.
F.H. Bell

TABLE 34. Origins of Catches by Depth of Water

	Catch				Percentage of Section Total		
Section	**1-50f**	**51-100f**	**100+f**	**Total**	**1-50f**	**51-100f**	**100+f**
Cape Spencer to Cape St. Elias	600	2,320	1,420	4,340	14.0	53.0	33.0
Cape St. Elias to Kodiak Island	3,200	7,570	4,940	15,710	20.0	48.0	32.0
Kodiak Island to Shumagin Islands	3,810	6,560	5,130	15,500	25.0	42.0	33.0
West of Shumagin Is. not including the Bering Sea	680	280	—	960	71.0	29.0	—
Total West of Cape Spencer not including the Bering Sea	8,290	16,730	11,490	36,510	22.8	45.8	31.5

less than 300 to 400 fathoms have been explored by the more curious halibut fishermen whose vessels now possess the sounding equipment to do so. None of these seamounts have been found to possess any notable quantity of halibut. It is doubtful that any will be found to harbor more than a few halibut, nor will any have any sustainable production potential.

WORLD CATCH TRENDS OF HALIBUT

Much may be learned about present trends and future prospects of the Northeast Pacific setline halibut fishery by observing what has happened to halibut resources elsewhere in the world as the supplies available in other countries have had an impact upon the market for halibut in the United States and Canada.

Unfortunately, most of the published tables of the world production of halibut are defective. Yearbooks of the Food and Agriculture Organization (FAO) usually fail to report any figures for Japan, and those recorded for USSR fail to indicate that, at least since 1967, they include a preponderance of other flatfish species of *Atheresthes* and of *Reinhardtius.* For example, the USSR is credited by the FAO with taking 40

TABLE 35. Landings of Halibut by Oceans and Countries of Major Importance in Millions of Eviscerated, Heads-off Pounds

	ATLANTIC									PACIFIC					
	NORTHEAST					NORTHWEST				INCLUDING THE EASTERN AND WESTERN BERING SEA					
YEAR	NORWAY	UNITED KINGDOM	WEST GERMANY	OTHER EUROPEAN COUNTRIES	TOTAL	CANADA	UNITED STATES	OTHER	TOTAL	CANADA	UNITED STATES	JAPAN	USSR	TOTAL	WORLD TOTALS
1959	9.6	8.1	1.9	7.7	27.3	5.3	.3	*	5.6	30.8	40.4	4.0		75.2	108.1
1960	10.0	8.3	2.5	7.4	28.2	5.5	.3	*	5.8	33.5	38.0	11.4		82.9	116.9
1961	7.2	6.3	2.5	4.4	20.4	4.9	.2	*	5.1	29.4	40.0	18.4		87.8	113.3
1962	8.8	5.8	1.3	3.6	19.5	3.9	.2	*	4.1	34.6	40.2	17.8		92.6	116.2
1963	7.0	5.2	1.3	3.0	16.5	3.6	.2	*	3.8	37.1	34.1	16.9		88.1	108.4
1964	6.7	4.9	1.4	2.9	15.9	3.3	.2	*	3.5	33.5	26.2	5.9		65.6	85.0
1965	6.2	5.0	1.3	3.1	15.6	3.4	.2	*	3.6	32.9	30.2	4.9		68.0	87.2
1966	4.6	4.4	1.1	2.4	12.5	2.6	.2	*	2.8	31.9	30.1	5.3		67.3	82.6
1967	4.7	3.4	1.4	2.6	12.1	3.1	.2	*	3.3	25.5	29.7	9.5		64.7	80.1
1968	4.3	3.3	.9	2.8	11.3	2.6	.2	*	2.8	29.4	19.2	8.2		56.8	70.9
1969	4.5	2.8	1.1	2.1	10.5	2.4	.2	*	2.6	33.5	24.8	8.5		66.8	79.9
1970	4.4	2.4	.8	2.9	10.5	2.2	.1	.3	2.6	29.2	25.8	10.8		65.8	78.9
1971	3.9	3.0	.7	2.9	10.5	2.2	.1	.6	2.9	25.4	21.2	12.7		59.3	72.7
1972	3.1	2.3	.7	2.4	8.5	2.3	.1	.6	3.0	27.5	20.4	6.9	.8	55.6	67.1
1973	3.3	2.1	.6	2.5	8.5	2.3	.1	.7	3.1	14.4	17.3	6.2	.5	38.4	50.0
1974	2.6	1.5	.6	3.0	7.7	2.1	.1	.9	3.1	7.4	13.9	3.0	.2	24.5	35.3
1975	2.4	1.5	.6	3.0	7.5	2.2	.1	.9	3.2	11.4	16.3	2.0	.3	30.0	40.7
1976	2.5	1.1	.5	3.1	7.2	2.1	.1	1.2	3.4	12.0	15.5	1.2	.1	28.8	39.4
1977	2.0	.6	.7	3.4	6.7	2.2	*	1.3	3.5	8.8	13.1	.8	.1	22.8	33.0

*Under 100,000 pounds

million pounds of *Hippoglossus* in the North Pacific in 1968. The true amounts annually were probably well under 10 million pounds, dressed weight. Such defective data along with other economic parameters of equally doubtful validity are reported by the United States Department of Commerce, National Marine Fisheries Service, in *Basic Economic Indicators No. 6128* (June, 1973) and in other United States publications.

World catch figures of halibut, including only the genus *Hippoglossus* caught by the major producing countries in the North Atlantic and North Pacific and Bering Sea, are shown in Table 35. The Pacific and Atlantic data have been corrected as far as possible for the erroneous inclusion of *Reinhardtius* and *Atheresthes,* the arrow-toothed flounder. No correction has been made for the unreported Pacific halibut taken by the domestic trawlers and the foreign trawlers and setliners. In the case of Japan and the USSR such unreported poundage was of significantly large proportions both in the Northeastern Pacific and Bering Sea. There are also strong indications that much of the Japanese poundage reported from the eastern and even western portions of the Bering Sea actually originated in Northeastern Pacific waters.

The magnitudes of the unreported catches are discussed elsewhere in this book.

From the early 1960s to the mid-1970s there has been a drastic 60% reduction in the world supply of *Hippoglossus.* All of the major halibut producing countries in Europe, including the United Kingdom, Norway and West Germany show the same pronounced downward trend. A decline of the same magnitude is also evident in Canadian Atlantic production, and the United States annual yields in the Northwest Atlantic have reached almost inconspicuous levels. The yields in the Atlantic and the Northeast Pacific waters in the 1970s are much below the average maximum sustainable yields for 1961-1966 as estimated by J.A. Gulland of the FAO, which were 67.5 million net weight pounds for the Pacific and 31 million for the Atlantic.

Large increases in the bottom trawling effort during the same period in all regions of the Northern Hemisphere resulted in declines in catches of halibut and not any increases. The immediate casualty is the population of adult halibut upon which the Pacific setline fishery of Canada and United States had depended for the past 75 years. Those stocks have been driven down to levels where the fishery is in

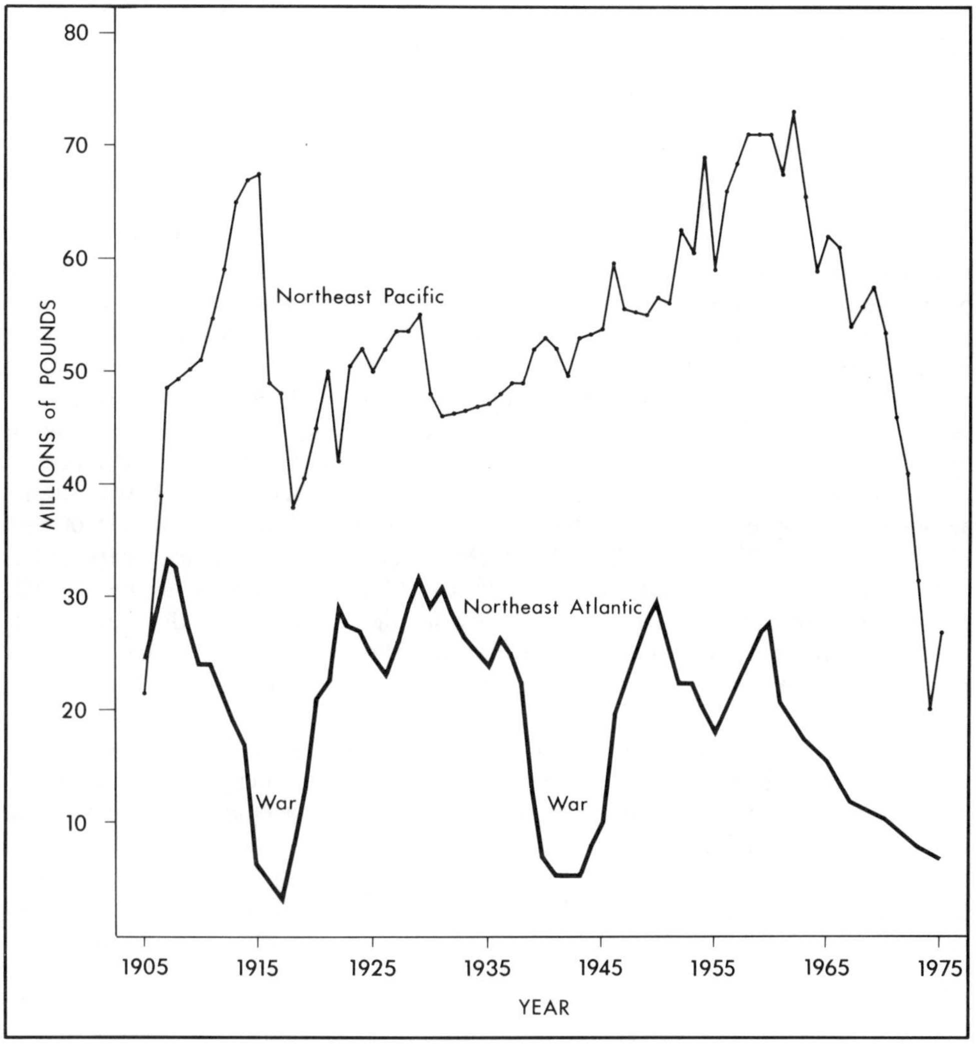

FIGURE 71. Eastern North Pacific and European halibut landings from 1905 to 1975 in eviscerated, heads-off weights.
F.H. Bell

jeopardy and its survival as a single-species fishery doubtful. The level at which the destruction by trawling can be held will decide whether the species will become merely an incidental but declining component of the catch of trawlers, as in the North Atlantic, or whether there can be a setline halibut fishery per se of some modest proportions.

Figure 71 extends that of one published by Bell and Pruter (1958). During the 70-year period there had been no controls imposed upon the European fishery except for two reductions in fishing caused by World Wars I and II. Autonomic economic controls have also been largely lacking in the European scene down through the years. This has arisen from the fact that the greatest proportion of the European halibut catch has been taken by trawlers whose fares include only a relatively small proportion of halibut, about three percent. In such a fishery the stocks of halibut can be driven to very low economic levels of yield as exists for example in the North Sea. On the other hand, vessels using line gear such as the British "liners" and the regular Pacific Coast setline halibut vessels depended largely upon the halibut in their fares for their earnings. They must cease to fish whenever the halibut stocks fall below the profit level.

As with the Pacific fishery prior to 1930, the fleets in Europe had brought all areas into production in a vain effort to offset the decline in yield from the older grounds. By 1929, in both the Pacific and European Atlantic, the fishery was covering the entire range of the commercial stocks in their respective regions. Thereafter the total yields commenced to decline in both regions. In the Pacific fishery the decline in total catch was accelerated by economic conditions in 1930 and 1931 which reduced the amount of fishing on some parts of the coast. With the reduction in fishing effort being continued after 1931 by the regulations the trend of catch rose steadily to the level of the maximum sustainable yield by about 1960.

In Europe the trend of the total halibut catch after 1930 was downward, in spite of a rising demand as shown by the increase in the average price of halibut in England whose market absorbs over 75% of the European production. The rise in price of competing species showed lower percentage increases.

The downward trend in the European production was temporarily halted in 1936 by the development of a short-lived net fishery for spawning halibut in Norwegian fjords in which 2,500 fishermen participated during the winter of 1936-37 (Devold, 1938). After the respite in fishing due to World War II, the trend of the European catch rose as the fleet was rebuilt and fishing was undertaken on the stocks accumulated during the war.

The rise after the end of World War I in 1918 and subsequent decline after 1930 has already had its counterpart after World War II; and the decline has been more abrupt than during the 1925 to 1935 period, because the Greenland-Labrador regions did not sustain the total catch to the extent that they did from 1925 to 1935. The uncontrolled European production did not level off at about 15 million pounds annually as predicted in 1958 (Bell and Pruter, 1958). Instead by the mid-1970s it was one half that amount, 7.7 million pounds in 1974 and still declining. This is a marginal production caught by trawl gear and consisting chiefly of juvenile fish under five years of age. Due to the great complexity of trawl fisheries, particularly those of North Europe, it is unlikely that any practical measures can ever be taken to restore halibut productivity.

In the Pacific halibut fishery, retention of trawler-caught halibut was prohibited. This led by about 1960 to the attainment of a maximum sustainable yield in the vicinity of 70 million pounds annually consisting almost exclusively of adults ranging from 6 to 25 and more years of age.

This utopian state, first time achieved in an exploited marine fishery, was soon to come to an abrupt end. A sharp increase during the 1960s in the catch of trawler-caught species by the large fleets of foreign trawlers, commenced to levy their toll upon both the adult and juvenile populations of Pacific halibut. The decline of the Northeast Pacific resource has been more rapid than was the case in Europe.

Some sections of the Northeast Pacific such as the Goose Islands grounds with a very considerable bottom trawl fishery, show a pattern of decline similar to that experienced by the halibut of the North Sea and the grounds off New England. In those two regions where once there was a very considerable production of halibut, the species has become almost a rarity.

At Georges Bank, off New England, once an important halibut producing area in early days, halibut production totaled in 1930 only 1,168,461 pounds of which 708,023 pounds were taken by line gear and the remainder by bottom net trawls of various sizes. By 1968, the total halibut caught was 17,530 pounds, all by bottom net trawls.

In the heavily trawled North Sea the catch of halibut had declined by the 1970s to a level of about 500,000 pounds annually, only about 10% of what it was in the mid-1920s, and even farther below the level that prevailed prior to World War I in the late 1890s and early twentieth century (Figure 72).

Goose Islands grounds off British Columbia that had been a high producing ground since it was first fished in 1902 was, in the 1970s producing only about 1 million pounds of setline-caught halibut annually and in some years less. As recently as 1955, production was at the 6 million-pound level. It is to be observed that during the succeeding 20 years domestic trawling increased sharply and the efficiency of the vessels at least doubled.

Grounds off Iceland provide a good example of what happens to a halibut resource exposed to a trawl fishery. That region of the Northeast Atlantic has accounted for a very large proportion of the total Northeast Atlantic halibut catch during the past 70 years. It is fished by a number of nations, which tends to buffer any effect that undue fluctuations of one nation's catch might have upon the overall production trends. Halibut has been caught by both setline and trawling in varying proportions over the years. However, unlike the Pacific Coast, the taking of halibut has been subject to no controls except those of two world wars and general economic conditions.

As the halibut fishery by trawlers and setliners off Iceland developed in the twentieth century, the catch by all nations in 1907 reached a total of nearly 9,000 metric tons (19.8 million pounds). Despite an intensified fishery effort, halibut production fell 50% by 1913 to about 4,200 metric tons (9.3 million pounds). After World War I, during which fishing had practically ceased (Kriegschonzeit), the rebuilt fishing fleets caught 6,600 metric tons (14.5 million pounds) in 1923 from the recovered stocks. Again, under the pressure of massive increases in fishing effort, the total catch had declined to about 2,600 metric tons (5.7 million pounds) by 1938. Granted another respite by World War II, the stocks again recovered, and by 1950 the fully rebuilt fleets of Northern Europe were able to catch 5,900 metric tons (13 million pounds) of halibut off Iceland.

With no further "war holidays" as yet, the overall catch of halibut by all nations off Iceland has

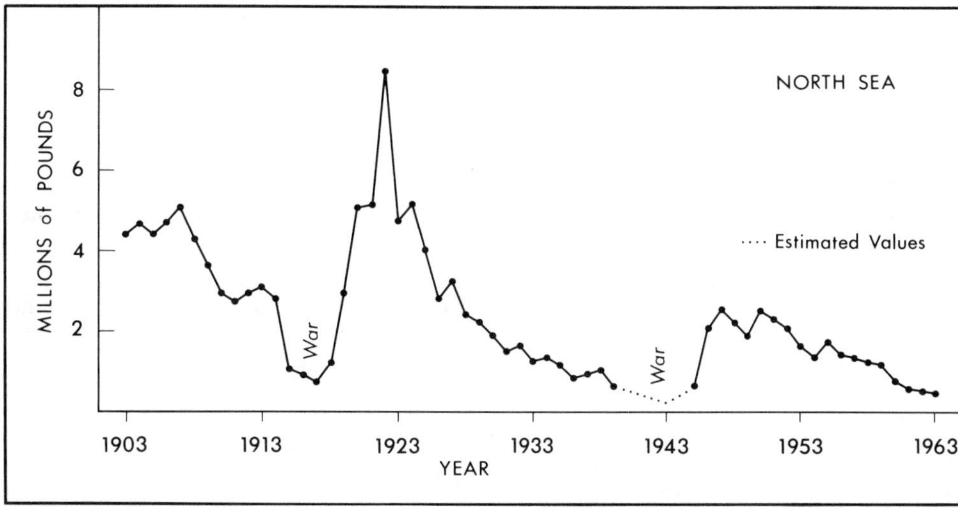

FIGURE 72. Landings of halibut from the North Sea from 1903 to 1975.
F.H. Bell

undergone another dismal decline over the past two decades. The 1,900-metric-ton (4.2-million-pound) level of the mid-1970s represented a drastic 70% decline from that prevailing after World War II.

The Northeast Pacific halibut fishery seems to be following the same course that prevailed in the Northwest Atlantic. From earliest times Gloucester, Massachusetts, was the most important halibut port in New England with Boston in low second place and Portland, Maine, a very low third. Practically all the United States catch of Atlantic halibut was landed in the early years in Maine and Massachusetts except for small amounts in New Hampshire, Rhode Island and Connecticut.

TABLE 36. Average Annual Landings of Halibut in New England in Eviscerated, Heads-off Pounds

Period	Landings*	Period	Landings
1879	10,650,000	1930-39	1,770,000
1880-89	8,300,000	1940-49	415,000
1890-99	9,100,000	1950-59	278,000
1900-09	2,360,000	1960-69	208,000
1910-19	2,450,000	1970-75	166,000
1920-29	2,820,000		

*From April 1918 through 1921 landings include some by Canadian vessels under reciprocal agreement between Canada and United States.

From historical information at hand the above annual average landings for the decade 1890-99 should probably be lowered about two million pounds annually as rail shipments from the Pacific Coast were reported along with the ex-vessel landings in some New England ports in those years, particularly in Boston. Also, prior to about 1915 an average of about one to two million pounds of the reported production of halibut were probably salted fletches, many of which were taken in the distant waters of Davis Strait and Iceland.

The statistics in the Table 36 beginning with the 1920s are a reasonably accurate estimate of the declining trend of ex-vessel landings of iced halibut in New England. The average rate of decline from one decade to the next is over 40%.

It is usually stated that the declining supplies of Atlantic halibut prompted the more venturesome to look to the Pacific in the late 1880s. While this was true to some degree, there was also a great increase in demand with the rapidly expanding population in the industrial states of the East. Such near grounds as Nantucket, No Man's Land and even offshore George's Bank could no longer supply the expanding market, and many New England vessels had to proceed to the more distant banks off Nova Scotia and Newfoundland to maintain production.

During the first 30 years of the twentieth century, United States production of Atlantic halibut remained relatively stable except during the late years of World War I. In the second 30 years of the century the scene changed radically as can be seen in Table 36. With the rapid expansion of domestic trawling on all the well-known halibut grounds, the overall production of Atlantic halibut fell precipitously. With the expansion of trawling in the late 1920s the halibut caught by United States vessels on the East Coast in the mid-1970s is less than 10% of what it was in the early 1930s.

There is no reason to believe that the Pacific halibut stocks are not as vulnerable, although there are probably many who will say that it "could not happen here." While extrapolating the New England experience to the Pacific scene, including the sharp decline in halibut, it can be seen that production is fraught with the same hazards. If the destruction of young recruits by trawling is not stemmed or stayed promptly the extinction of the Pacific halibut fishery will be more traumatic and more rapid than was experienced by the halibut stocks and fishery off New England.

It cannot be proven beyond all reasonable doubt that trawling for other species of demersal fish has been the prime cause of decline in halibut yields in all parts of the northern oceans, but to disregard the consistency and timing of the circumstantial evidence would border on the irresponsible. To credit the declines to some long-term changes in survival or recruitment conditions in all parts of the Northern Hemisphere but with each part having its own time scale is even more untenable.

The Halibut Commission reports of the 1970s seem to lean to the idea that some of the decline in the stocks in the 1960s is due to a long term decline in recruitment. This of course takes some of the responsibility off the foreign and domestic trawl fisheries.

7 INDUSTRY ORGANIZATION

Most industries are organized into various action groups. The fishing industries of Canada and United States are no exception. While there are a few organizations exclusively concerned with the Pacific halibut fishery, there are many others in which the halibut fishery is but one of many concerns of the particular organization. Since most of the industry groups, from time to time, have played varying roles in relation to the development and management of the Pacific halibut fishery, the origins and objectives of the more important associations will be briefly described.

NATIONAL FISHERIES INSTITUTE INC. (NFI)

This institute is a United States trade association that includes all types of firms engaged in the fresh and frozen fish business including producers, processors, wholesalers, brokers, importers, exporters and members of allied supportive industries. The prime purpose of the NFI is to achieve better understanding of the mutual problems, communication with governmental agencies, with the consumer and other organizations.

There are several major activities and functions of concern to NFI members including:

Market Promotion
Technology
Special Hotel Rates at Conventions
Communication and Public Relations
Group Insurance
Government Relations

Membership fees vary with the type of business and is proportionate to the volume of the firm's business. The national headquarters of the institute are in Washington, D.C. NFI also maintains a promotion and advertising office in Chicago.

The National Fisheries Institute was formed in 1946 and was to some degree a successor to the defunct United States Fisheries Association. Charles E. Jackson, former United States Member of the Halibut Commission, was the first general manager. With a budget of only $75,000 and the invaluable assistance of Ray Steele and Mel Xavier, a first-year membership of 600 was achieved against difficult odds. Organizing an industry with so many diametrically opposing interests was a major feat of the early executive committees and their small staff under Charles E. Jackson.

With the right leadership in each part of the country a strong sense of working for the common good emerged, and NFI soon became a powerful and effective force in the industry. As the years went on the incumbency of each annually elected president was characterized by a project or theme that gained the support of all members, such as wartime price ceilings under the federal Office of Price Administration (OPA), imports, plant sanitation, membership drives, federal school lunch programs, purchase of surplus fish supplies for European relief, lamprey control in the Great Lakes, federal support of state fishery research, promotion of fish sticks and fish in general, recipes in national magazines, and many other objectives of high priority to the entire fishing industry.

Participating in these projects were many prominent members of the Pacific Coast halibut industry. Harold J. Synnesvedt Sr. of the New England Fish Co. was an early national president (1950) and Eugene (Gene) A. Ruthford, of the San Juan Fishing and Packing Co., who later became national president (1966) gained early recognition as the "most sincere man" at the 1947 annual convention. Charles D. Alhadeff, of the Whiz Fish Co., a major halibut buyer, was national president in 1950.

Further recognition was given the Pacific halibut

industry by the elections of Harry J. Tillman in 1949, of Harold J. Synnesvedt Sr. in 1953, William C. Eardley in 1958 and Eugene (Gene) A. Ruthford in 1963 as NFI Man of the Year. In addition many other members of the Pacific halibut industry played active roles as members of various committess of the NFI.

As of 1970, the NFI consisted of an executive committee, including the president and other officers, and six regional vice presidents representing each section of the nation. There were 19 standing committees through which much of the business of the NFI was conducted.

Charles E. Jackson, after 17 years as general manager, retired in 1962. Subsequent managerial functions were carried out with the assistance of a professional management firm.

CANADIAN FISHERIES ASSOCIATION

The Canadian Fisheries Association was organized February 15, 1915, through the efforts of James J. Harpell, publisher of the *Canadian Fisherman*, an industrial journal that commenced publication in January, 1914. Major objectives were to improve transportation, to obtain more favorable freight rates for fishery products, to raise the fisheries in the federal government to ministerial rank (finally achieved in 1935 after 20 years), to obtain government product inspection by competent inspectors, and to begin joint industry-government promotion of fishery products with more practical and less academic research by the Biological Board of Canada.

The following persons who were also major factors in the Pacific halibut industry played very prominent roles in the association: W.H. Barker of the B.C. Packers, Vancouver, a founding vice president; A.L. Hager, 1920-23, of the Canadian Fish Co. and the New England Fish Co., Vancouver; T.H. Johnson, 1929-30, and J.W. Nicholls, 1938-44, both of the Canadian Fish and Cold Storage Co., Prince Rupert, presidents. For the latter three persons, halibut was the major concern of their respective companies.

FISHERIES COUNCIL OF CANADA

By 1941 emerging problems indicated the need for local or regional organization, so the Fisheries Council of Canada was formed on January 29, 1945. Council membership consisted of representatives from most of the major fishing industry groups in Canada including cooperatives. It functions under an elected board of directors. Under Clive Planta, the first secretary-manager, the council continued many functions of the disbanded Canadian Fisheries Association. Captain F.W. Wallace, author and consulting editor of the *Canadian Fisherman* was an early honorary president.

The council deals with a very broad spectrum of concerns of the Canadian fishing industry including internal problems and those involving both the federal and provincial governments. It maintains close liaison with the National Fisheries Institute, its counterpart in United States. It functions on a modest budget derived from the dues of member organizations.

FISHERIES ASSOCIATION OF BRITISH COLUMBIA

The Fisheries Association of British Columbia was a development of the Canned Salmon Negotiating Committee, an industry group primarily concerned with the contract fish prices to be paid union fishermen other than for halibut. Shore workers' wages and working conditions generally were also involved.

As the unions and Native Brotherhood became more effective and the infrastructure of the industry more complex, it became necessary to give more stability and continuity to the committee and thus the association came into being. One of the early executive secretaries was James Sinclair, former Federal Minister of Fisheries.

As most of the British Columbia halibut dealers were members of the association by reason of their salmon, herring or other interests, the association played a much lesser role in halibut matters. It is a member organization of the Fisheries Council of Canada.

HALIBUT ASSOCIATION OF NORTH AMERICA (HANA)

This is an organization of Canadian and United States halibut dealers whose object is to promote the sale and broaden the market for Pacific halibut.

An organizational meeting was held March 15, 1961, in Seattle, with E.A. Ruthford as temporary chairman. It was incorporated in the State of Washington March 21, 1961. The first officers and their affiliations at the time were:

E.A. Ruthford, President, San Juan Fishing and Packing Co., Seattle.

H.M.D. Lambe, Vice President, British Columbia Packers Ltd., Vancouver, B.C.

William S. Gilbert, Secretary, New England Fish Co., Seattle
Donald S. Johnston, Treasurer, Seattle Seafoods Inc., Seattle

The Board of Trustees and affiliations at the time included:

Joseph E. Antonelli, Prince Rupert Fishermen's Cooperative Association, Vancouver, B.C.
H. Borden Smith, Canadian Fish Co., Vancouver, B.C.
Malcolm McCollum, McCollum Sales Ltd., Vancouver, B.C.
James G. Pinkerton, Ketchikan Cold Storage Co., Ketchikan, Alaska
Elton E. Engstrom, Juneau, Alaska
Norman E. Weitkamp, Washington Fish and Oyster Co., Seattle
W.B. Hall, Booth Fisheries, Seattle
W.C. Eardley Jr., Whiz-Eardley Fisheries Inc., Seattle
John W. McCallum, McCallum-Legaz Fish Co. Inc., Seattle

The stated purposes of the association in the Articles of Incorporation were:

To engage in the promotion of the use and consumption of halibut products and to undertake research, public education, and advertising programs calculated to increase the use and consumption of halibut products by the consuming public.

The presidents of HANA and their industry affiliations as of 1979 have been the following:

Two Year Term
1961-1962 Eugene A. Ruthford: Icicle Seafoods Inc., Petersburg and Seattle
1963-1964 Palmer G. Olson: New England Fish Co., Seattle
1965-1966 William B. Hall: Retired
1967-1968 William S. Gilbert: Washington Fish and Oyster Co., Seattle
1969-1970 Hugh C. Lavan: Cal-East Foods, Oakland, California
1971-1972 Rufus A. Littlefield: Retired
1973-1974 David G. Roy: Booth Fisheries, Seattle
1975-1976 James G. Ferguson: Pelican Cold Storage, Seattle
1977-1978 William J. Kelliher: New England Fish Co., Seattle
1979-1980 William J. Kelliher: New England Fish Co., Seattle

Promotional advertising is contracted to an advertising agency, and it accounts for about 75% of the annual budget. The advertising and other activities are funded by an assessment on the member companies. The organization functions with an extremely low overhead.

The assessment was initially 1/10 of 1¢ per pound for Pacific halibut marketed in United States. In 1963 it was raised to 2/10 of 1¢ to accumulate a reserve, reverting to 1/10 in 1964. It was raised to 2/10 in 1976 and ¼¢ in 1977. To provide a fund for members when sales of halibut in United States are low, there is a minimum annual fee of $50.

Initially, membership accounted for 90 to 95% of the Pacific halibut marketed in the United States. By the late 1970s with the very low total production, membership declined to represent only 60 to 70%.

In addition to general promotion, it produced and exhibited a 16mm film titled *Halibut—Treasure of the Deep* that was shown from 1964 to 1975 to a wide range of audiences in North America.

To combat in the courts and legislatures the false advertising of imported turbot as "Greenland halibut" required a major allocation of funds by HANA from 1967 to 1969. William S. Gilbert of the Washington Fish and Oyster Co. played a major part in success of this project.

HANA also worked closely with the United States Food and Drug Administration in working out a plan for dealing with any heavy metal problem with respect to halibut. Through membership in the National Fisheries Institute, it has brought its influence to bear on many questions facing the fishing industry as a whole. It has also supported various industry-related projects such as the Seattle World's Fair, the Halibut Fishermen's Wives' Association, and the conservation measures of the Halibut Commission.

FISHING VESSEL OWNER'S ASSOCIATION (FVOA) OF SEATTLE

This organization, incorporated on the 30th day of December, 1914, in Washington State, was destined to be one of the most effective and durable industrial fishery organizations in the country. The incorporators were all owner-operators of halibut vessels. Much of its effectiveness has stemmed from the continuity and perspicacity of its management over the past 50 years. On March 13, 1924, Harold E. Lokken (still active in 1981) assumed the position of general manager succeeding a brother who had held the position for the previous two years.

Provisions of the Articles of Incorporation were well chosen, and their fulfillment accounts for much of the organization's success. They advocated maintenance of an office and meeting place, a fish exchange, aids to navigation, standards of vessel safety, new markets and the conduct of any other business for the mutual benefit and profit of members.

These mandates have been pursued with vigor. Orderly collective bargaining has been carried out with union groups and disputes have been settled with reasonable dispatch. Effective presentations on behalf of the membership have been made to regulatory and legislative bodies, and at all times conservation has been a guiding principle. The organization has been called upon to lead the fleets, both owners and fishermen in Canada and in the United States, in many projects of mutual concern. The Seattle Fish Exchange has been successful in creating a free market not only in Seattle, but has assured fair prices up and down the coast. A branch office was opened in Ketchikan in 1928 at a time when Seattle vessels were landing many trips in nearby Prince Rupert.

Much of the business of the association was carried out in committee with full opportunity for minority dissent. It organized cargo and hull insurance pools and other activities as the need arose. It was successfully involved in the cooperative marketing of vitamin bearing livers and viscera and carried out many functions of value to its membership, including protests of unnecessary bureaucratic controls.

FISHING VESSEL OWNER'S ASSOCIATION (FVOA) OF VANCOUVER

This association carries out many of the same functions as those carried out by the Fishing Vessel Owner's Association in Seattle, but it is of recent origin. The British Columbia group was established January 3, 1935, and incorporated March 25, 1938, with a registered office in Vancouver.

The British Columbia organization has a much broader membership base than the Seattle association that consists primarily of owners of setline halibut vessels. The constitution of the British Columbia group covers independent owners of vessels in halibut fishing, salmon seining, herring and pilchard fishing, trawl fishing and vessels used as tenders or packers.

The British Columbia organization formed a successful vessel insurance pool and took an active interest in the management decisions of the Halibut Commission and fulfilled many other functions for the benefit of its members. It was the bargaining agency for the owners in dealing with the fishermen's unions for agreement on terms of employment and compensations.

HALIBUT FISHERMEN'S WIVES' ASSOCIATION

This association was organized in 1963 by wives of halibut fishermen and vessel owners operating out of Seattle. Its general purpose was to promote the welfare of the halibut fishery upon which so many depended for their livelihoods.

One of its first objectives was to vigorously protest the "Bering Sea Giveaway" as a result of which Japan was no longer required to abstain from halibut fishing in the eastern Bering Sea. The effective manner in which the association acted in the matter drew national notice and the organization immediately gained considerable influence in government circles.

All sections of the industry aided the organization in its many projects that promoted the marketing of halibut and the economic welfare of the industry. Leading business establishments and some financial institutions in the community provided support and cooperation. State and federal agencies assisted where they could.

The association was thus able to publish and distribute throughout North America many thousands of booklets containing selected recipes for cooking halibut. Annual festivals were organized and displays shown in leading retail stores, at fairs and exhibitions. Television and the printed media featured the cooking of halibut. The association established the Golden Halibut Award given annually to an individual who had made a significant contribution to the conservation of the resource or the welfare of the industry.

Each year the governor of Washington officially proclaimed Halibut Week, and each year the senior United States senator of Washington State had Congress request the president to call upon the people of United States to observe the week.

Delegations or individuals from the association attended many legislative sessions and hearings and testified on questions of concern to the fishing industry. The success of the association inspired the establishment of similar groups connected with other Pacific Coast fisheries. On occasion the several groups would develop a common front along the coast on matters of general concern to the fishing industry.

While many women contributed to the success of their organization, the Halibut Fishermen's Wives' Association owed much to the uncommon tenacity and dedication of the late Mrs. Gus (Fern) Molvik.

FIGURE 73. Governor Daniel J. Evans of Washington signing the 1975 Proclamation of Halibut Week. Behind him are members of the Halibut Fishermen's Wives' Association. *From left:* Mrs. Fern Molvik, Mrs. Ragnar Ness, Ingrid Ness, Alice Thomassen, and Helen Johnson.
F.H. Bell

FISHERY COOPERATIVES

Canada

Fishermen's cooperatives have been a long-standing feature of the British Columbia fishing industry. One of the earliest and ultimately most successful of such associations was that of a salmon-trollers' marketing group organized early in the 1920s on the west coast of Vancouver Island out of Kyuquot. Subsequently other producing groups of varying but modest magnitudes organized cooperatives in British Columbia. These groups included the Prince Rupert Fishermen's Cooperative Association (1931) and the North Island Troller's Cooperative Association (1934). Activities of the cooperatives were restricted mainly to the production and primary marketing functions. They used the established wholesale and retail channels for the profitable disposal of their catches, which were chiefly salmon and occasionally incidentally caught halibut.

In 1939 the North Island Trollers amalgamated with the Prince Rupert Fishermen's Cooperative with the latter being the surviving association and continuing as such to the present. By 1941 British Columbia fishery cooperatives handled 1.5 million pounds of products with halibut a major part of the total and assets of about $500,000.

In 1944 the Fishermen's Cooperative Federation became the central sales organization for the fishermen's cooperatives of Prince Rupert, Masset, Sointula, Kyuquot and the United Fishermen. The marketing of its products in United States was effected through the Fishermen's Federation, Inc., mutually owned with a United States group, the Halibut Producer's Cooperative.

Growth was rapid and the cooperatives became major factors in the British Columbia salmon, halibut and other foodfish industries and in the short-lived but prosperous vitamin A liver oil market. A large segment of the Canadian halibut fleet thus became a "captive" group, no longer disposing their individual catches on the open market.

During the period from 1945 to 1949 the cooperative had an annual production of between 25 to 30 million pounds of foodfish. By 1950 the non-depreciated value of their plant was about $2 million. Also in that year the Kyuquot, Sointula and United Fishermen amalgamated to form the Fishermen's Cooperative Association to control operations in southern British Columbia.

During the succeeding 15 years the large Prince Rupert association prospered and expanded into other fishery activities. Under sound management it became an important factor in the competitive fishery economy of the province. In addition to good management and sound fiscal policies the early success of the venture was undoubtedly aided by the extremely lucrative market for Vitamin A-bearing materials, the livers and viscera of halibut and other species. This provided a strong capital base upon which to build.

Some early cooperatives in United States were organized in times of great economic stress which greatly reduced their chances of success. Also they were sometimes directed by some vessel owner rather than by a professional manager. A "high-line" fisherman does not necessarily make a good manager of a shoreside operation. In Prince Rupert the cooperative in 1941 employed Kenneth F. Harding, a technologist of the Fisheries Research Board of Canada and graduate of Queens University, Ontario, to be manager of their large liver-oil business. In 1947 he was elevated to general manager of the cooperative and continues in that capacity in 1980. Harding has held prominent positions in the Cooperative Union of Canada and has been member of the Fisheries Price Support Board and of the Industry Advisory Board of the Fisheries Research Board of Canada. He also was a member of various industry advisory groups established by the Halibut Commission.

Several of the vessel-owner trustees of the cooperative have served as Canadian Members on the Halibut Commission, including Harold S. Helland, 1953-1963, Martin K. Eriksen, 1963-1973, Jack T. Prince, 1974-1976, and Peter C. Wallin, 1977-present member.

In the new competition the cooperative provided and conversely in the competition it encountered in the rest of the industry, lies the strength of any free economic system. In 1965 it possessed non-depreciated plant assets valued at nearly $5 million, foodfish sales of nearly $10 million, and sold about $2 million of fishmeal and oil from its herring reduction activities. Consumer functions involved retail sales to members of nearly $3 million in 1965.

The association operates modern fish receiving, freezing, cold storage and shipping facilities at railhead ports and possesses a coastwise system of many fish collecting stations, some of them with supply stores and other service facilities for member fishermen.

In addition to fish production and sales the association has provided a growing number of ancillary functions and items for its consumer and fishermen members including ship chandlery, food, home fuel, and a credit union. A cooperative vessel insurance program was also conducted.

With the increasing magnitude of both the producer and consumer functions of the association and their essentially different requirements the two divisions—the Producing and Consumer—were divorced on April 1, 1965, to become separate but closely coordinated associations.

United States

In the United States several attempts were made in the 1920s to develop fishermen's marketing groups, but they were not successful as some were undertaken in the times of general economic adversity of the early 1920s and early 1930s, both unfavorable environments to initiate any new enterprise. While the halibut fishery out of Seattle, both vessel owners and fishermen, had strong organizations and excellent leaderships, it was much more difficult to achieve unanimity on such matters as cooperative marketing than in Prince Rupert or Southeastern Alaska ports. Two attempts to negotiate minimum price agreements with the dealers were unsuccessful.

Nevertheless, the Seattle fleet successfully created the Halibut Liver Oil Producers (HLOP), a cooperative that under good management became the major producer of high potency Vitamin A-bearing oils in North America. The cooperative did handle some other fishery products on a very limited scale over a short period. For a number of years the Petersburg halibut fleet carried on a successful cooperative venture with halibut as their main product. Alaska salmon trollers also maintained an effective cooperative marketing arrangement. It involved the small amounts of halibut that are frequently taken incidentally when salmon trolling. The cooperative has been managed for many years by Donald E. Reinhardt and markets its products out of Bellingham, Washington, through the Fisherman's Federation, Inc.

NEW ENGLAND FISH COMPANY

Halibut and NEFCO are almost interchangeable terms. No story about halibut could be complete without an accounting of the New England Fish Co., which was usually referred to by the familiar acronym, NEFCO. Conversely, no account of NEFCO is possible without mentioning halibut. For the first 50 years of the existence of what was a century-old company, it traded in halibut almost to the exclusion of all other species of fish, including salmon. It was the dominant factor in the halibut trade in the new world for over a hundred years.

NEFCO was formed in Boston on March 9, 1868, by 11 wholesale fish dealers after the end of the Civil War. Headquarters were established in Gloucester where it acted as a cooperative marketing agency for Atlantic halibut. As a result, order quickly replaced chaos in the market place and other dealers soon joined the consortium. The more stable market attracted vessels from various ports and the agency widened its scope. In November, 1877, operations were extended to New York where seven of the wholesale dealers of the Fulton Fish Market joined the organization.

About this time two events occurred that not only are of some historic significance, but also demonstrate how fate so often plays a part in the affairs of men.

On May 30, 1877, Freeman W. Emery, president of the New England Fish Co., paid one year's rental to the recently formed New England Telephone and Telegraph Co. for a telephone in his home and for one in that of this brother, James. This was the first payment ever made for telephone service in the United States, and halibut, as well as NEFCO, were involved in this historic event. It may be said in passing that the brothers lived only across the street from one another!

In July, 1879, an incident occurred bearing upon the halibut fishery. President Emery had occasion to threaten suit against the Western Union Telegraph Co. for transmitting an error in prices from Boston to Gloucester for the purchase of a trip of halibut by NEFCO. In settling the matter out of court the struggling telegraph company, short of money, offered NEFCO the telephone and telegraph franchise for a large section of Massachusetts for $1,500. By declining, NEFCO probably missed becoming a major factor in what subsequently became a multibillion dollar communications industry. A century later this telephone and telegraph company was one of the creditors of a bankrupt NEFCO.

In 1884, NEFCO moved headquarters from Gloucester to T-Wharf in Boston where the company continued to prosper despite the fast falling supplies of Atlantic halibut. In view of that decline, the company by 1890, taking official cognizance of the growing interest of others in the halibut potential of the Pacific Northwest coast. Some shipments were already reaching the East Coast from the Pacific. In January, 1894, NEFCO took the definitive step of employing Grier Starratt of Vancouver, British Columbia, to represent the company on the Pacific Coast. This decision shifted the emphasis of the company to the Pacific Coast as a major source of supplies, to meet the needs of the growing market in the eastern United States, as well as to offset the decline in yields from Atlantic waters.

Grier Starratt, who had managed the Haney office of the New Westminster and Port Haney Freezing Co., was instructed by NEFCO directors to organize and supervise a halibut operation out of Vancouver, British Columbia. This city was chosen because the Canadian Pacific Railway offered more favorable terms and service than the Northern Pacific Railway out of Tacoma, Washington. The latter road was not willing (or permitted) to attach refrigerator cars to the faster passenger trains. Moreover, the C.P.R. was willing to cancel transportation charges if delivery was delayed, provided NEFCO would not claim for loss of fish due to any ensuing spoilage. The lack of similar support of the Northern Pacific Railway out of Tacoma had been the primary reason for the failure to establish a halibut fishery out of Puget Sound ports five years earlier.

Vancouver had the further advantage of possessing an established halibut operation. The S.S. *Capilano* and S.S. *Coquitlam,* two small freight and passenger vessels of the Union Steamship Co., had already been chartered in previous winter months to various companies to service halibut fishing camps on the British Columbia coast but with mediocre success. Furthermore it is doubtful that a United States company such as NEFCO operating other than from a British Columbia port and with carriers other than of British registry would have been permitted to maintain a fishing camp on the British Columbia coast even under the *modus vivendi* of the 1888 treaty.

The S.S. *Capilano* was chartered from the Union Steamship by NEFCO and left Vancouver on September 22, 1894, with a 9-man crew, 20 fishermen and a cook to establish a fishing camp on Banks Island with Captain Anderson to be in charge until the arrival of Captain H.B. Joyce from Boston. The *Capilano* returned to Vancouver on October 4 with 90,000 pounds of halibut caught in three days' fishing. The load counted out 2,066 fish with a high boat share of $67.00 per man and a low boat share of $20.75 per man, each fisherman at the camp being paid by the count of fish he caught. The landing of this successful trip of halibut on October 5, 1894, can be regarded as the beginning of NEFCO's commitment to the Pacific Coast. This area was destined to dominate its activities at least for the next 90 years during which time NEFCO would usually handle more Pacific halibut than all other companies combined.

While a number of people had contributed to the early success of NEFCO on the Pacific Coast, the name of Captain Hanson Benjamin Joyce must be remembered. Captain Joyce was a native of Swan Island, Maine, and had been the "high liner" in the New England vessel fisheries in 1874, 1876, 1877, 1879, 1880, 1884, and 1887. He formed a team with Captain Absalom Freeman that assured success for the company's camp operations where other companies had failed. He was involved in the design of the S.S. *New England*, the first steamer constructed expressly for halibut fishing along the Pacific Coast, and he introduced many innovations in dory fishing. He made production records during the years he was captain of the S.S. *Kingfisher.* His name is commemorated in Joyce Island near Brown and Edye passages in Hecate Strait.

NEFCO's first season on the Pacific Coast ended

with the arrival of the *Capilano* in Vancouver on February 26, 1895, with its thirteenth and last load of halibut. It also towed down the camp scow. Production totaled 900,000 pounds of which 750,000 pounds were shipped out on the Canadian Pacific Railway, 29 cars to the east and 2 south as far as Portland, Oregon. The remainder was disposed of locally. Profits from the first season's operations seemed to be marginal and there was some concern as to whether the fishing operation would continue the next winter. There were probably start-up costs, and some of the operational changes made at the camps to increase efficiency had not been in effect the whole season. Also, the relatively high freight rates and pending petitions to the railroad for relief probably prompted a certain amount of reluctance to admit publicly to any profit from the operation. In addition, there is a long and continuing tradition in the fishing industry that no one ever makes any money and certainly never admits it.

Freeman, captain of the *Capilano,* and Joyce, camp superintendent, made what were important and probably vital changes in the camp operation during their first season. New charters with the Union Steamship Co. now not only freighted supplies to camp and packed the fish caught to Vancouver, but also permitted the steamer used, while waiting for a load of fish, to tow the fishing boats to the fishing grounds and tend them during the day. This allowed the boats to go much farther afield and overcome the problem of dwindling fish stocks near camp. And as less time was consumed rowing to the grounds, more time was provided each day to fish. A fish scow, especially designed, built and equipped for the purposes at hand, was towed to the camp the first season, and six new fishing boats, about 22 feet long and better adapted to the operations than those used heretofore, were built by NEFCO.

During the first winter season NEFCO was not alone in the fishery. It was in competition in Vancouver with another Boston company, The American Fishing Co. The latter chartered the S.S. *Coquitlam*, practically a sister ship to the *Capilano*. Under the command of Captain Foulton it departed on September 29, 1894, seven days later than the S.S. *Capilano.* Instead of a specially constructed fish scow, they towed the hull of an old schooner, the *Delaware,* to anchor at Goschen Island as a receiving station. They had only eight boats fishing of which three were independent, i.e., not company-owned. Operations were terminated March 6, 1895, having landed about 675,000 pounds of halibut. However, the firm encountered financial difficulties and withdrew from the scene.

In addition to the American Fishing Co. in Vancouver there was competition out of Victoria where the Victoria Fishing and Trading Co. had the S.S. *Thistle* chartered under Captain W. Langley. It operated out of an old hulk, the *Isabel,* anchored off Goschen Island. Despite its not leaving on the first trip until November 11, 1894, the *Thistle* made eight trips before terminating the season March 10, 1895, and landed about 700,000 pounds. Two of the trips were shipped east via the Canadian Pacific Railway out of Vancouver, three trips were landed in Tacoma and shipped on the Northern Pacific Railway along with three others that had been transshipped from Victoria to Tacoma on the freighter *City of Kingston.*

The development of the camp fishery on the British Columbia coast has been discussed in some detail as it demonstrates why NEFCO had been able to survive the many vicissitudes to which the Pacific Coast fishing industry has been exposed over the past 100 years. Probably the most important factor in the durability and success of NEFCO has been the caliber of people that it sought to employ. This was evident in its earliest venture on the Pacific when it employed such experienced and capable persons as Grier Starratt, Absalom Freeman, and Hanson B. Joyce. Each of these men were already of high reputation and have later become almost legends in the Pacific halibut fishery. In its subsequent development NEFCO apparently continued to place great emphasis upon the people, as well as on the physical assets that it has acquired down through the years.

Whether the first season was profitable or not, or whether it was New England determination and a sound financial structure, NEFCO, with start-up costs behind them, did recharter the S.S. *Capilano* for its second season of 1895-1896. With the failure of the American Fishing Co., the S.S. *Coquitlam* became available, and it also was chartered by NEFCO on November 1, 1895. On the first trip northbound the S.S. *Coquitlam* freighted coal to Juneau for the Alaska Commercial Co., then southbound, picked up a load of halibut and made alternating trips thereafter with the S.S. *Capilano* from the old camp at Freeman's Pass and from a new one at the northwest end of Banks Island in Hecate Strait.

Before the 1896-97 season NEFCO made provisions for a major change in fishing strategy. It had six dories built *that could be readily hoisted on board the tending vessel.* This was another major development. It greatly extended the range of operations and also eliminated the necessity to return each night to camp. This clearly was the forerunner of the dory-fishing steamer era.

For the winter season of 1897-98, NEFCO had three

vessels under charter. The S.S. *Thistle,* which had operated for a Victoria company for several years, was chartered along with the S.S. *Capilano* and *Coquitlam.* Fishing was conducted from Butler's Cove on Stephen's Island and the season's catch was about the same poundage as the previous year. This was NEFCO's fourth and last year of camp operations using chartered vessels.

The next development was to build a vessel especially designed for halibut fishing on the Pacific Northwest Coast with dories. The fjord-like topography and many productive grounds in close proximity to the rocky shores, precluded the use of sailing vessels such as were used on the East Coast. Also, speed was essential in order to bring the fish as fast and fresh as possible to railhead for the long 3,000-mile journey to the main markets on the East Coast.

NEFCO incorporated the collective experience of its operating personnel in what was to be the first vessel of a type that would dominate the Pacific halibut fishery for the next 20 years. It was appropriately named the S.S. *New England.* During its first winter fishing season, 1898-99, the vessel landed 42,000 halibut for a total weight of about 900,000 pounds. During the remainder of the year the vessel engaged in other revenue-producing activities. It possessed the unique distinction of being not only the first but also the last vessel of its class to fish halibut on the Pacific Coast.

By carefully adjusting production to market demand NEFCO steadily expanded its halibut operations. To supplement the catches of the *New England,* the S.S. *Capilano* was also chartered in the winter of 1899-1900, and in the winter of 1900-01, the *Thistle* replaced the *Capilano.* The next winter, 1901-02, the *Capilano* was again chartered.

In 1902 NEFCO bought the S.S. *Kingfisher,* a new vessel, 141 net tons and built in San Francisco by Captain Baker of Boston. It was the same length and breadth as the 70-net-ton S.S. *New England,* but had a somewhat deeper draft. Arriving in Vancouver in mid-August, 1902, it departed on its maiden voyage to the halibut grounds on August 30, 1902.

In 1903 another United States bottom vessel was purchased, the S.S. *Manhatten,* 134 net tons. This vessel was built in Philadelphia. Late the same year the *Saga,* a United States steamer which had been chartered earlier, was purchased by NEFCO, refitted, renamed the S.S. *Columbia* and transferred to British registry.

The fleet operated by NEFCO in 1904 consisted of the *New England, Kingfisher, Manhatten,* and *Columbia,* and despite a crippling strike in the fall, these four vessels accounted for about 95% of the 12 million pounds of halibut that were landed in British Columbia that year. The S.S. *New England,* for which detailed records were still available, alone landed 3,363,000 pounds that year. Both the *Kingfisher* and the *New England* landed several very large trips of 240,000 pounds each.

Other Canadian companies were again active in the halibut fishery but British Columbia Packers Association had their own vessel, the *Unican,* fishing, as well as the chartered schooners *Athena* and *Kilkenny.* It used as a base the cold storage of the Cleeve Canning and Cold Storage Co., a subsidiary of the B.C. Packers Association, at New Westminster, British Columbia. The St. Mungo salmon cannery near New Westminster, with its cold storage, purchased some halibut and the Western Canadian Fish Co. used their cod fishing schooner, the *Blakely,* to bring in a load of halibut for processing at their cod-curing plant at Barnett, near Vancouver. After the heavy halibut production in 1904 and with the depression the next year, only the *Kingfisher* and *New England* fished in 1905. These two vessels accounted for most of the 7.2 million pounds of halibut landed in British Columbia that year, the S.S. *New England* alone landing 3,290,000 pounds.

During the first decade of NEFCO's Pacific Coast operations its halibut production had increased from 900,000 pounds in the first winter season of 1894-95 to 7 million pounds by 1905. The catches of vessels owned or chartered by the company had also been augmented by purchases from independently owned boats and vessels. In the first year of camp operations about 50% of the 900,000 pounds had been taken by independent fishermen who used their own boats and gear. They were afforded accommodation and food on the scow and were paid 37.5¢ per fish compared to 25¢ paid the company fishermen. This practice continued during the period that the camps operated. NEFCO also was usually able to accept the catches of any small boat that landed in Vancouver. In later years with the passing of the company-owned or chartered steamers, the catches of independently owned vessels became the primary source of supply for the company. Also many of their salmon seiners were fitted for halibut fishing prior to the beginning of the salmon season. Their catches were bid upon in the open market unlike in the days of the company steamer.

This greatly increased production necessitated larger quarters and in August, 1904, NEFCO completed the construction of a two-story building of 21,000 square feet on the Vancouver waterfront. It was close to the Vancouver Ice and Cold Storage Co.

where NEFCO could freeze up to 500,000 pounds of halibut for winter sale. Until that time NEFCO continued to rent docking and other facilities from the Union Steamship Co., from which it had chartered its first vessel, the S.S. *Capilano* in 1894.

In 1906 NEFCO contracted to buy the catch of the S.S. *Celestial Empire*, a British flag ship which had been purchased in England by the Canadian Fishing Co. This company had been formed on April 30, 1906, by three Vancouver drugstore owners, Messrs. J.M. Atkins, H.H. Watson, and L.G. Hendersen, along with Captains Absalom Freeman and Johnson. The company maintained its own vessels and sold its catches under contract to distributors. NEFCO continued to operate its own vessels, the *New England, Kingfisher,* and *Manhatten,* all United States bottoms.

In 1907 NEFCO again operated its own vessels and contracted for the catches of the *Celestial Empire,* except for 1 million pounds of halibut that the latter vessel had contracted to deliver to the B.C. Packers in New Westminster between April 15 and October 1, 1970. NEFCO also contracted for the catches of the S.S. *Flamingo*, another vessel of British registry bought in England by a Vancouver company. The company later went bankrupt and NEFCO bought the vessel. In 1908 the Canadian Fishing Co. was incorporated as the Canadian Fishing Co. Ltd.

In 1907 NEFCO also commenced the construction of a cold storage plant at Ketchikan. The plant had been in the planning stage the year before. In 1908, NEFCO probably made what was probably its most notable acquisition when it secured the services of Alvah L. Hager in connection with its purchase of the Northwestern Fisheries Co. of Boston. Mr. Hager, who had organized the latter company to assure a supply of Pacific Coast salmon for eastern dealers, brought to NEFCO a breadth of vision and business acumen that in the next 40 years was to carry the company to a position of preeminence in the North American fishing industry. His influence was to carry on long after his passing in 1948, in the further expansion of the company under members of his family and others who had worked with him.

Grier Starratt, who had effectively guided NEFCO through the initial and often difficult stages of his early development on the Pacific Coast, left NEFCO and continued his own notable career. He became manager in charge of design and construction for the Canadian Fish and Cold Storage Co. at Prince Rupert. This company, established in 1907 in Hull, England, was to build in Prince Rupert what was to be the largest fish cold storage plant in the world.

In 1911 NEFCO's Vancouver fleet of United States vessels — the *Manhatten, Kingfisher,* and the *New England* — produced 9,874,535 pounds, about equally divided between each vessel. In addition, the *Prospector,* a recently acquired power schooner with a trip capacity of about 90,000 pounds of iced halibut, is known to have landed about 1,322,000 pounds in Ketchikan in 1911. Over another 1 million pounds were bought from independent vessels at the same plant.

The S.S. *Flamingo* and S.S. *Celestial Empire* of the Canadian Fishing Co., now a subsidiary of NEFCO, are estimated as having together produced about 4 million pounds. Thus, despite the entry of other companies into the Pacific halibut fishery with large steamers, both in Canada and the United States, NEFCO vessels produced over 80% of the halibut landed in Canada in 1911, and NEFCO's Pacific Coast halibut production exceeded that of any individual company in both United States or Canada.

In 1912 NEFCO expanded its halibut operations to northern British Columbia where it acquired the Atlin Fish Co. of Prince Rupert to engage in handling both halibut and salmon.

NEFCO's almost exclusive concentration on halibut continued until 1918 when its subsidiary, the Canadian Fishing Co., built in Vancouver a salmon cannery known as the Home Plant. Five years later, in 1923, it began a soundly financed acquisition of a large number of well-established salmon canneries on the British Columbia coast commencing with those at Butedale, Shushartie, Margaret Bay, and Lagoon Bay. The latter plant had been already under lease by the Canadian Fishing Co. With further acquisitions CANFISCO soon became a major factor in the British Columbia salmon industry. The parent company NEFCO also had salmon operations in Alaska and the Pacific Coast states.

There is no magic formula for the prolonged dominance of NEFCO in the competitive fishing industry. The strength of the company lay in several areas. Its fiscal policies obviously had been sound as it weathered many serious economic and financial crises in the country for over 100 years. However, surviving the Great Depression of the early 1930s strained its reserves to the utmost. Nor was it financially comfortable at that time when the bank in Vancouver with which it had dealt for many years withdrew its support and called their loan. Fortunately another bank took advantage of the opportunity and gained an important customer. NEFCO was also favorably situated in that its ownership was then very closely held. In a cyclic industry such as the fisheries, there is much in favor of being independent of the demands of outside stockholders or financial forces such as occurs with publicly held companies.

Initially NEFCO specialized in one commodity, halibut, and when it finally did enter the more volatile and cyclic salmon canning business, it did so with a strong financial structure built upon half a century of trading in halibut. This permitted NEFCO to acquire proven canning properties at realistic valuations during the depression at the end of World War I and in the Great Depression of the early 1930s. Consolidations, with their attendant cost savings, had been an important element in NEFCO's production strategies, but the major asset of the company seems to have lain in the managerial talent that it had recruited down through the years, either directly or by the technique of making acquisitions that included such talent.

A large number of circumstances came together in the late 1970s and reached a climax by 1980 to the detriment of NEFCO. The company was declared bankrupt in United States Federal Court in Seattle on May 2, 1980.

Thus ended an enterprise that had played the leading role in the North American fishing industry since shortly after the end of the Civil War. While the dissolution may be attributed by some to internal management problems, exogenous considerations seem to have been more important.

Extensive penetration of the domestic fishing industry by Japanese interests during its last 10 years reduced domestic control of the industry. An overdependence upon the recently booming Japanese market for some United States fish products further reduced such control. Consequently, the collapse of that market due to unduly high prices was disastrous for many in the United States fisheries. These market conditions occurred during a time of disastrous inflation brought about primarily by irresponsible fiscal policies of central governments. The resultant chaotic changes from month to month in the cost of working capital made corporate planning an almost impossible task particularly in cyclic industries such as the fisheries.

A large company like NEFCO, requiring large amounts of working capital, would have greater difficulty. It is also possible that the main bank involved, having recently undergone some large losses in another industry, was unduly prudent. Although NEFCO's annual earnings had not been spectacular the previous five years, the company was far from being insolvent. Despite liquidation during a pronounced recession and a sharp downturn in the fishing industry, NEFCO stockholders, let alone its creditors, are expected to realize a substantial return, an unprecedented event for a corporation forced to take bankruptcy.

BRITISH COLUMBIA PACKERS LTD.

The British Columbia Packers (B.C. Packers) is one of the largest firms in the Pacific Northwest fishing industry. Most of its production facilities are centered in British Columbia, but its markets are throughout the world.

Since it was first conceived in eastern Canada early in the 20th century, it has undergone a number of changes in corporate structure and direction. It was initially chartered in the state of New Jersey in 1902, probably for tax purposes, then in British Columbia on March 9, 1910, and later on April 1, 1914, under the Dominion Companies Act. Later on May 18, 1928, a holding company of the present name was formed to acquire the Gosse and the Millerd cannery properties. More recently it became a member company of Westons Inc., a large Canadian conglomerate with world-wide interests. Despite some changes in the name of the firm down through the years, it will be referred to in this book for simplicity as the B.C. Packers.

The firm has a background and a character that is both similar and at the same time in sharp contrast to the Canadian Fishing Co. which was the Canadian subsidiary of NEFCO. Both parent corporations came into being as the result of the consolidation of a number of independent entities. B.C. Packers arose from the unification of 42 British Columbia salmon canneries. Similar consolidations in the salmon industry had occurred on the Columbia River and in Alaska with the resulting Columbia River Packers Association (CRPA) and the Alaska Packers Association (APA) respectively. NEFCO was formed by the pooling of 18 fish brokers in New England. NEFCO, like the B.C. Packers, first operated as an American company in British Columbia but in October, 1909, NEFCO bought the Canadian Fishing Co. The head office of NEFCO remained in Boston until July 1, 1931, when it was transferred to Seattle, Washington.

The initial objectives of the two firms were very different. B.C. Packers at the outset was exclusively concerned with salmon and NEFCO with halibut. With the acquisition of the Cleeve Canning and Cold Storage Co. Ltd. of New Westminster in 1902, the B.C. Packers ventured slowly into the halibut field. NEFCO had been practically exclusively concerned with halibut for the first half century of its existence, from 1868 to 1918. In both firms it seems conditions caused by World War I attracted each company into broader fields. B.C. Packers expanded its halibut activities in the war and prewar years. NEFCO went into salmon toward the end of World War I and in its economic aftermath was able to acquire some good salmon canneries in the early 1920s.

The corporate structure of the two firms was also very different, one was a public company, the B.C. Packers, but NEFCO remained privately held until many years later. In this probably lies some of the explanation why B.C. Packers went through a number of reorganizations and financial difficulties over its 75-year history. It is not unlikely that of the 42 canneries that formed the original promotion some could have been more of a liability than an asset to the original association. In addition, in a cyclic industry such as the salmon fishery, management has enough problems without being burdened with complaints of stockholders, criticisms of financial analysts and maintaining a public image. Also, dividends may have to be paid when the funds should go into reserves for the "bad years" which are sure to come. Annual reports make your business everybody's business, which in the highly competitive fishing industry, could be disastrous. Also, publicized profits one year can make for labor and stockholder demands the next.

NEFCO probably entered the salmon fishery with reserves from 50 years of successful trading in halibut. It was able to "pick and choose" in its acquisitions during depressed conditions both after World War I and during the early 1930s. The public did not know what NEFCO's stock was worth, and NEFCO did not have to maintain a "public image" in that respect.

When B.C. Packers entered the frozen and fresh fish business with the Cleeve plant, it had the distinction of shipping the first carload of frozen fish to eastern Canada. All earlier carload shipments from British Columbia had been to the United States' East Coast. The auxiliary powered vessels *Unican, Athena* and *Kilkenny* fished for the company at that time. In 1910 the cold storage facility was relocated on land leased from the city of New Westminster and called the Columbia Cold Storage. Again displaced in 1914, the B.C. Packers built its own Columbia Cold Storage at Steveston. It also used a cold storage at its Balmoral Cannery on the Skeena River and at Claxton.

At this time the company was contracting for halibut from the then independent Canadian Fishing Co. with its two vessels the S.S. *Celestial Empire* and the S.S. *Flamingo.* In 1907, B.C. Packers bought 1 million pounds from the *Celestial Empire.* These sources terminated when NEFCO bought out the Canadian Fishing Co. in October, 1909, along with its two steamers.

In January, 1910, the B.C. Packers purchased the S.S. *Roman* with a capacity of 220,000 pounds of iced halibut from the Imperial Steam Fishing Co. of Hull, England, and brought it to Vancouver under the command of Captain Bernard L. Johnson. In January, 1914, the S.S. *Onward Ho*, also with a 220,000-pound capacity, was purchased in England arriving in Vancouver in July. Both vessels were converted to oil from coal burners. As recounted elsewhere the *Onward Ho* was lost with all hands in January, 1916, and the *Roman* wrecked in November, 1916. These were severe blows to the company as both vessels had been successful in supplying the company with needed halibut. Their replacement was not possible due to wartime conditions. Also, the steamer era was starting to fade due to production costs relative to what had to be paid for fish purchased from smaller independently owned vessels. The company built or acquired 10 smaller 65-foot vessels with gas engines that along with the independent fleet, could keep it supplied with halibut. Such 65-foot vessels could also be used in the salmon fisheries as packers or seiners.

With the reorganizing of the firm in 1931 the fresh and frozen fish division was transferred to Messrs. Edmunds and Walker, two B.C. Packers executives. They moved to New Westminster operating from the public cold storage plant located on the Fraser River in that city.

In 1939, the Edmunds and Walker firm was purchased by the B.C. Packers and moved to Vancouver where it became the fresh and frozen fish division of the company. Messrs. Edmunds and Walker rejoined the B.C. Packers, the latter becoming president in 1956.

The company proceeded to expand its halibut business, which integrated well with the increasing proportion of salmon being frozen rather than canned. In 1941, the company bought the Victoria Cold Storage in Victoria and many halibut fares were bought on the Vancouver Fish Exchange for delivery to the Victoria plant. In 1945 the company which had withdrawn from Prince Rupert in the early 1930s purchased the Canadian Fish and Cold Storage Co., probably the largest halibut handling and storage facility on the Pacific Coast. In 1968 the firm moved its head office from Vancouver to Steveston at the mouth of the Fraser River, the location of its Imperial Plant, the largest and most diversified fishery operation on the Pacific Coast.

In this post-World War II period the B.C. Packers was in possession of strategically located plants equipped for handling fresh and frozen fish in volume, including halibut. The very large Prince Rupert plant in northern British Columbia, the Namu, with a freezing facility on the coast between Vancouver and Prince Rupert, the Victoria Cold Storage in Victoria on Vancouver Island, and Vancouver and

Steveston on the lower mainland provided coastwise coverage. The recently built Bellingham Cold Storage plant was also used as back-up capacity to Vancouver. The firm also had arrangements with commission buyers in Western Alaska.

Physically, the firm was now in a position to recapture some of the importance in the halibut fishery it possessed 30 years previously when it operated the large steamers *Roman, Onward Ho* and *B.C.P.* from New Westminster and Steveston.

Previous to the foregoing developments there had been major changes made in the firm after the traumatic events of the late 1920s and early 1930s. In 1930 alone, after packing over 1 million cases of salmon, nearly 50% of the provincial pack, a $1,266,058.38 loss ensued. As a viable enterprise the firm had reached a very low ebb.

In 1933 Harvey R. McMillan, who had been a director since 1930, was recruited from the lumber industry and made president of B.C. Packers. Forthwith the cumbersome corporate structure was streamlined, unprofitable properties liquidated, costs cut, and plans made to diversify at the appropriate time. In 1935 John M. Buchannan became general manager. His accounting background, managerial expertise and imaginative leadership had long been needed by the firm. By 1946 he became president, and the future of the firm was assured. H.R. McMillan continued as chairman of the board.

By this time the firm was in possession of what were probably its most valuable assets for its expansion into the fresh and frozen fish field, namely, experienced and capable personnel. Walker and Edmunds, who had successfully maintained the firm's original fresh and frozen fish business during the difficult 1930s, returned to the B.C. Packers. Roy M. Winslow, one of the most knowledgeable persons in Canada on merchandising quality frozen fish, including halibut, and Colonel G.W.E. Nicholl, who had been comptroller of Canadian Fish and Cold Storage Co., both joined the B.C. Packers on its acquisition of the Prince Rupert firms.

The company also maintained close liaison with the Nelson Bros. Fisheries Ltd. a tight-knit family company whose members were thoroughly versed in all aspects of the fish business. Successful as salmon trollers on the west coast of Vancouver Island after World War I, they packed their fish and those of others to the ultimate market, namely the United States, by-passing the secondary distributors in Vancouver. Many times in the mid-1920s their packers *Sundown* and *Western Chief* would be seen unloading on the Seattle waterfront.

Later, taking advantage of the depressed conditions after 1929, the Nelson Bros. Fisheries Ltd. acquired other properties, including a fish reduction plant on the west coast of Vancouver Island, a floating saltery and a salmon cannery on the Fraser River. As the economy moved ahead so did Nelson Bros. Fisheries Ltd. A large fishery development was initiated at Port Edwards on the Skeena River as well as a U.S. based salmon operation in Bering Sea. Their larger fishing vessels, when not occupied in the herring and salmon fisheries, were also active in the halibut fishery. Mr. Richard Nelson, president of the company which had been incorporated September 14, 1929, was a member of the Halibut Commission, serving 11 years, 1953 to 1964, and also serving on the Fraser River Salmon Commission.

Continuing the family tradition his son Richard I. Nelson, formerly manager of their Port Edwards plant, became president of Nelson Bros. Fisheries Ltd. in January, 1966, when Richard Nelson Sr. became chairman of the board. Several years later Richard I. Nelson was named president of the B.C. Packers. He later became chairman of the board when Bruce Buchannan, son of John M. Buchannan, was named president of the firm.

8 ECONOMIC HISTORY OF THE PACIFIC HALIBUT FISHERY

Although the fishery has usually reacted to the major fluctuations that have occurred in the North American economy over the past 100 years, economic changes have also taken place that were largely restricted to the halibut fishery.

The late nineteenth century was a period of discovery for the halibut fishery, of exploring the markets and developing production procedures. After the turn of the century the fishery enjoyed a profitable 15-year period of expansion with an almost geometric rise in total catch in the first 7 of those years. Fishing was being conducted on primeval stocks on nearby grounds close to railhead ports. These two elements, close in and abundant supplies with a growing market in the East, almost assured high returns.

While prices were low compared to present standards, only 2¢ per pound in Seattle in 1902 rising to 4¢ by 1904, the net returns in stable dollars were high during most of the first 15 years of this century. Such earnings generated investment capital to build more vessels, both large and small. Also, more firms entered the industry both in Canada and in the United States.

By 1915 there were marketing problems brought about by the overly rapid increase in production and other dislocations partly due to the onset of World War I. Prices had remained level at about five cents per pound since 1907, but costs on the large company-owned fishing steamers had increased sharply. Manpower for crews was scarce. Later a number of the large steamers were wrecked and the high wartime shipbuilding costs precluded their replacement.

The growing fleet of independently owned owner-operated vessels with their smaller crews prospered during the later years of World War I. Prices paid in Seattle for number 1 halibut had doubled by 1916 to 10¢ per pound and reached 16¢ by 1920; demand for all fish was good, and prolific new halibut grounds west of Cape Spencer provided a profitable winter fishery for the larger halibut vessels. The exploitation of those grounds was furthered by the opening of Prince Rupert in 1915 to landings by United States vessels for shipment in bond through Canada to the eastern United States markets. During this period many crew members accumulated earnings that later led to their acquisition of shares in the ownership of new or of older vessels.

The impact of the postwar depression of 1921-1922 with a temporary price decline was partly offset by the installation of the more economical diesel engine in many of the vessels. Halibut prices soon recovered to the predepression 16¢-level and there was an almost boom in vessel building, particularly those of the schooner type. Ship chandlers to the fishing industry supplemented the earlier savings of former crew members to provide some of the construction capital.

Except for some short-lived and unsuccessful efforts at cooperative marketing by the fleets, the decade of the 1920s was another period of expansion in the fishery as it was for the whole North American economy. However, prices paid for first class halibut remained at an even level of about 16 to 18¢ per pound in Seattle from 1923 to 1929. By the end of that decade the "cost/price squeeze" began to be serious as the items purchased by the fleet had risen steadily in cost.

With the onset of the Great Depression halibut prices like those of most other commodities declined precipitously, 62% between 1929 and 1932. The independent and resolute manner in which the halibut fleets attempted to cope with the disaster is discussed elsewhere in this book.

As the general economy recovered from the world-wide depression, halibut prices rose in general conformity with the consumer price index. Coupled with the improving halibut stocks now under the control of the commission, the halibut fishery prospered. Vessels were attracted to it from other fisheries and a few new ones were built for the fishery. The fishing seasons became shorter in length despite some increases made in the catch limits as allowed by the commission. A strong demand arose for the high potency Vitamin A oil of halibut livers and later of the viscera. Proceeds from such materials began to represent an important addition to earnings of the fleet.

Also, a successful fishery cooperative was launched in British Columbia that contracted for the catches of a large segment of the Canadian fleet. This withdrawal from the open market led the independent buyers to equip and generally further the development of a large fleet of small one- or two-man boats to setline for halibut out of camps which the buyers maintained to service the troll salmon fishery. The halibut fleet continued to engage in a voluntary lay-in program designed to spread the permitted catch over as long a period of the year as possible.

World War II added new dimensions to the fishery. There were mandatory price controls, allocations of the permitted annual catch among the buyers based on their respective historical shares of the catch, and many other dislocations directly or indirectly attributed to the war.

After termination of hostilities, the fishing seasons continued to shorten. The increased abundance of halibut attracted more vessels to the halibut fishery chiefly from less profitable pursuits. Although increases were made in the catch limits they could not offset the fleet increase. The fleets did not resume their voluntary lay-in program that was discontinued during the war. However, halibut prices maintained their postwar level and were well over parity until 1955.

In 1955 halibut prices dropped rather sharply to a level about 50% lower than that in the postwar period. Also, the price level relative to what the fleet had to pay for their requirements deteriorated to a level considerably below parity. Notwithstanding this severe but temporary setback, the fleet possessed the resilience and fortitude to withstand it. They generally resisted the efforts of a few of their members to relinquish their independence and petition for government subsidies, price supports, embargoes and the like. All such measures of government assistance with the possible exception of low cost loans and anti-dumping laws were usually self-destructive.

As it had done more than once in the past the fleet proceeded to solve its own problems. Crew earnings were sustained by reductions in the size of crews on many vessels. Technological improvements were made in the vessels and in their operation and the fleet set in motion arrangements to reinitiate the between-trip lay-in program that was developed at the time of the Great Depression of the early 1930s. In 1956 such a lay-in program was started.

During this rather critical period of about three years a number of economists developed an interest in the fisheries. With no research upon which to base their findings, they expressed the belief that the halibut management program had been an "economic failure" and of "doubtful biological advantage." They were concerned about the common property character of the fisheries generally and of the Pacific halibut fishery in particular. They espoused limited entry and felt that biologists "did not function in the real world." Very few facts were adduced and most of their conclusions seem to have been based on hearsay.

The United States subsidized a study of the economic effects of the regulations of the Halibut Commission on the Pacific halibut fishery. It had been represented by the sponsors of the study that the halibut industry was faced with economic uncertainties that "endangered the stability of the fishery and indirectly affected adversely the national economy and more severely the economies of Alaska and the Pacific Northwest States," a rather portentous statement, to say the least.

The study appraised the economic status of the fleet fishing out of Seattle. While a fairly representative sample of vessels was used it was of a segment of the Pacific fleet that was in a state of incipient decline. It also covered one of the short but temporary periods of economic decline in the fishery that in no way reflected the conditions that had prevailed over the long period that the fishery had been subject to commission management. Furthermore, none of the dire predictions made at the time were borne out in the succeeding years.

The basic theme of that report was advocacy of bureaucratic control of entry into the fishery. To examine the merits of such entry controls within the framework of a free (but regulated) enterprise system of a domestic fishery is beyond the scope of this book. However, to apply them in an internationally regulated fishery such as that of Pacific halibut would pose insurmountable difficulties.

The report did conclude that increases in the size of the fleet tied up resources that could be used to better advantage elsewhere and that uncontrolled entry and the low level of vessel earnings arising from an inequitable share system inhibited technical progress. While the latter may have been partly true, the condition did not seriously hinder the progress of the fishery. The fishery subsequently declined for other reasons.

There was little evidence of fleet increases tying up resources to any consequential degree. Vessels constantly moved from one fishery to another from season to season or from year to year. When the halibut season was very short, over 95% of the vessels engaged in other gainful pursuits after closure of the halibut season. Also most of the increases in fleet size arose from an influx into the halibut fishery from other fisheries not from the building of new tonnage.

While the share system may have had some inequitable features as far as the vessel share was concerned, it did not inhibit the installation of any proven technological device that would increase safety or profitability. This is evident in the changes made in specific vessels such as the M.V. *Polaris* discussed elsewhere in this book as a type.

After the short and depressed episode of the early 1950s the fleet generally experienced an acceptable level of profitability. It managed to cope with the various negative as well as positive forces that prevail in any private enterprise system. Imports of lower cost halibut particularly were becoming an increasing market problem. It was intensified to the point of being disastrous by the fact that most of the halibut imported from Japan was being taken from North American stocks that the halibut fleets of Canada and United States had sacrificed so long to conserve. It was also an added blow when in 1962 Canada and the United States "legalized" such depredations in the eastern Bering Sea by removing the halibut therein from abstention by Japan under the Tripartite Fisheries Convention between the three countries.

To compensate for the catches made by Japanese trawlers and setliners, it was necessary in the 1960s to sharply reduce the catch limits provided for the domestic setline fishery, and the earnings of that fleet suffered accordingly. While this slowed the decline in the stocks, any more radical reduction in catch limits would have been self-defeating if not disastrous as far as Canadian and United States interests were concerned. The foreign fleets, which were under no restraints, would have benefited more from a lessening of the rate of stock decline.

With some restraints placed upon the foreign fleets under the bilateral agreements and the 200-mile limits in the early 1970s, the possible benefits of holding a "close rein" on the domestic setline fleets may be realized at least in part by the domestic fleets making the sacrifice. Furthermore, the lowered production and reduced imports have brought above-parity price levels for halibut that may make the continuation of a modest single species setline fishery for Pacific halibut by Canada and United States a possibility. Much will depend upon how Canada and United States resolve the problems occasioned by their respective 200-mile exclusive fishery limits.

The impact on the halibut fleet of the changed conditions in the fishery brought about by the regulations of the commission, such as the shortening of the fishing season, and the increase in the abundance of the stocks, is evident in Table 37.

By 1965 all productivity measures per crew member had about doubled over what they were immediately prior to regulation. The average size of crew had been reduced 35% by 1965, largely made possible by technological improvements in the vessels and to some extent in the fishing gear as well as changes in operational strategies. The gains to the vessels themselves were not as great as for crewmen, although by 1965 the catch per season and the catch per day were generally higher than in the unregulated fishery. Not evident in Table 37 is the fact that the vessel's share in the gross stock was slightly increased over pre-regulation years.

The gains in production represent even greater gains in net dollar earnings. The fares, being larger, are consequently fewer for the same season's production, thus reducing the proportion of the revenue consumed for expenses. Also, there is less running time per pound. Furthermore, under regulation a much larger proportion of the catch has consisted of the more valuable large fish over 60 pounds.

TABLE 37. Productivity Measures for About Twenty Nearly identical United States Vessels Fishing Largely in Area 3 For the Preregulation Years 1930 and 1931 and the Post-regulation years 1954, 1957 and 1965

Year	Season Length* (days)	Catch per Man in pounds			Catch per Vessel in pounds			Average Crew Size
		Per Trip	Per Season	Per Day	Per Trip	Per Season	Per Day	
1930	272	2,966	34,620	127	27,785	323,700	1,190	9.3
1931	224	3,566	31,203	139	33,101	287,600	1,280	9.3
1954	99	6,900	29,097	301	47,217	206,300	2,080	6.9
1957	153	6,816	35,198	231	46,397	240,200	1,570	6.8
1965	173	8,600	57,000	329	59,000	346,000	2,000	6.0

*This is not the fishing season. It is the number of days from date of clearance from port on first trip to date of landing of last trip.

Gordon (1953) and other economists believed that the building of larger and faster boats to secure the largest possible share of the limited catch has not resulted in any net gain from the improved stocks but dissipated them through higher costs. Size and speed in themselves should not add to total costs but should detract from them. Within reasonable limits man's individual productive capacity has been greatly augmented by increasing both the size and speed of production machinery.

Without becoming further involved in an extremely complex and probably futile analysis of economic considerations, it would seem from Table 37 and other facts of the fishery that management of the resource by the Halibut Commission has not been the disaster that many economists claimed it to be. Similar data for the Canadian and Seattle vessels fishing in Area 2 south of Cape Spencer show results comparable to the above for the United States vessels fishing in Area 3 west of Cape Spencer.

EFFECTS OF WARS ON THE FISHERY

The Pacific halibut fishery has not been immune to the forces that were unleashed and restrictions that were imposed in the wars in which Canada and United States participated. In addition to the general effects felt by all, there were a number of developments that were more or less restricted to the halibut fishery.

Spanish-American and Boer Wars

These two wars around the turn of the century had little impact directly or indirectly upon the early, but developing, fishery. There was not an increased demand nor a general rise in prices as is often the case in time of war. Probably the only close encounter was the narrow escape of the newly-launched S.S. *New England* from being commandeered by the U.S. Navy for war service. After launching in Camden, New Jersey, in 1897, the vessel made a forced run to the Virgin Islands to avoid United States ports when en route to Vancouver via the Horn.

World War I

At the time of the Declaration of War in 1914, the Canadian (British registry) fleet of company halibut steamers was of some proportions. One or two of the steamers as well as their captains were members of the British Naval Reserve and forthwith returned to Britain to serve as minesweepers and other naval duties. The recruitment of men into the armed services in Canada and in United States after 1916 resulted in a scarcity of capable persons to crew the growing fleet of company-owned steamers. Also, the replacement of vessels that were wrecked or lost was impossible as shipyards were fully occupied in a desperate effort to replace the merchant ships lost to German U boat action.

Although halibut prices first lagged behind production costs because of overproduction in 1914 and 1915, the halibut fishery became quite profitable in the later years of the war. This resulted in the accumulation of capital by crew members and owners that later aided the growth of the fleet in the early 1920s.

Another of the positive values of the war was the closer cooperation in fishery matters between Canada and the United States than had prevailed during the previous 100 years. Measures were enacted and restrictions were removed in both countries to facilitate the joint war effort. Discussions of a joint fisheries commission were also undertaken at this time.

World War II

World War II had a greater impact upon the Pacific halibut than any military conflict heretofore. However, since the Northeastern Pacific was the scene of only limited enemy action, there was no *Kriegschonzeit,* "wartime holiday," such as prevailed in the European Atlantic fisheries during both world wars. The effects of the war were more indirect than direct.

Shortly after the beginning of hostilities in 1939 the Royal Canadian Navy Fishermen's Reserve was organized in Canada, and about 40 units of the fishing fleet including a number of halibut vessels served as emergency picket boats and coastal patrol vessels. They were later replaced by vessels built for the purpose.

The entire United States halibut fleet was surveyed for potential use as emergency transportation of cargo to Alaskan military bases. Several of the large schooners were converted to coastal patrol vessels. One in particular, the *Foremost*, served out the war with the United States Coast Guard, which in wartime becomes a part of the U.S. Navy. One halibut schooner, the *Sitka,* was engaged in supplying halibut directly to distant military bases in Western Alaska.

As hostilities moved into the eastern Pacific and the far-western Aleutian Islands were invaded, the United States Navy closed areas west of Trinity Island to halibut fishing. Vessels fishing west of Cape Spencer were formed into Navy-supervised convoys at Ketchikan. Each convoy was required to have a radio

FIGURE 74. The M.V. *Helgeland*
Ray Krantz

telephone and the boats were instructed to fish where possible within reasonable proximity to one another. On return they were to report any sightings of enemy mines, submarines and similar intelligence. The United States Navy also requested the fleet to abandon their voluntary between-trip lay-ins to expedite the taking of the catch limits and removal of the fleet from the area.

While there were no confirmed losses of the fleet to enemy action, the large halibut schooner *Helgeland* (Figure 74) of Seattle and its crew of 11 under Captain Louis Holm, disappeared without trace when fishing in the Gulf of Alaska in 1942. Several of the most seaworthy vessels of the Seattle halibut fleet organized a prolonged search for the vessel but without success. There were rumors during the war that the vessel had been seen in Adak and other places occupied by the Japanese. Others claim it was seen in Japan toward the end of the war. True or not, the *Helgeland* and its crew joined the many brave men that have gone down to the sea and were lost while engaging in their chosen pursuit in life. Brave, too, were those who manned the search vessels.

The Canadian halibut fleet was augmented during the war by the Norwegian fishing vessel *Kaare* that escaped from Nazi-occupied Norway and made the long journey to British Columbia.

Wartime demand occasioned a significant rise in prices paid for halibut but production costs also rose. Changes had to be made in the fishing gear as Italian hemp for ground line was no longer available and had to be replaced by a succession of other fibers. National plans were made in the United States for the allocation of scarce goods, supplies and fuel so the production of food would not be hindered. The U.S. Army Quartermaster Corps placed priority orders for large amounts of halibut to be delivered under specified conditions to certain markets.

The mobilization of civilian manpower along with the Selective Service in the United States affected the fishery. Men were drafted, others were exempted, and although good crews were sometimes scarce, production was not hindered. Several captains of the fleet served as pilots for supply and other vessels moving in the poorly chartered waters of the Aleutian Islands.

Probably the greatest wartime dislocation in the fishery was caused by the emergency price controls in United States under the Office of Price Administration. While controls are almost mandatory for a nation engaged in all-out war, the repercussions provided an excellent demonstration of what happens when the free market is suspended. There was confusion as inequitable prices were set between ports and black markets begun. The catch limits of the commission were allocated between buyers based on their historic interest in the fishery with all the attendant problems — questionable allotments, protesting sellers and buyers and costly bureaucratic controls.

By June 30, 1946, price controls elapsed under one law, but allocations continued under other legislation until July 22. However, the Office of Price Administration reinstated price controls on July 29 that were largely ignored by the vessels freezing their catches or landing in ports that ignored the ceiling prices. Finally, on August 15, 1946, open market sales were reinstated and crews which had frozen their catches disposed of their holdings in the regular channels.

World War II also had a major impact upon the Halibut Commission. Despite sharply increased costs of operation the commission received no increase in funds from either country. Field investigations became impossible, and at one time the scientific staff was reduced to three persons: the director, assistant director and a biologist. Although the stocks were rebuilding, the fishery was in a ferment of change. There was an urgent need for information to guide the commission in its management of the halibut resource.

LABOR RELATIONS IN THE HALIBUT FISHERY

The Pacific halibut fishery has been regarded as a "closed shop" industry since early in the century. Although remuneration of the fishermen has always been on a share basis, the share structure, or "lay," and the terms of employment have been negotiated through collective bargaining between fishermen's unions and either vessel-owning companies or associations of individual vessel owners. A few of the labor disputes that involved significant principle will be briefly reviewed. The development of unionism in the halibut fishery in British Columbia will also be compared with that in Puget Sound.

One of the first labor disputes in the halibut fishery occurred in Vancouver in October, 1904, when New England Fish Co. decided to pay the fishermen by the pound rather than by the piece. The fishermen went on strike, and the company recruited a mixed crew of shoreworkers and took the S.S. *New England* to sea. The strikers chartered the eight-dory *Dauntless* under Captain A. Dawe and fished while negotiations were under way. The strike was over in about three weeks with both sides making some concessions but no closed shop.

The conditions for union growth and collective bargaining in the halibut fishery differed widely between Vancouver and Seattle. The present Deepsea Fishermen's Union in Seattle was organized in 1913 as a branch of the long-established and powerful Sailors Union of the Pacific. The union later became independent of the Sailors Union of the Pacific. Considering the diversities that prevailed in the halibut fishery the union has maintained a remarkable continuity and cohesiveness over the past 65 years.

In 1913 the union successfully struck to have the owners' share of the gross stock reduced from 25% to 20%. While there was a further adjustment in 1916 which provided for the boat share to bear a portion of the cost of worn gear, it was the reduction to 20% in boat share that was the most meaningful change that was to be made for the next 50 years.

An earlier union in Puget Sound, the Pacific Halibut Fishermen's Union, was organized in Seattle in the winter of 1907. It organized the 72 dory fishermen of the United States vessels that were fishing out of Vancouver at that time, the S.S. *Kingfisher,* S.S. *Manhatten* and S.S. *New England.* On June 18, 1909, the crew of the *Kingfisher* struck when a crew member, Paddy Shean, became a *cause celebre* by refusing to join the union. This set the stage for a test of the closed shop principle. The crews of the other two vessels also struck. However, the importation of 20 men from Nova Scotia and the defection of some of its own members resulted in the union capitulating after about a month. Some of the original strikers got jobs in the salmon fishery.

The foregoing disputes and their settlements show some of the differences that prevailed between British Columbia and Puget Sound with respect to union organization in the halibut fishery. Puget Sound and Southeastern Alaska possessed a fast-growing fleet of independent owner-operated vessels and a few large company-owned steamers. This provided varied employment opportunities for the fishermen. There was also alternative employment in the vessel fishery for cod in Bering Sea and in the nearby salmon seine fishery in Puget Sound. This provided a broad employment base in vessel fisheries which furthered the fishermen's bargaining position in any dispute.

In British Columbia the fishery was initially dominated by one company using large company-owned steamers. Later there were two more companies, one in New Westminster and one in Prince Rupert. There was no independent halibut fleet or vessel cod fishery, and the salmon fishery was conducted by gill-net boats. The halibut fisherman was in a sense locked to the large company steamer. He had to bargain with a vessel owner and a distributor at the same time, unlike with the independent vessel crewman who bargained with only an owner. The existence of a large reservoir of experienced men in the Nova Scotia and Newfoundland fisheries also weakened bargaining power.

The companies in British Columbia were beset with their own problems. Unlike those in Puget Sound, they lacked the boxed halibut received from Southeastern Alaska and the larger catches of small, owner-operated vessels in Seattle in the winter. The fish from these sources were sold in the open market at prevailing prices. In British Columbia the dealers were dependent almost exclusively upon the company-steamer landings which had fixed costs. The costs could be relatively high at times so that the profits lay in selling the fish at higher prices in eastern markets and not in producing the fish in the West. One company, however, was particularly successful due to its well-organized distribution system in the East. This led at times to abrasive situations that did not help relations with labor. The production side of the business was under constant pressure from the head office and stockholders to carry its share of the cost load and not depend only on the distributor level for a profit.

The company-vessel fishermen had many problems. They were constrained by seasonal quotas on the size of trip that could be taken. This and unduly quick turn-arounds in port were points of complaint. Furthermore, pressure on the companies to cut costs increased as the stocks declined. The steamers were more susceptible to stock declines than were the smaller independent vessels. Access to old company cost accounting records has indicated that, except in the early years, it was cheaper for the company to buy on the open market from the independents than produce the fish on company vessels, because of their high fixed costs including 11 to 13 salaried persons on the crew. The main value of the steamers lay in the volume that they could produce pending the expansion of the independent fleet.

By the winter of 1912-13 with the support of the Halibut Fishermen's Union of Seattle under P.B. Gill, Vancouver fishermen prevailed upon the British Columbia companies to raise the 1904 settlement of 1¢ per pound to 1.25¢ and afforded union recognition. One company again imported men from Nova Scotia and offered them 1.5¢ per pound without union recognition but finally agreed on March 31, 1913, to the general settlement of 1.25¢ with union recognition. The closed shop finally prevailed.

On November 4, 1915, the Vancouver branch of the Halibut Fishermen's Union affiliated with the Vancouver Trades and Labour Council.

A strike in 1916 had little impact on the Canadian fishermen because it applied only to the independent owner-operated vessels which were few in British Columbia. They did support the United States fishermen and discouraged United States owners breaking the strike by landing in British Columbia ports.

Union membership grew in British Columbia to about 200 in 1917. When the Deep Sea Fishermen's Union of Seattle, which had an office in Prince Rupert, finally closed in 1931, the Canadian halibut fishermen formed the Deep Sea Fishermen's Federal Union of British Columbia and affiliated with the Trades and Labour Council of Canada.

By this time unionization of the halibut fishery in British Columbia had firm footing. In later years another strong group entered the scene—the United Fishermen and Allied Workers Union, a horizontal organization that crossed into all sectors of the fisheries of British Columbia to include the shoreworkers. This union encompassed the halibut fleet, and in the 1960s a prolonged and involved jurisdictional controversy arose between the two unions in Prince Rupert.

Labor-management relations in the Pacific halibut fishery over the past 80 years can be classed as benign compared to most industrial or craft situations. While there have been many differences that were resolved, only three ended in what could be considered a strike of any major proportion. Despite there being several unions in an area, there was only one jurisdictional dispute of consequence and that involved a special situation. There have been no lockouts per se, but a number of tieups have occurred from disputes between buyers and boat owners and fishermen.

There are several reasons for the relatively tranquil conditions. The early resolution of the closed shop principle and the limited number of absentee vessel ownerships, with most owners being former fishermen, have contributed to the peaceful situation. It also helped that disputes with buyers and distributors were fewer with open market pricing than with yearly contract prices. (With yearly contracts prevailing in the salmon fishery, a steady state of dispute exists between fishermen and processors.) In the halibut fishery both owners and crews have been represented by generally stable and conservative leaders.

The relative infrequency of strikes has also facilitated the management of the resource for the Halibut Commission.

THE SHARE SYSTEM OF COMPENSATION

From the earliest days, the proceeds of halibut fishing have been divided according to a certain lay—an arbitrary measure of each person's share of revenue. Daily wages or monthly salaries are not known to have been used at any time as a system of compensation for the fishermen in the Pacific halibut fishery. The non-fishermen members of the crew on company boats were salaried, some with a bonus based on size of catch.

In the very early years of the Puget Sound fishery the International Fish Co. of Tacoma, which was the major producer of halibut on Puget Sound until about 1905, paid the fishermen on their steamers 25¢ per fish over 34 inches in length and 12.5¢ for those 28 to 34 inches. Those below 28 inches were rejected at sea as undersized. Such a steamer fishing on some sections of Goose Islands grounds had been known to reject during a trip as many as 10,000 undersized fish.

At the start of the fishery in Vancouver in the mid-1890s, when it was conducted from winter shore stations in Hecate Strait, company fishermen were paid 25¢ for halibut over 10 pounds. Independent fishermen with their own boats and gear received 35 to 37.5¢ per fish. Fish under 10 pounds could be taken at half price but only at the captain's discretion. Company men were given double price for those over 50 pounds. This was not a company rule but was done on the captain's own initiative in order to maintain morale and discourage the fishermen from releasing the large fish. When the S.S. *New England* first fished in 1899 and shore stations were abandoned, the basic 25¢ per fish over 10 pounds continued, but the discretionary double price for fish over 50 pounds was dropped.

Because night fishing often made accurate fish counts impossible, New England Fish Co. proposed paying for the fish by the pound. The use of weight instead of count as a measure of payment was an advantage to the company as the average size of the fish was declining. The fishermen went on strike.

The strike was settled with the establishment of a straight price of 1¢ per pound for fish over 10 pounds (later changed to 1.25¢ for fish over 20 pounds) and 1¢ for those 10 to 20 pounds. The fishermen were no longer required to load the railroad cars after unloading the vessel without extra pay. The proceeds of the trip were still prorated among the fishermen according to their estimated dory counts. This settlement initiated the grading of fish by size at the dock which has continued to the present in most agreements.

Later increases were made in the price per pound paid on the company steamers, and prorating the proceeds by individual dory counts was abandoned. By 1925 on the S.S. *New England* the 22 fishermen on board divided a straight price of 5¢ per pound for all halibut landed. Expenses such as grub, bait, gear, and fuel were paid by the company. The captain and crew members were paid a monthly salary. (The captain received a bonus of $2 per 1,000 pounds of halibut landed.)

For the past 60 years the independently owned vessel crews have divided the gross proceeds from a trip as follows:

Gross Stock (Gross Sales Revenue) included all income of every kind from the fishing operation—minus

Gross Stock Expense included cost of lost or stolen gear, brokerage, fish exchange, customs fees, watchmen's wages and other incidentals. This gave the Net Gross Stock.

Total Boat Share—20% of Net Gross Stock. The owners usually assign 10% of Boat Share to the captain as a bonus. The latter is usually an owner or part owner.

Man Share the remaining 80% of Net Gross Stock less the Crew Expense divided by the number of men including the captain who is not infrequently also an owner.

Crew Expense includes cost of grub, bait, fuel, grease, engine room supplies, dish towels, broom and mop for living quarters, medicine chest, some radiophone charges and other incidentals.

The foregoing is a typical lay that has prevailed in the fishery down through the years, although there have been many relatively minor modifications from time to time. Some of the modifications required varying amounts of negotiation between the principals, namely the unions and the vessel owners' associations. The less substantive adjustments, such as whether to charge the cost of a floor mop to crew expense or to gross stock expense, were usually settled with reasonable dispatch. Other differences, such as the question of the size of the total boat share, have on occasion been difficult and their negotiations more extended.

In 1966 the Fishing Vessel Owner's Association of Seattle opened their contract with the Deep Sea Fishermen's Union on February 14, requesting an increase of 6.5% in the boat share. This represented a considerable increase in the basic 20% that had been in effect for 60 years since 1917. (There had been two 1% increases, one in 1946 and the other in 1958, to cover United States social security and state unemployment insurance taxes.) There were some differences for Alaska-based vessels because of the differences in the state unemployment tax.

One of the items in the 6.5% increase was 1% for the 1966 rise in the United States Social Security tax. The union conceded this increase since payroll taxes only benefit the crew. The remaining 5.5% covered owner cost claims for maintaining items that were not involved in 1917 when the basic 20% boat share was established. Those items included hold refrigeration, radar, life raft, galley refrigeration, loran, radiophones, hydraulic steering, gurdy and winch, extra fathometers, engine filters, stabilizers and greater electrical supplies. The owners claimed that the crew was the beneficiary of 80% of the increased earning capability that the above installations generated and that it should contribute accordingly. The union regarded the demands of the owners as a cut in pay.

After considerable negotiation a new agreement was ratified by the union on April 28 and by the owners on April 29. A 1% increase was provided for vessels with hold refrigeration and the aforementioned 1% for Social Security tax increase. On this occasion the negotiations did result in the United States fleet from Seattle being unable to participate in the early fishery in the Bering Sea and in Area 38.

These negotiations were recounted in some detail as the settlement involved a departure from the basic 20% boat share that had been in effect for over one half century. For the first time the principle of an incentive scale for capital investment in technological investments by owners was brought into the lay. It seems that in recent years the crews have benefited disproportionately from the various technological advances made in the vessels.

To cover more than some of the highlights of the share system in the fishery and the lesser changes therein is beyond the scope of this book. Generally speaking the labor management relations have been relatively tranquil in the Pacific Coast setline halibut fishery with few strikes or "lockouts" during the 90 years since the fishery began.

HALIBUT PRICING

The modern Pacific halibut fishery has been conducted upon an open market basis wherein the buyers or dealers and the vessel owners and crews have controlled the disposal of the individual catches by a system of competitive bidding. In many other fisheries pricing has been on the basis of a preseason negotiated contract price for the whole season or parts thereof.

Open market daily pricing has been achieved through fish exchanges in the major ports and prices arrived at by the auction process. In ports without exchanges the price was usually based on what prevailed at some nearby port with an exchange. Halibut landed by small boats at outside buying camps were usually bought at a fixed price for the season, although in some instances competition between buying camps of different companies led to price changes or the granting of some perquisite such as free ice.

The Seattle Fish Exchange was initiated by the vessel owners dissatisfied with the practices of some buyers. It was operated by the dealers in the 1920s. Later, due to restraint of trade indictments resulting from the exclusion of some buyers, the exchange was brought under the joint control of the vessel owners and the unions with its rules subject to the concurrence of the dealers.

In later years with the dispersion of landings away from the railhead ports and the improvement of ship-to-shore communications, many trips were sold directly to a particular buyer. In the securities market such sales could be classed as "over the counter" sales.

The fish exchanges were usually operated by both the sellers and the buyers and financed by a buyer's small membership fee and a jointly shared assessment on the poundage landed by the selling vessel. As with any commodity or securities market there were agreed-upon rules. Buyers had to have a safe place for vessels to unload, unloading had to be done expeditiously, and the buyer had to be financially able to pay for their purchase without any delay. The vessels were required to list or hail their fares by a certain hour of the day, and after the sale they were to be at the buyer's dock before a certain hour. Weighing methods were specified and grades, according to size and quality, were defined in the Rules of the Exchange. The buyer could assess a price penalty if the weight and size composition of the catch deviated beyond certain limits from what was hailed or listed on the exchange.

Down through the years, the incorporated fish

exchanges in Seattle, Vancouver and Prince Rupert and less formal ones in Ketchikan and Petersburg have collectively sold many hundred million pounds of halibut. Often unusual accumulations of fares in port—500,000 to slightly under 1 million pounds—were sold on a single day through Seattle or Prince Rupert exchanges.

Compared to some commodity and securities markets, halibut has been bought and sold down through the years under relatively healthy conditions. There were some occasions when there might have been collusion in bidding or restriction of membership on a fish exchange. Such suspicions often drew the attention of the respective federal agencies in Canada and the United States concerned with restraints of trade. Generally, it has been a competitive market, and all parties have benefited accordingly.

Various segments of the fleet from time to time have proposed contracting with the distributors for minimum prices. One objective was to reduce price fluctuations. They also briefly attempted to regulate the weekly flow of fish by scheduling arrivals. While the fleet did not succeed at this the voluntary between-trip lay-in program was conducive to a more orderly marketing of the permitted catch with some reduction in price fluctuations from week to week.

Independent of the auction bidding process, there have been, however, quite consistent differences between firms in their grading and weighing practices, and some vessels refused to sell to certain companies. While both grading and weighing were conducted under the close surveillance of the vessel owner or crew delegate, or both, unscrupulous firms had many ways to manipulate the processes without detection.

The matter of grading and weighing was a concern of the Halibut Commission insofar as the catch limits set for an area were based on the weight of salable halibut as taken from the invoices of sale or the dock tallies. However, based on personal observation over many years and some informal comparative tests of the weighing and culling process, it is apparent that most companies endeavored to be fair and were so known to the fleet. This was not always easy as a few less scrupulous competing firms gained a trading advantage in the secondary market.

Pricing in the Pacific halibut fishery is subject to a number of factors including the primary economic ones of supply and demand. There has been little meddling by governments with the process except for the conservation necessity of controlling the annual catch. The halibut conventions required that the maximum physical yield be secured from the resource. At times the halibut fleets did voluntarily control the rate at which the permitted catch was landed during the season in the interest of more orderly marketing and more stable prices.

Supplies at any one time consist of holdings of unsold stocks, domestic landings as permitted by the Halibut Commission and imports. Demand is subject to the availability of alternative food items of equivalent attractiveness to the consumer and many other considerations. The prevailing wage scale will affect what the consumer will buy. The sharp increase in the share of the food budget being spent in restaurants creates demand for the scarce and more expensive species such as crab, salmon and halibut. Increase in export demand for some species has also caused a sharp rise in price.

Since a comprehensive economic study of the Pacific halibut fishery is outside the purpose of this book, only a few facets of the question will be briefly discussed. They are designed to bring out principles

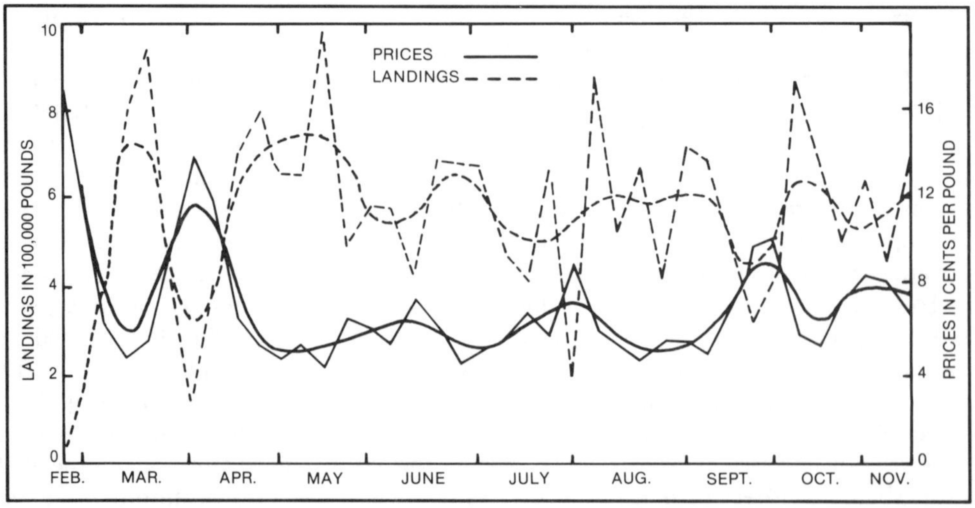

FIGURE 75. Halibut landings and average prices of medium halibut by weeks in Seattle, 1932 (lines are smoothed by moving 3's).
F.H. Bell

that prevail in the fishery. An example of how supply and demand functions at the port level is suggested in Figure 75.

The obvious inverse relation between weekly volume and weekly prices portrays the open market conditions that have characterized the fishery. It is the reason why the fleet by their voluntary lay-in program endeavored to reduce the weekly volume and thereby lower the amplitude of the price fluctuation. The program would not have necessarily raised the average price to the consumer.

Most studies on Pacific halibut prices have used the average price paid for the total coast catch without correcting for the relative amounts of the catch that are landed in the several ports each of which have their price levels and usually based on the cost of getting the product to the ultimate consumer. On account of the historical shifts of landings between ports of different price levels, price trends based on the total catch will also tend to reflect the shift of landings rather than the price level alone.

Price trends over the longer term also show some relation to the trends of the domestic catches. In Figure 76 halibut prices in stable dollars (1939 = 100) in Seattle from 1946 to 1977 show an inverse relation to the size of the Pacific Coast catch.

There are some indications that since 1971 there has been a sharp rise in price over parity as the volume fell below 40 million pounds. With the sharp decline in North American halibut production since 1960 ex-vessel prices have shown a sharp increase, an increase greater than those for most fishery products. With world halibut production also declining sharply during the same period imports have not been available to compensate for the short domestic supplies.

A strong above-parity price structure for halibut may make it possible for the fishery to survive and regain some of its previous importance to the fishery economies of Canada and United States. However, very high prices could be disastrous for the setline halibut fishery and the resource. It will greatly increase pressure from the domestic trawl fleet to participate in the halibut fishery.

The average ex-vessel price paid for eviscerated heads-off halibut in Seattle annually since 1920 is shown in Figure 77.

The relationship between available supplies and the continental market is indicated in Figure 78.

As holdings rise, prices decline and vice versa with about a five-month lag in the effect on prices. The 20- to 40-pound category is used as it is the best indicator of the market level. The holdings are plotted in Figure 78, five months ahead of the prices to emphasize the inverse but delayed relationship between the two factors.

MARKETING OF PACIFIC HALIBUT

Pacific halibut is marketed throughout Canada and the United States and there is an export market of varying dimensions in Europe, largely in Great Britain.

The continental distribution of halibut has over the years been one of the major features of the Pacific North American fishing industry, eclipsed only by the similar trade in canned salmon and canned tuna. In past years many carloads of fresh, iced halibut were dispatched almost daily from the railhead ports of Prince Rupert, Vancouver and Seattle. They were destined for such redistribution centers as Chicago, Boston, New York and points of lesser consequence. The transportation charges were an important source of railroad revenue to the western roads and the large volume of business stimulated many improvements made in the design and operation of railcar and truck refrigeration.

As the halibut fishing season shortened and retail merchandising shifted more and more to frozen foods and to supermarket chains, an increasing proportion of the halibut catch was shipped and marketed frozen. This trend was furthered by the improvements in the technology of freezing and storage. (The closing or shortening of the fishing season was a secondary factor.) The various proportions of the catch frozen by decades has been approximately as follows: 1920s, 20%; 1930s, 40%; 1940s, 75%; 1950s, 85%; 1960s, 90%; and 1970s, over 90%. Although the fishing season lengthened sharply after the mid-1950s the increasing proportion of the catch landed in Western Alaska, where freezing was a practical necessity, further increased the overall proportion frozen.

From the consumer marketing standpoint halibut is a most versatile species. Its large size permits it to be cut into transverse slices or steaks or to be sliced longitudinally as boneless fillets or flitches. It can also be sold in chunks, or roasts, and in measured portions. It can be successfully frozen whole or in pieces, and has excellent keeping qualities, both in the fresh and frozen state.

From a dietetic standpoint, halibut is very low in fat and sodium content, sharply contrasting with what prevails in the flounders which have been marketed misleadingly as halibut.

Thurston and MacMaster (1960) have shown that except for the very thin layer of dark meat under the

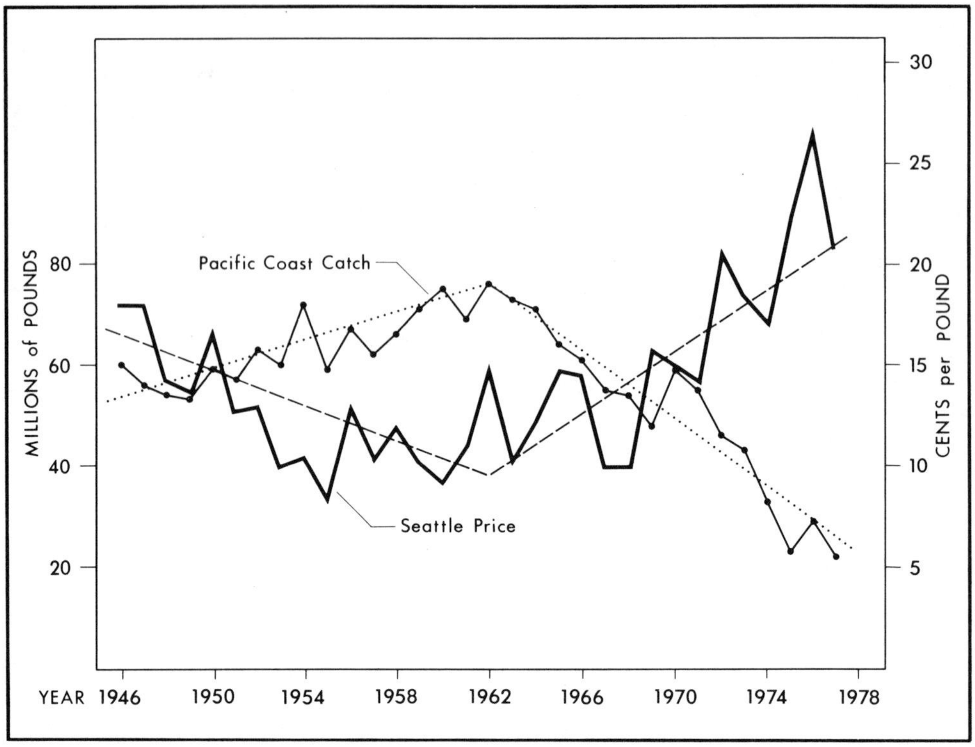

FIGURE 76. Ex-vessel price of halibut in Seattle in stable dollars (1939) and the total Pacific Coast landings.
F.H. Bell

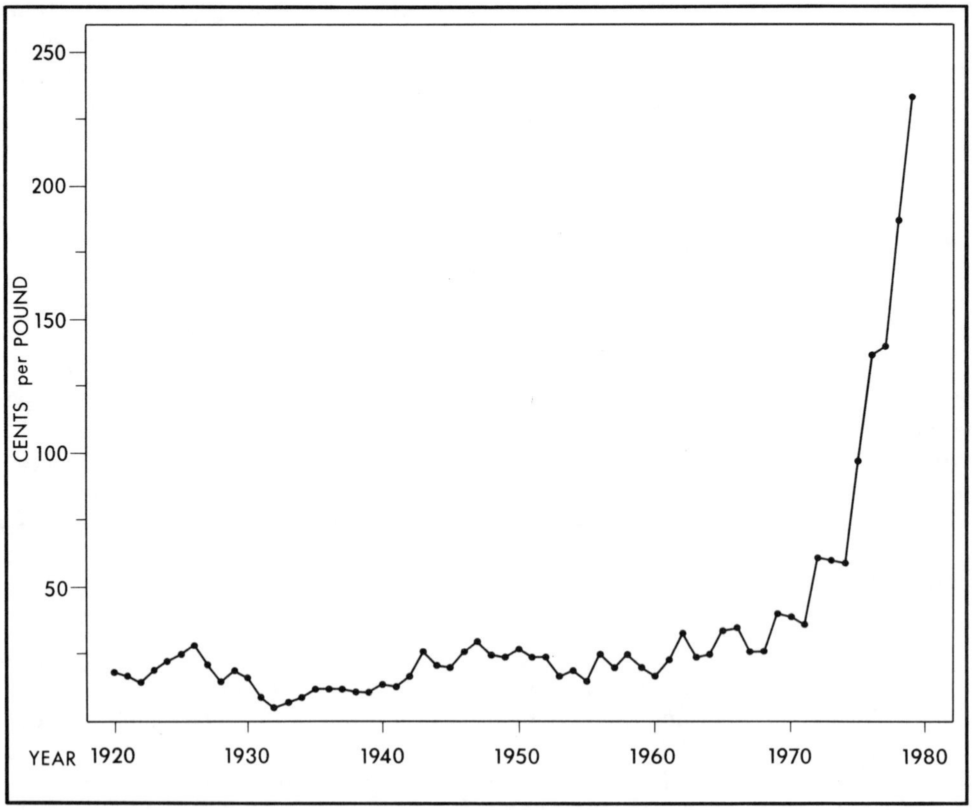

FIGURE 77. Average annual ex-vessel price per pound of eviscerated, heads-off halibut in Seattle from 1920 to 1979.
F.H. Bell

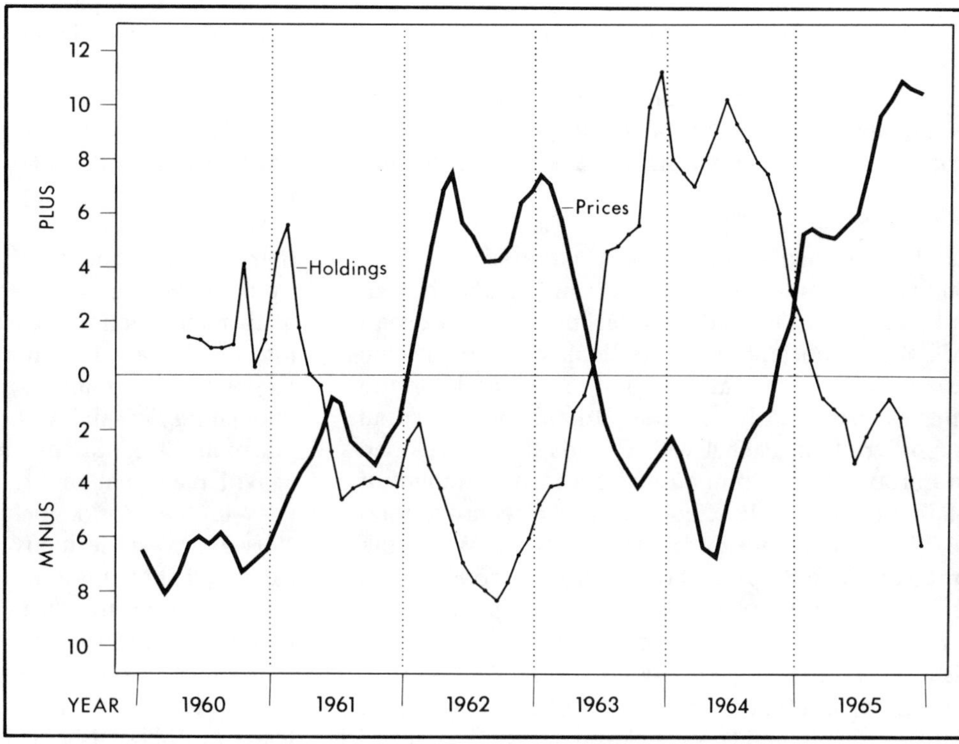

FIGURE 78. Deviation in cents per pound of frozen 20-40-pound halibut in New York and in millions of pounds of Canadian and United States holdings of frozen halibut from six-year monthly averages.
F.H. Bell

TABLE 38. Approximate Analysis of Halibut and *Reinhardtius* (Turbot)

	Raw Fish			
	Per 100 Grams, Edible Portion		Per One Pound Flesh, as Purchased	
	HIPPOGLOSSUS	REINHARDTIUS	HIPPOGLOSSUS	REINHARDTIUS
Water (%)	76.5	74.5	—	—
Protein (gr.)	20.9	16.4	94.8	74.4
Fat (gr.)	1.2	8.4	5.4	38.1
Ash (gr.)	1.4	1.0	—	—
Phosphorus (mg.)	—	—	957	953
Thiamine (mg.)	—	—	.29	.05
Food Energy (cal.)	100	146	454	662

skin along the center of each side of the fish, the average fat content of *Hippoglossus* was less than 1% based on a very large sample of Pacific halibut.

In contrast, in the far northern latitudes *Reinhardtius* has been used by the aboriginal people not only for food but also as a substitute source of oil to heat and light their dwellings when the other sources such as seal oil were in short supply. Currently in certain European countries including the USSR, much of the catch is smoked because of the high fat content.

It is a commentary on the "truth in advertising" myth to find the following promotional copy on retail packages of turbot *(Reinhardtius)* in the supermarket:

> From the ocean . . . Boneless Turbot Fillets . . .
> Sweet, Mild, Flavor . . . Low in CALORIES!

Turbot has seven times the fat content of halibut *(Hippoglossus)* and 25% less protein—hardly low in calories!

EXPORT AND IMPORT TRADE IN HALIBUT

Canada and United States, partners in the conservation of the Pacific halibut for over a half century have different trading positions in halibut—Canada with a smaller population has been a net exporter of halibut, and United States has been a large importer, chiefly from Canada. This situation has prevailed from the outset when halibut was first shipped from New Westminster to Boston in the early 1890s.

However, it should be recalled that much over 50% of the halibut landed in British Columbia up to recent years consisted of tonnage landed by United States vessels. This commenced with the S.S. *New England* in 1898 as provided in Order-in-Council of November 8, 1897. In 1915 a similar provision was applied to United States vessels landing in Prince Rupert. Such poundage was shipped through Canada in bond and is not classed as exports from Canada or imports into United States.

In early years of the fishery, at least 90% of the catch by Canadian vessels was exported to United States. With the rebuilding of the Pacific halibut stocks in the 1930s and their decline in the European Atlantic, importers in Europe looked to the Pacific Coast to supply what was a firm demand for the species, particularly in Great Britain. A substantial import trade in halibut with Canada then developed in Europe. It rose from an initial recorded amount of 32,000 pounds in 1932 and attained a recorded total of 5,685,000 pounds by 1939. After World War II the trade dropped because of the increased production in the European Atlantic from the stocks replenished as a result of the wartime "holiday" and the respite from fishing on the fishing grounds off Europe.

After World War II imports into United States from Iceland, Greenland and Scandinavian countries reached meaningful proportions totaling 3 to 5 million pounds in a year. They were often quoted in New York at prices that the Pacific Coast industry had difficulty in meeting. Also, by the early 1950s amounts reaching 1 million pounds were imported from Japan at equally low prices. By 1962 imports from Japan had escalated to 8 million pounds and had a demoralizing effect upon the North American market. For the remainder of the 1960s imports from Japan remained at an estimated minimum of 4 to 6 million pounds annually. In some years there were large exports from Japan into the European market, particularly into Great Britain. In some years total exports from Japan exceeded their reported receipts even after allowing for carryovers.

In the late 1920s, when the stocks of halibut in the Northeast Pacific were at a low ebb, imports from Japan were encouraged by North American companies. The following amounts were imported; 1925, 860,000 pounds into Vancouver; 1926, 234,000 pounds into Seattle; 1927, 500,000 pounds into New York; 1929, 321,917 pounds into Seattle and small amounts thereafter. The quality was poor.

However, the major international trade in halibut has always been the imports into the United States from Canada. Some of these were landed by Canadian fishing vessels in United States ports. From 1956 to 1975 such landings averaged 1.25 million pounds annually. In the past there were brief periods when Canada embargoed such ex-vessel exports. Fresh and frozen imports of halibut from Canada into the United States averaged 27 million pounds annually from 1955 to 1970. With the great reduction in the total Pacific catch in the 1970s, the level of imports from Canada has been greatly reduced.

With respect to United States and Canadian trade there was in some years a substantial flow of halibut from United States to Canada amounting to as much as a million pounds or so. This arose from possible internal imbalances in the holdings of some companies having branches in both countries as well as fluctuating demand and supply problems at different places in the continental market.

With major halibut producing and consuming countries in the world, including the United Kingdom, Norway, West Germany, United States, Canada and Japan, all showing a drastic reduction in halibut production during the past decade, the volume of international trade in halibut will continue at low levels but probably at premium prices.

An exhaustive analysis of the import and export trade in halibut is not feasible nor appropriate in a book of this type. Anyone contemplating such a study should be forewarned of various deficiencies in the official figures available. Most agencies in arriving at the total volume of imports will add the amounts of fillets or steaks to the whole fish without converting to a common basis. This sometimes causes large errors as the recovery of fillets from whole fish is about 55% and of steaks about 70%. Also, in some United States tabulations salmon and halibut fillets are grouped into a single category and more often than not, what is recorded as halibut fillets are actually halibut steaks.

Furthermore, in United States import statistics of halibut *(Hippoglossus)* the importation of a large flounder *(Paralichthys)* from countries such as India, Taiwan, Canary Islands, New Zealand, Australia and other countries are classed as halibut and included with the true halibut. Also, turbot *(Reinhardtius)* is often included under the halibut category.

Many years ago the United States Federal Trade Commission, investigating the duty status of halibut for a committee of Congress, was prepared to send an investigator to Canton and Formosa, Taiwan, to inquire into the cost of production of halibut in that area. An amazing amount — 3 million pounds — was indicated in United States import statistics as originating in the China Sea!

9 NET FISHING FOR HALIBUT

Since halibut is a bottom or demersal species, it has been assumed for a long time that an effective method of catching halibut would be to use some form of bottom net. Based on such a premise, there have been a number of attempts to develop a net fishery on the Pacific Coast with halibut the target species.

Probably the earliest and more significant of these efforts was in 1892 by the S.S. *Iona,* built and operated by Fader and Co., Vancouver. The vessel was in the charge of Captain Arthur Edgett and fishing captain Mitchell. It landed about 30,000 pounds of halibut in Vancouver on February 26, 1892, from virgin fishing grounds off the Queen Charlotte Islands. The vessel was equipped with a 50-foot North Sea-type, 25-inch mesh old-style beam trawl, not the otter trawl currently in use. Before the vessel was secured under debtor's liens, as recorded at the Canadian Customs, it landed several trips of halibut caught with a beam trawl. The fish were shipped to eastern markets. This use of a beam trawl by the S.S. *Iona* in 1892 was the first and last attempt to use a beam trawl for the specific purpose of catching halibut.

Some years earlier in 1885 the *Carrie B. Lake*, a small vessel of 36 net tons, had fished a 40-foot beam trawl off the Washington Territory coast between Cape Disappointment and Shoalwater Bay. The catch, mostly flatfish with some halibut, was delivered to Portland, Oregon. Whether halibut was intended to be the primary target of the Oregon operation, as in the case of the *Iona,* cannot now be ascertained.

In 1914, with interest developing in the halibut potentialities off the Oregon coast an Aberdeen, Washington, company equipped the auxiliary schooner *Zillah May* with a 96-foot otter trawl. It fished off Yaquina Bay where Seattle dory vessels were making moderately successful halibut trips. The trawl operation was not successful in catching much halibut.

The general use of beam trawls or bottom nets had been the object of concern for many years. In 1892, at a hearing conducted in Vancouver by a Royal Commission, the presiding officer, Commissioner Wilmot, commented that the beam trawl had a reputation in Europe of "harrowing the bottom and caught immature fish" and that the halibut "demanded the husbandry care of the government." Mr. Wright of the B.C. Fishing and Trading Co., owner of the *Iona,* stated that the beam trawl "never caught young halibut and that the small ones can be released alive."

The harrowing of the bottom, or "destroying the grounds," and the taking of small fish have been longstanding accusations leveled against trawling with bottom nets at least as far back as 1376 A.D. (Fulton, 1911).

> In the Parliament Roll *(f)* of 51 Edward III., 1376 A.D., we find the *first mention of trawling.* The Commons petition the King, complaining that, where in creeks and havens of the sea there used to be plenteous fishing, to the profit of the kingdom, certain fishermen for seven years past have *subtily contrived an instrument which they call "wondyrchoun,"* made in the manner of an oyster dredge, but which is considerably longer, upon which instrument is attached *a net so close meshed* that no kind of fish, be it ever so small, which enters therein can escape, but must stay and be taken. And

> that the *great and long iron* of the wondyrchoun *runs so heavily and hardly* over the ground when fishing that it destroys the *flowers of the land below water there,* and also the spat of oysters, mussels, and other fish upon which the great fish are accustomed to be fed and nourished. By which instrument in many places the fishermen take *such quantity of small fish* that they do not know what to do with them; and that they *feed and fat their pigs* with them, to the great damage of the Commons of the realm and the destruction of the fisheries, and they pray for a remedy.
>
> In consequence of this petition a commission was appointed to inquire into the matter (g). The commissioners met at Colchester, and reported that the net called wondyrchoun was three fathoms long and ten of men's feet wide, and that the net had a beam ten feet long, at the end of which were two frames formed like a closrake; that a leaded rope weighted with many great stones was fixed on the lower part of the net between the two frames, and that another rope was fixed with nails on the upper part of the beam, so that the fish entering the space between the beam and the lower net were caught. The net had meshes (maskes) of the length and breadth of two thumbs. They say that the net ought to be used in the deep water, and not in the waters of Colne and Pont, or in any other like places.
>
> No legislation appears to have taken place in consequence of this report, but we find at later times trawling, which is evidently the mode of fishing above described, was prohibited in many private fisheries.

With regard to destroying "the flowers of the land below water" there has been very little research conducted on what, if any, irreparable damage to the bottom is caused by trawling. For that matter "furrowing" the sea floor could be restorative by releasing nutrients otherwise trapped in the overlying sediments. What little evidence there is suggests that bottom damage is minimal (Ketchen, 1947). Such is not the case for the catch of small halibut. There is abundant evidence that bottom trawling tends to capture "such quantity of small fish" when such small fish are present.

Some of the above remarks of the 1892 Royal Commission are rather prophetic. That the halibut "demanded the husbandry care of the government" has become more urgent than ever in view of the present both purposeful and inadvertent taking of halibut by the massive foreign trawl fisheries in the North Pacific and Bering Sea.

After the 1892 beam trawling venture of the *Iona*, the growing British Columbia halibut fishery in the 1890s depended first upon the use of small hook and line boats operating out of inshore stations close to the fishing grounds and somewhat later upon large dory-fishing steamers. The halibut fishery situated out of Puget Sound ports was also based upon dory-fishing vessels that included both large and small vessels.

In the first decade of the twentieth century a new production method for halibut appeared upon the Pacific scene—the otter trawl using two large, heavily weighted doors so rigged that when towed through the water, they spread the mouth of the net laterally, the doors acting as underwater kites. The vertical spread is accomplished by buoying up the top or head rope and weighting down the foot rope so that the net will remain on or in close proximity to the bottom. The width of the net's mouth can be extended far beyond what is feasible with a rigid beam, which, beyond a certain length, cannot be handled safely or effectively upon a vessel. The otter trawl design increased the efficiency of bottom trawl net gear materially. Later modifications, in the early 1920s, such as the Vigneron-Dahl or V.D. gear, and other advances, further extended the effectiveness of the otter trawl by as much as 50% to 200%.

In the spring of 1906 the Canadian Fishing Co., a new Vancouver company, brought the *Celestial Empire* from England, fully equipped with otter trawl, to fish halibut. This first use of the otter trawl for the taking of halibut is described in some detail from notes made 45 years ago during one of many interviews with Captain Absalom Freeman who had participated in initial use of the otter trawl for halibut fishing on the Pacific Coast.

Shortly after arrival in Vancouver, the S.S. *Celestial Empire* with a crew of 15 under Captain W. Johnston, including 5 experienced Scottish trawlermen, departed for Cape Scott. Three drags were made at Lanz Island, one of which was good, but unfortunately the net burst and only 4,000 pounds of halibut were saved. The fish were small so the vessel moved to the Flats 12 miles south of Goose Islands and dragged in 22 fathoms for but a few fish. They shifted to the northeast corner of Goose Islands grounds, secured several good hauls and moved to the northwest corner where they had one haul of 25,000 pounds. The vessel departed for home after two-and-one-half days on the grounds with 100,000 pounds for which they received 1.5¢ per pound from the New England Fish Co. The catch would have been larger if the crew had been more experienced in dressing and icing down halibut. Also the deck was not "checkered"—compartmented—and with big hauls operations had to stop while the crew dressed and stowed the fish in the hold, otherwise they would slide dangerously around on the deck.

Captain Freeman was on this trip of the *Celestial Empire* as an observer for his employer, the New England Fish Co., to compare the relative effectiveness of otter trawling and dory fishing.

The second trip was only one-half the size of the first and was two days longer despite there being twice the number of fishermen in the crew. About 8 to 10 trips were made using the otter trawl, each averaging about 35,000 to 40,000 pounds. It was said, however, that Captain Johnston was a deepsea man who possessed little fishing experience and had practically no acquaintance with the grounds being fished. The fishermen's union looked with disfavor upon trawling, fearing it would supersede dory fishing and thus reduce employment as well as deprive certain men of their jobs. The result was that the men would give Captain Johnston "all the trouble they could and at no time cooperated like they should have."

Accommodations for a larger crew were installed and by late fall of 1906 the *Celestial Empire* proceeded to northern Hecate Strait with six dories but with the otter trawl equipment still on board. It was now in the command of Captain A. Freeman who had left New England Fish Co. and invested in the Canadian Fishing Co., which sold most of its production under contract to New England Fish Co. The vessel made a few trawl hauls around Rose Spit and off Masset, but not many fish were caught and dory fishing was resorted to for the rest of this November trip. Thus ended the first effort to establish an otter trawl fishery in which Pacific halibut was the target species. This did not terminate the attempts, but subsequent ones, after the pioneer attempt of the *Celestial Empire*, would hardly start before collapsing for one reason or another.

Another Vancouver company that had fishermen as shareholders brought the S.S. *Flamingo* from England fully equipped with otter trawl, arriving in Vancouver in 1906, a few months later than the *Celestial Empire.* It fished otter trawl under Captain Jack Gosse but with indifferent results. Late in the fall the vessel was converted to the use of 12 dories under Captain A. Dawe. The following spring the Canadian Fishing Co. acquired the vessel in bankruptcy proceedings.

An overriding consideration at that time for deciding in favor of dory gear, according to Captain Freeman, was the fact that the trawl "seemed to be quite efficient for small fish but not for large." This observation made in the first decade of this century still prevails and has been supported since by a large body of scientific data acquired down through the years. Trawl gear of the type presently used in the commercial fisheries tends to overfish the smaller sized halibut, and probably underfishes the larger individuals, both conditions being inimical to maximizing the yield from the resource.

For the next few years there were no further significant attempts to use trawl gear in the halibut fishery in British Columbia. During that early period there was no experimentation with the gear by Puget Sound vessels.

In the fall of 1912 the over-promoted and much publicized B.C. Fisheries Ltd. (no connection with the B.C. Packers Ltd.) operated one of its two otter trawls brought from England, either the S.S. *Triumph* or the S.S. *Canada,* with halibut as the primary target of the operation. Dories were kept on board, but most of the halibut was taken by otter trawls. The stock-selling prospectus of the company had anticipated in glowing terms that it would soon have 20 large otter trawlers in the halibut fishery. Such grandiose predictions of course never materialized as the company soon foundered.

The well-organized Canadian Fish and Cold Storage of Prince Rupert also brought out three large North Sea-type trawlers from England in 1912, the *George E. Foster, James Carruthers,* and *Andrew Kelly,* each fully equipped for otter trawling. They never operated in the Pacific halibut fishery as trawlers, but were converted on arrival to dory fishing or longlining. In 1917, however, the *George E. Foster* and *James Carruthers* from time to time during 1917 to 1919 under the impetus of wartime demand and subsidized by payment by the Canadian Government of freight rates as far east as Winnipeg did engage in otter trawling for other demersal species which involved some incidentally caught halibut.

This early preoccupation with trawl gear as a means of catching halibut was a natural one as the gear had long been an effective way to catch bottom or mid-water species of fish. The gear is of great antiquity, being used in Egypt as early as dynasties VII-XIV, 2180-1640 B.C. (A.A. Aleem, 1972). Also, what some believe to be a recent development was the mesh of the early Egyptian gear that was graduated, large at the head and foot ropes and progressively smaller toward the cod end.

For the next thirty years otter trawling, specifically for halibut, lay quiescent until it was abruptly revived out of Seattle. No attempts to trawl for halibut had ever been made before from Puget Sound ports. In this attempt no large steel trawler was involved, as was the case many years ago in British Columbia.

Early in 1943 a resourceful owner-operator of the Seattle setline halibut vessel *Soupfin,* 67 feet long, successfully used the otter trawl with halibut as his prime target species. The vessel had a crew of seven including Captain Sverre Jangaard. This event was a milepost and for that reason will be recounted in detail with the permission of Captain Jangaard, owner, who later became prominent in the California tuna industry.

The first 15 days of the historic venture that commenced April 26, 1943, out of Seattle involved a succession of frustrating mishaps that would have discouraged a less motivated person. First, the winch broke off Cape Flattery and later the *Soupfin* collided with a Canadian fishery patrol vessel. After a return to Seattle for repairs, the boat was again under way when the net became tangled off Cape Flattery. Hauls off Destruction Island were unsuccessful. The net snagged again, now on 40-mile Bank near Cape Flattery. Two more snagged hauls were made off Cape Scott at the north end of Vancouver Island. Finally, 16 days after the first departure from Seattle, the *Soupfin* made on May 10 four net hauls on the Chicken Patch of the Goose Island grounds for 15,000 pounds of halibut. The next day 10,000 pounds were taken in four hauls. The third day they were crowded off the spot by setliners but some halibut were taken on the northwest edge along with sole and rockfish. Another 10,000 pounds of halibut were taken at Chicken Patch on five hauls before returning to Seattle with a substantial load of 40,000 pounds iced halibut and 3,000 pounds of frozen fillets. Considering the size of the vessel and the condition of the halibut stocks at that time, it was an excellent catch.

This event brought the matter of the retention of halibut by trawl gear up for immediate decision by the Halibut Commission. The question had been under review by the commission due to the increasing amount and proportion of halibut caught while trawling for other demersal species. The reported coastwise total of halibut landed by trawlers had increased from 115,000 pounds in 1941 to over 600,000 pounds in 1943. Also, there was a growing interest in halibut by certain individual vessels and the increased proportion of halibut in the trawl catches reflected the shift of the trawl fishery to areas possessing larger stocks of halibut. A survey of the industry also revealed that a broad expansion of trawling was imminent, as evidenced by the orders for trawl winches for vessels previously engaged in other pursuits including the halibut fishery. As a result of these developments the Halibut Commission in 1944 extended its earlier 1938 prohibition against the use of set nets for the capture of halibut, to include all types of nets.

The prohibition in 1938 against the use of set nets was to forestall their imminent introduction to the Pacific halibut fishery. Recent experiments in Norway (Devold, 1938) had indicated that they were undesirably selective of the large spawning individuals and thus inimical to a balanced use of the population.

The action with respect to trawl net gear was based upon the biological evidence then at hand regarding the nonselectivity of commercial trawl net gear with respect to halibut which made it capable of catching a much larger proportion of the smaller sizes than in the case of the setline gear currently in use. The decision continued as a long established policy of the commission. Comparisons of the growth and natural mortality rates had shown that a net gain in yield from the Pacific halibut stocks would accrue by allowing small halibut to grow to a reasonable size before capture. To achieve this objective the Halibut Commission had enacted a number of earlier measures:

(1) Since 1932 the commission has closed certain well defined nursery areas.

(2) Since 1932 the commission has controlled the catch in order to reduce fishing mortality and thereby increase the average size of the fish in the population.

(3) Since 1935 the commission has prohibited the use of dory gear by reason of its tendency to be used on grounds characterized by small fish.

(4) Since 1944 the commission applied a size limit below which halibut could not be retained. This was implemented by the commission despite the fact that it did not have such authority until the 1953 convention.

The prohibition against trawl retention of halibut was an extension of the above program.

The tendency of trawl gear to catch smaller halibut than setline gear did had been long recognized both on the Pacific Coast and in Europe. While the evidence was limited and some was circumstantial, conditions required prompt action to prevent the development on the Pacific Coast of a vested situation. In the succeeding years extensive research in the North Pacific and field observations in the North Atlantic have shown that trawl gear catches have a much higher proportion of small halibut and a lower proportion of the large sizes than is the case with setline gear. This is indicated in Table 39, showing the

TABLE 39. Size Composition of Trawler and Setline-Caught Halibut

Location	Weight Category: Under 10 Pounds Percentage		10-60 Pounds Percentage		60 + Pounds Percentage	
	Trawl	Setline	Trawl	Setline	Trawl	Setline
Western Bering Sea	87	45	13	54	—	1
Eastern Bering Sea	90	9	10	80	—	11
Gulf of Alaska	80	11	18	62	2	27
British Columbia	73	30	27	69	—	1
Iceland	68	—	27	64	5	36

FIGURE 79. Comparison of weight composition of trawl and setline recoveries of tagged halibut.
F.H. Bell

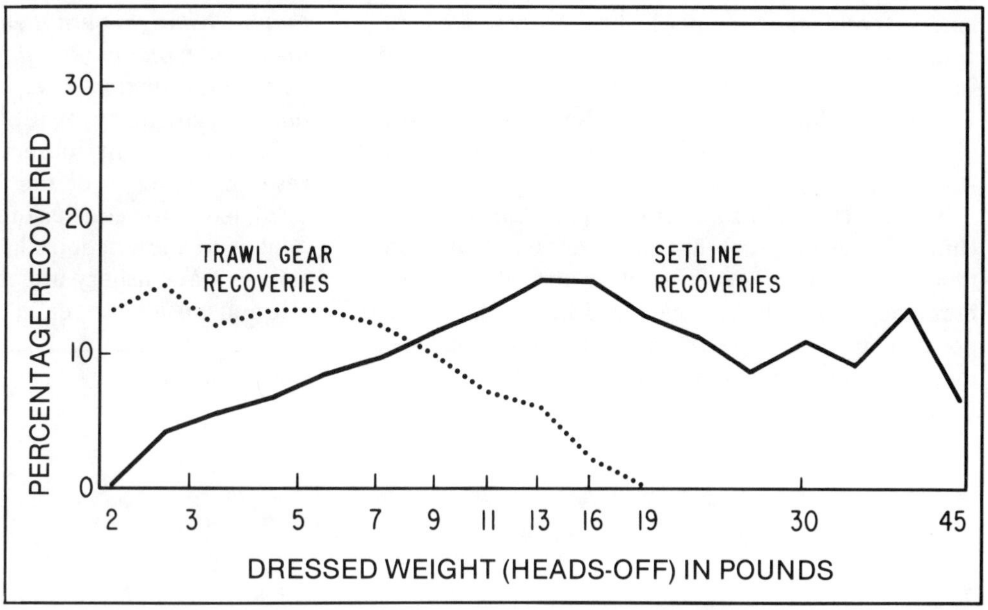

percentage in each weight category by areas taken by commercial trawl and setline gear under controlled conditions. On the basis of numbers in each category, the comparison is even more unfavorable for the trawl gear.

The different selection properties of the two gears in regard to halibut is also shown in Figure 79.

This is a very rigorous test as the two gears are fishing on the same marked population. Both the above figure and table indicate that trawl gear takes a larger proportion of the smaller and younger fish below the optimum harvesting size of halibut, and also tends to under-fish the larger and older fish. Both conditions are inimical to the attainment of the maximum sustainable yield of halibut.

By restricting the halibut fishery to hook and line gear, the commission had capitalized on the fact that halibut have a much greater length range than all other demersal species. Consequently, with the proper choice of gear, halibut catches can be largely isolated from the remainder of the demersal fish community and accorded the specific management controls necessary for attainment of maximum sustainable yield.

Also, hook and line gear of the type used in the North American halibut fishery is selective, not only in the size of halibut caught, but in a catch that is nearly all halibut. There is no conflict of interest between species such as occurs in a multispecies trawl fishery, which makes that type of fishery practically impossible to manage rationally.

Ketchen and Forrester (1966) concluded a comprehensive treatise on the petrale sole on what they referred to as a gloomy note that any effective regulation of the multispecies Canadian and United States bottomfish fishery presents problems "of almost insurmountable difficulty."

That halibut should be isolated from a multispecies demersal trawl fishery is concurred by the world's outstanding experts on resource management. McCracken (1958) concluded that with respect to Atlantic halibut, "increasing the age at first capture would increase the sustained yield . . . to design gears which would take other groundfish and not small halibut is probably impossible . . . that returning small halibut to the water seems an obvious solution."

Beverton and Hodder (1962) in their critical analyses of the condition of the fisheries of the Northwest Atlantic, noted that trawl-caught halibut were much smaller than those taken by longlines due to the selective properties of the respective gears, and that the optimum harvesting size for halibut at first capture was probably from 24 to 30 pounds, much above the average weight of 12 pounds of trawl-caught halibut in the Northwest Atlantic. Since the mesh required to achieve such an optimum average harvesting size for halibut would be useless for the capture of other species, they concluded that the only way of improving the halibut fisheries is to discard the inadvertently taken smaller halibut. Losses, of course, would tend to be minimized if retention and sale of such individuals were prohibited.

Novikov (1964) stated that "in Bering Sea the most efficient and suitable gear for halibut (speaking for the USSR) will be longlining."

Needler (1965) in support of a joint Canadian and United States proposal that all vessels employing any type of trawl net gear shall return to the sea immediately any halibut captured in the Bering Sea pointed out that, "to *further conservation,* it had for a long time been found necessary to prohibit North American fishermen from retaining halibut incidentally caught in trawls."*

The incompatibility of net gear and halibut is long standing. The coastal hook and line fishery for halibut in Japan began in the various provinces at approximately the indicated dates: Ochiishi, 1893; Nemura, 1920; Rausv, 1900; Abashiri, 1891; and Kushiro, 1910. The decline in all prefectures by the mid to late 1920s was "estimated about 10% per annum, *due to the fact that net fishing is permitted in the areas frequented by the halibut, and the small fish, or "Hirame" or flat fish as they are known in Japan are taken in very large numbers*" (Langley, 1930).*

The foregoing observations, along with the research findings of the Halibut Commission since 1925, led to the conclusion that to relegate the Pacific halibut to the uncontrollable vicissitudes of a multispecies trawl fishery intentionally would be indefensible and outside the realm of reason.

---

*Author's italics

10 FLEET PROGRAMS FOR ORDERLY MARKETING

The Pacific halibut fleets have been faced with many problems during the 93-year history of the industry. The economic depression of the early 1930s was one of the more critical. Although the stocks of halibut had ceased to decline first due to economic conditions that caused the fleet to reduce fishing, and subsequently, due to the controls of the Halibut Commission, they were still at very low levels. Market prices had plummeted. These two factors, low stocks involving smaller trips, combined with a precipitous decline in price, spelled economic ruin for the fleet with owners unable to meet their fixed costs, and unemployment for fishermen.

During the period of generally declining abundance in the 1920s any economic distress resulting from the decline in the stocks was largely averted by the steady rise in prices for halibut during that decade. For the larger vessels the situation was also aided by the introduction of the more efficient diesel engines and the extension of the fishery to some new and prolific, if not virgin, grounds in Western Alaska.

However, in the early 1930s all factors were negative, and to paraphrase the jargon of the Space Age, all systems were "no go." Prices were low, fish scarce, and there were no new grounds to exploit. The fleet on its own initiative attempted to cope with the economic emergency in the same independent and resolute manner that it had faced past problems and would probably continue to do so in the future. Since Seattle was the home port of the largest segment of the Pacific halibut fleet and had the ably managed Fishing Vessel Owner's Association and a long established Deep Sea Fishermen's Union, it was here that the leadership lie. In later years the effectiveness of this self-imposed control was greatly enhanced in Canada by the strong United Fishermen's and Allied Workers Union. This union, with jurisdiction over fishermen in both halibut and other fisheries as well as shoreworkers could strengthen compliance particularly in the unorganized small-boat fleet fishing out of camps.

The primary objective was to stave off financial ruin by ameliorating some of the effects of the extreme declines in commodity prices including halibut. It was quite obvious that the halibut fleet like all fisheries would have to fend for itself. This was particularly true in the United States where fisheries had received little meaningful economic assistance from government in times of stress. It might be said that the Pacific halibut fishery had survived not because of, but in spite of, many actions of the federal bureaucracies down through the years. In Canada more concern was directed to the fishing industry but even there it has been a "poor relation" compared to the disproportionate attention directed to Canadian agriculture.

1932 TO 1942

In 1932 the Seattle vessel owners and fishermen formed the Halibut Control Board, an unincorporated committee whose immediate purpose was to provide for the more orderly flow to market of the annual catch limits set by the Halibut Commission. While the Halibut Commission could not, under the convention, participate in any part of the program, its services as a public body were always available to the fleet. It conducted many statistical and other analyses of the various measures that the control board might have under advisement. There was only one restriction. The commission could not reveal any information dealing with individual enterprise without first securing the written release of the individuals concerned.

There was no attempt made by the fleet to *curtail* production in 1932 but, notwithstanding, the whole program came to be labeled with the word *curtailment* and was almost universally referred to as such. In the United States this drew from time to time the attention of some over-zealous persons in the Department of Justice looking for a conspiracy to restrain trade in violation of provisions of the Sherman act. Also, an occasional vessel owner not willing to conform to the majority will of his associates in the industry would threaten action under the same act. However, both official suspicions and individual threats came to naught. It was quite clear that the only real restraints on the fishery and on production were those imposed under law by the Halibut Commission. This included setting catch limits, closed seasons and the application of other restrictions in accord with the Halibut conventions.

The fleet believed that price fluctuations could be reduced and the general level raised if the bunching of port landings could be reduced, particularly at the beginning of the season and, in addition, if the total production permitted by the Halibut Commission could be spread more evenly over about a nine-month season. Another and rather different type of objective was to secure a more even distribution of the catch between vessels. This could be regarded as a radical departure from the free enterprise system. Right or wrong, it was very little different from the graduated income tax or excess profits tax both of which penalize individual efficiency and productivity.

In order to reduce the bunching of landings, the United States fleet was divided into two groups with the first half of the alphabet commencing to fish on the opening day of the season and the second half, seven days later. The two groups alternated from year to year and the system of split departures, with various modifications to guard against its abuse, continued for 10 years through 1941. To further spread the catch and slow down the rate of landing, the vessels in 1932 were required by their control board to undergo a number of between-trip lay-ins of varying lengths. All vessels were required to take two six-day lay-ins in March to mid-April, 14 days between April 15 and June 15 and a seven-day tieup in September.

The Canadian fleet, which at that time was mainly centered in Prince Rupert, and some of the Alaska vessels did not subscribe to the 1932 program. Canadian participation in the halibut fishery then was very low, accounting for only 15% of the total Pacific Coast catch. The Canadian share was also in a period of sharp decline, undergoing about a 10% drop annually from 1928 to 1932. It is doubtful whether adherence or non-adherence to any control plan by the Canadian fleet from 1932 to 1934 would have significantly affected the rate of landings. The very low prices of the times made the United States tariff on Canadian fish even more restrictive, and rather than fish halibut many Canadian boats went trolling for salmon. There also seemed to have been more than the usual number of casualties from fire and other mishaps with the usual suspicions that arise in depressed economic times.

Whether the foregoing voluntary measures in 1932 were efficacious in improving the economic condition of the fleet is a moot point and one that cannot be resolved conclusively as there is no basis upon which to make a valid judgment. It can be said, however, that prices might have been much lower without the program despite its shortcomings. It is also indicative that the participants proceeded with determination in the following year to build on those first efforts.

In 1933 two new ingredients were added to the fleet's program—a maximum catch per crew member per trip, and advance schedules of arrival dates set for each vessel. The limit of catch-per-man-per-trip was expected to reduce the rate of landings and also improve quality by the shorter and smaller trips. The advance assignment of each vessel with an arrival date was designed to keep weekly landings in the then major halibut ports of Seattle, Prince Rupert and Ketchikan, in balance with market demand, and also to improve quality by the shorter and probably smaller trips. In addition to the foregoing measures, the fleet limited the total catch to 35 million pounds, which was 11 million below that provided by the Halibut Commission's regulations, the goal being to hold prices to a minimum of 6¢ per pound for No. 1 fish and 4¢ for No. 2's. Whether such limitations on production below the amount legally permitted invoked the attention of the Department of Justice in the United States was never tested as the projected price levels were achieved in the open market without any self-imposed reduction in the total annual catch allowed by the commission.

The catch limits per-man-per-trip in Area 2 lying south of Cape Spencer were set at round figures of 3,000 pounds, and 4,000 pounds in Area 3 west of Cape Spencer. They were determined by the Halibut Commission at the request of the control board from catch records in 1932. In addition to the foregoing primary features of the voluntary control program, there were many administrative provisions in the plan designed to make the rather complex program functional and enforceable.

During this time the world-wide Great Depression was at its worst. In the United States economic conditions had reached an almost chaotic stage, with numerous bank closures, excessive deflation of all values, widespread foreclosures on mortgages and other debts, and nationwide unemployment mounting to dangerous levels. Drastic measures were undertaken at the federal level, including the establishment of many federal corporations and administrations designed to stabilize the economy along with the temporary bank holiday to avert catastrophic panic. Two of the major stopgap stabilizing agencies were the National (Industrial) Recovery Administration (NRA) and the Agricultural Adjustment Administration (AAA or Triple A). Fisheries problems were shuttled back and forth between the two agencies but usually ended up with the NRA, whose promotional logo was a stylized Blue Eagle.

The Fishing Vessel Owner's Association of Seattle prepared a Divisional Code of Fair Competition, to be implemented and given the force of law under the National (Industrial) Recovery Act of 1933. A formal hearing was conducted upon the proposed code in Seattle on May 3, 1934. The plan included two basic provisions, a catch-per-man-per-trip for Area 2 of 3,000 pounds for fares landed in Seattle and 3,300 pounds in British Columbia and Alaskan ports and 4,000 pounds for vessels fishing Area 3 regardless of the port of landing. It omitted the pre-assignment of arrival dates that had been in the 1933 voluntary plan but included the staggering of departures at the beginning of the season to even out the landings. There were some secondary provisions for special situations, such as the effects of tidal cycles in some areas and breakdowns. There was general agreement among the fleet in support of the code, although the Deep Sea Fishermen's Union wanted assurance of representation on any governing body. The few larger vessels in the fleet took some exception to the 40,000-pound limit on size of fare allowed regardless of crew size.

Despite practical unanimity among the fleet in favor of the code and the proven feasibility of such an arrangement as indicated by the results in 1932 and 1933, it was not put into force by the NRA. The reasons for the rejection were not evident, but it is not unlikely that dealers and distributors may have had some negative input as there was nothing in the plan that would have been of any direct benefit to them. There were also features of the plan such as the limits on the catch-per-man-per-trip and the maximum trip allowed for any vessel, that may have been inadmissible under the National Recovery Act.

The fleet proceeded to put into effect in 1934 on a voluntary basis essentially the same plan that was in the proposed code. Although the program lacked the force of law it would have possessed under the NRA, it did remain free from the almost inevitable interference of a cumbersome, inflexible, and frequently inefficient bureaucracy. Furthermore, the voluntary program, despite its deficiencies, lasted long after the welcome demise of the "Blue Eagle" NRA, which from the outset was frequently shot at in United States courts on constitutional grounds. In 1935 the United States fleet attempted to develop a price stabilization scheme to be financed by a loan from the Reconstruction Finance Corporation (RFC). The program was not consummated, but it delayed the start of fishing about six weeks after the legal date of opening of March 1 set by the Halibut Commission. The 1935 Control Program provided a sharp reduction from the 1934 level in the allowed catch-per-man-per-trip in pounds, which was then maintained at about the 1935 level through 1941 as shown in Table 40.

TABLE 40. Trip Limits in Pounds Per Man Per Vessel by Ports and Areas

	Area 2		Area 3
Year	Seattle	British Columbia and Alaska	All Ports
1933	3,000	3,000	4,000
1934	3,000	3,300	4,000
1935	2,500	2,800	3,500
1936	2,500	2,800	3,500
1937*	2,500*	2,800*	3,500*
1938	2,500	2,800	3,500
1939	2,500	2,800	3,500
1940	2,700	2,800	3,500
1941	2,700	2,800	3,500
1942	3,000	3,000	4,000

*Limits reduced on first two trips.

In 1942 all control by the fleet in Area 3 was terminated as of June 29 and in Area 2 by the end of the year. Due to wartime conditions, the United States Navy wished the fishing season to be closed as soon as possible. There was also increasing difficulty in maintaining an adequate level of voluntary compliance.

In 1934 the Canadian fleet had participated in the program but on a reduced scale. Between-trip tieups were 4 to 5 days compared to the 10-day and longer tieups for United States vessels. The reduced level of participation was predicated on the basis that the Canadian fleet was in a more depressed state and also had to contend with a 2¢ per pound United States tariff. At prevailing prices this was an extremely high *ad valorem* rate, approaching 100%.

In March, 1935, the Canadian fleet requested that the Dominion Government establish a Halibut Marketing Board under the Natural Products Marketing Act of 1934. In June of that year the request was implemented by an Order-in-Council. The legally supported program provided between-trip tieups ranging from 7 to 10 days in Area 2, depending upon the size of vessel, and 10 days in Area 3. Area 2 boats were limited to 3,100 pounds per-man-per-trip and 4,000 pounds in Area 3.

In 1936 the Dominion Act was invalidated by the Supreme Court of Canada and also declared *ultra vires* by the Privy Council in London as independent milk producers in British Columbia had contested the Dominion Marketing Act. Subsequently the halibut program was placed under a British Columbia Provincial Marketing Act. Despite the fact that the program in Canada had the force of law, infractions of the scheme were no less than with the United States fleet under the private or voluntary plan. The strategems used in both countries for circumventing the program were manifold and in most cases incapable of being detected by ordinary surveillance methods.

This 11-year-long episode in the history of the fishery which has been recounted in some detail had a marked effect upon some features of the fishery including the distribution of landings in the several ports. It also revealed some characteristics of the fishery that would otherwise have been overlooked. Very few, if any, industries have been able to exercise such self-control over its members and for such a long time as did the Halibut Control Board from 1932 to 1942 inclusive. These were also difficult years. The negotiation and implementation of such a complex voluntary program of individual restraints is a monument to the self-reliance of the fleet. It also reflected credit upon the leadership of the dozen or so organizations involved and particularly Harold E. Lokken, manager of the Fishing Vessel Owner's Association in Seattle. This group represented the largest segment of the fleet. It initiated most of the programs and bore its full share of self restraint.

To secure the required unanimity, many compromises had to be made by all in order to reconcile extremely diverse elements in the industry. While at times this left the plan with shortcomings, it was always a workable, if not perfect, arrangement. Admittedly, it is impossible to prove that the economic condition of the fleet was benefited by the program. On the other hand, however, it cannot be denied that the program did reduce the great concentration of landings and also lengthened the fishing season. These actions promoted a more orderly flow to market of the permitted production, and the longer season had a definite conservation value.

Since the program was basically voluntary, except for a short period in Canada where it was under the sanction of the Marketing Act, there was an increasing number of infractions. The commission was precluded from policing the program by the provisions of the conventions from which derived its authority. The infractions were chiefly restricted to individual vessels violating the poundage limitation per-man-per-trip. They reached serious proportions as the years went on and not only threatened the program but posed major threats to the integrity of the commission's statistical system. As has been recounted elsewhere, both the fleet and the commission made concerted and repeated efforts to have the Halibut Convention amended to permit the commission aid in implementing the program.

Although all initial goals may not have been attained and despite complications for all concerned, this 11-year effort of self-help left the fleet still independent of the government handouts such as price supports, subsidies and other illusory benefits so widely indulged in by agriculture.

1955 TO 1976

After the end of World War II, programs to improve the economic state of the fleet were discussed occasionally by the fleet, but it was not until the mid-1950s that there was any concerted coastwide effort to reinstate any plan. The fishing seasons had become extremely short with a great concentration of landings and ex-vessel halibut prices remained at a relatively low level compared to those of other commodities. Action by the fleet was precipitated by the sharp price decline in 1955 despite only a moderate total production. A program was developed for the 1956 season consisting of between-trip lay-ins of 7 days for the regular halibut vessels and alternating open and closed periods of 7 days for small boats operating out of fishing camps, except at the outset of the season when there was an open period of 14 days. The plan was relatively simple compared to that in effect in the 1930s. There were no allowances for breakdowns and other concessions except when a breakdown held a vessel in port over 7 days. Those days in excess of 7 could be deducted from subsequent lay-ins. Compliance was enforced through fishermen's unions one of which in Canada also controlled the shoreworkers. Control of shoreworkers made it possible to secure compliance from the small boats fishing out of camps not belonging to the fishermen's unions.

Halibut vessel owners and fishermen from various sections of the coast decided first to seek minimum halibut prices for the 1956 season. It was proposed to the buyers that there be minimum initial bids of 25¢ per pound in Seattle, 24¢ in Vancouver, 23¢ in Prince Rupert, 22¢ in Ketchikan and 21¢ for other Alaska ports. With the rejection of the minimum price proposal by the dealers, strike votes were taken in all ports, but the Deep Sea Fishermen's Union of Seattle achieved only a 58% affirmative vote instead of the two-thirds majority required by its constitution. This required the abandonment of the project in spite of an affirmative carrying vote in other ports.

Negotiations on a minimum price delayed final consummation of a lay-in plan forcing the fleet to voluntarily agree not to start fishing before May 20. The fleet had requested the commission on May 2 to delay the already approved opening date of May 12. While the commission could extend the fishing season on short notice, to delay an opening date was another matter. It required approval of the two governments, which at that time required the actual signature of the president of the United States and the enactment of an Order in Council in Canada. The fleet on several occasions failed to realize that the foregoing were both time-consuming procedures.

With the experience gleaned from the earlier post-World War II experiences the fleet carried out an effective program during the next 20 years. Again it required extensive annual negotiations to satisfy the increasing number of elements in the fishery, each with their own views and demands. Once agreed to, compliance with the plan was generally good with only a few vessels failing to subscribe to the plan. This occurred chiefly in Southeastern Alaska where several nonunion vessels fished. However, even in that area noncompliance involved under 5% of the landings from 1966 to 1968. In 1969 the proportion increased to 8.4%. Notwithstanding all shortcomings a more orderly fishery resulted, and the voluntary program had a salutary effect upon conservation as the resultant longer season assured a more balanced use of the stocks.

Despite some questioning to the contrary the Halibut Commission had always supported, and at times publicly advocated within the limits of its powers, the fleet's program of self regulation. In that respect it twice formally petitioned the two governments to broaden such powers but without success.

By 1977 the low annual catches and the increased number of small boats in the fishery, which were now difficult to control, resulted in the abandonment of the 20-year-old voluntary program. The commission established a series of openings and closings— multiple seasons whose need had been anticipated and provided for in the 1953 convention in order to distribute the permitted catch over a longer period of the year.

11 INTERNATIONAL AGREEMENTS

The high seas character of the Pacific Halibut fishery very early on required some form of agreement between the participating countries if this exclusive Canadian and United States fishery were to be conducted in an orderly and prudent manner. There were a number of abortive attempts at agreement down through the years, but it was not until 1923 that any meaningful understanding was reached by Canada and the United States for the conservation of the resource. The 1923 agreement was successfully negotiated and was ratified by both countries. Earlier attempts at agreement had failed to achieve either of these two stages. The contractual obligations of the 1923 convention were simple and were expressed in unequivocal terms.

1783 TREATY OF PARIS

The full import of the 1923 Halibut Convention can be appreciated best by reviewing some of the difficulties experienced by Canada and the United States in composing their differences in fishery questions. For over a century and a quarter, from 1783 to 1912, there was an almost steady friction between the two countries on fishery matters. Much of the exacerbation stemmed from obscurities in the fishery provisions of the 1783 Treaty of Paris that terminated the American War of Independence.

Article III of the Treaty of Paris accorded certain rights and liberties to United States fishermen.

> It is agreed that the people of the United States shall continue to enjoy unmolested the right to take fish of every kind on the Grand Bank, and on all the other banks of Newfoundland; also in the Gulf of Saint Lawrence, and at all other places in the sea where the inhabitants of both countries used at any time heretofore to fish. And also that the inhabitants of the United States shall have liberty to take fish of every kind on such part of the coast of Newfoundland as British fishermen shall use (but not to dry or cure the same on that island) and also on the coasts, bays and creeks of all other of His Britannic Majesty's dominions in America; and that the American fishermen shall have liberty to dry and cure fish in any of the unsettled bays, harbours and creeks of Nova Scotia, Magdalen Islands, and Labrador, so long as the same shall remain unsettled; but so soon as the same or either of them shall be settled, it shall not be lawful for the said fishermen to dry or cure fish at such settlements, without a *previous agreement* for that purpose with the inhabitants, proprietors or possessors of the ground.

There are a number of ambiguous and vexatious clauses and words in this passage. Many attempts were made during the next 100 years to reconcile differences in their interpretation. When it was agreed that the War of 1812 did not abrogate the provisions of the Treaty of Paris of 1783, the convention of October 20, 1818, attempted to redefine the rights and liberties of United States fishing vessels in Canadian coastal waters. This was only partially successful.

While a later Treaty of 1854 also attempted to be more definitive it was terminated in 1866 by the United States because of its provision for reciprocal abatement of customs duties. Canada, however, continued the privileges of United States vessels by a system of licensing until 1870. A new treaty was consummated in 1871 which lasted until 1885. Another agreement, the Bayard-Chamberlain Treaty negotiated in 1888 was not ratified by the United States, but the negotiators had provided a *modus vivendi* which was maintained. Further treaties of 1890 and 1902 to clarify the uncertainties and compose the abrasive situation also failed to be ratified by the United States again due chiefly to the inclusion of provisions for reciprocal freedom from customs duties. Notwithstanding these setbacks, the *modus vivendi* of 1888 continued to prevail.

However, with the exacerbations increasing, the original convention of 1818 was finally submitted in 1909 to a tribunal chosen from the members of the Permanent Court at the Hague. The arbitration award led to the agreement of July 20, 1912, that provided procedures for amicable and reasonably effective means of resolving disputes regarding the provisions of the 1818 convention.

During the protracted period of temporary successes and frequent failures, there was an underlying disposition to endeavor to exchange privileges that might be accorded United States vessels in Canadian waters or ports for the ready access to United States markets of Canadian-caught fish. However, such exchanges were constantly being thrown off balance or nullified by frequent shifts in the political climate in the United States with respect to trade protectionism and exclusionary tariffs. During the years from 1818 to 1919, a number of the fishery compacts had foundered or even failed to materialize on account of the inclusion of provisions for the reciprocal freedom from customs duties on fishery products.

The port privileges and other questions discussed above applied only to the Atlantic Coast as the Pacific halibut fishery did not exist. However, very soon after the start of the latter fishery in 1888, the problem of port-use privileges came to the fore. In 1894 the New England Fish Co. established its headquarters in Vancouver, and the Canadian government, acting unilaterally by an Order-in-Council of November 8, 1897, granted that company the privilege to land halibut in Vancouver caught by its vessels of United States registry. The company could ship this fish in bond to the United States and the vessels would be permitted to ship crews and outfit in Canadian ports. The privileges, which were continued from year to year, created no problem in themselves, but the concessions that were involved became a matter of major moment in the number of subsequent agreements between the two countries.

With the completion of the Pacific terminus of the Grand Trunk Pacific Railroad at Prince Rupert in 1908, and the subsequent building of fish-handling facilities at that port, a demand arose for similar bonding and other port privileges as prevailed in Vancouver for owners of American vessels landing in Prince Rupert. This was permitted by Order-in-Council of December 10, 1914. However, in Prince Rupert most of the American vessels involved did not qualify. They were not owned by Prince Rupert companies engaged in the purchase and distribution of halibut, with the exception of the *Sumner* and *Sitka*, as was the case with American vessels in Vancouver. Accordingly, the December 10, 1914, Order-in-Council was amended on March 9, 1915, to permit the sale in bond of American-flag halibut to Prince Rupert dealers licensed for the purpose, who themselves could then process, store and transship such fish in bond.

These port-use arrangements, including some subsequent modifications, had an important bearing upon the development and magnitude of the Pacific halibut fishery on different parts of the coast. Furthermore, due to competition between ports for the business generated by the halibut fleets these port privileges were also the object of a great amount of political maneuvering and agitation both at local and national levels. Notwithstanding, the agreements were maintained with few interruptions.

These year-to-year agreements were given a more stable basis by a Port Privilege Convention between Canada and the United States on March 24, 1950, which became effective on July 13, 1950, upon the exchange of ratifications. It provided for halibut fishing vessels of either country to land their halibut and sablefish in Canadian or United States ports without payment of duty, for their sale locally on payment of applicable customs duties, for their transshipment in bond to any port within the country of the vessel's flag or for export, and for supplies, repairs and equipment.

None of the foregoing agreements furthered the cause of conservation except in a very remote and vague direction. They did at times create a climate for discussion, and on a few occasions an atmosphere that could be mistaken for temporary cooperation. Nevertheless, it will be evident that only by the complete divorcement of trade and economic matters from conservation considerations did it become possible to come to any durable and lasting understanding regarding the preservation of the Pacific halibut fishery.

The divorcement process was a gradual one and there were a number of abortive attempts before the first effective conservation agreement was finally consummated. Table 41 provides a chronological list of the more important agreements between Canada and the United States that had some direct bearing on the conservation of the Pacific halibut fishery and the

TABLE 41. Agreements, Conventions and Treaties Concerning the Pacific Halibut

Dates		Common Name and Place of Signing	Plenipotentiaries		Joint Commission (Initial Members Only)	
Signed	Effective		Canada (Great Britain)	United States	Canada (Great Britain)	United States
September 3, 1783	January 14, 1784	Treaty of Paris, Versailles	David Hartley	John Adams Benjamin Franklin John Hay	—	—
(Notes Exchanged)	December 6, 1892	Agreement of 1892, Washington	Sir Julian Pauncefote British Ambassador	John W. Foster, Secretary of State	William Wakehouse	Richard Rathbun
(Notes Exchanged)	June 4, 1908	Bryce-Root Treaty, Washington	James Bryce, British Ambassador	Elihu Root, Secretary of State	Edward Ernest Prince	David Starr Jordan
(Not signed)	—	Draft Treaty of 1919, Washington	John Douglas Hazen, Chief Justice of New Brunswick, Member of Privy Council	Robert Lansing, Secretary of State	—	—
March 2, 1923	October 21, 1924	Halibut Convention of 1923, Washington	Ernest Lapointe, Minister of Marine and Fisheries	Charles Evans Hughes, Secretary of State	John P. Babcock William A. Found	Miller Freeman Henry O'Malley
May 9, 1930	May 9, 1931	Halibut Convention of 1930, Ottawa	W.L. Mackenzie King, Prime Minister and Secretary of State for Eternal Affairs	R. Reath Riggs, Charge d'Affaires in Canada	John P. Babcock William A. Found	Miller Freeman Henry O'Malley
June 29, 1937	July 28, 1937	Halibut Convention of 1937, Ottawa	W.L. Mackenzie King, Prime Minister and Secretary of State for External Affairs	Norman Armour, Envoy Extraordinary and Minister Plenipotentiary	George J. Alexander A.J. Whitmore	Edward W. Allen Frank T. Bell
March 2, 1953	October 28, 1953	Halibut Convention of 1953, Ottawa	James Sinclair, Minister of Fisheries Hughes Lapointe Minister of Veterans Affairs	Don C. Bliss, Charge d'Affaires *ad interim* William C. Herrington, Special Assistant for Fisheries to the Under Secretary of State	George R. Clark Harold S. Helland George W. Nickerson	Edward W. Allen J.W. Mendenhall Seton H. Thompson

resource. The Treaty of Paris 1783, the first agreement, has already been discussed at some length.

AGREEMENT OF 1892

The Agreement of the 6th day of December, 1892, between the United States and Great Britain with respect to Canada was the first significant attempt at formalizing joint conservation measures for fishery resources of common interest. The agreement was concerned with the prevention of destructive fishing in both the territorial and contiguous waters of the two countries and in the open seas, the prevention of pollution or obstruction of contiguous waters, the enforcement of closed seasons, and the preservation and increase of fish life by restocking with fish and other means.

It is timely to observe that in the public hearings on the agreement in Canada it was noted that the catching of halibut in beam trawls was regarded as destructive, as that gear had a propensity for the taking of halibut below the optimum harvesting size. Also, it is interesting that the other stated concerns of the agreement are still major ones today, 80 years later, namely pollution, closed seasons and aquaculture in the broad sense of the term.

After four years of delays, a report was finally submitted on December 31, 1896, which concluded that the facts "clearly demonstrated the inexpediency in attempting to regulate any of the fisheries herein." This early effort thus came to naught as far as any substantive accomplishment was concerned.

BRYCE-ROOT TREATY OF 1908

The Bryce-Root Treaty signed April 11, 1908, was also an important though unsuccessful attempt to consummate a fishery conservation agreement between the two countries. The convention was restricted to fisheries in waters contiguous to the common boundaries of the two countries and would thereby largely exclude such fisheries as that for halibut.

The International Fisheries Commission, appointed in 1909 under the treaty, proposed a number of conservation regulations for species other than halibut to be applied by concurrent legislative action of the two countries. Canada implemented all the proposed measures by Order-in-Council in 1910, but repeated attempts continuing until 1914 to secure passage of concurrent measures in the United States failed. There was strong opposition by Washington State interests whose concerns included the possible erosion of state's rights.

The British ambassador to the United States stated on October 19, 1914, "that the Government of Canada now purpose to resume their liberty of action in the matter of the administration of the fisheries in Canadian waters contiguous to international boundaries." Thereupon, the 1908 convention became non-operative and the International Fisheries Commission was dissolved.

Although the 1908 compact made no substantive contribution to the cause of conservation, it contained two important features. It was oriented toward conservation and also provided for a joint fisheries commission. A major weakness in the compact lie in the fact that its effectiveness depended upon concurrent legislation in the two countries to implement any regulation or changes therein rather than through executive action. In the fisheries field where time is often of the essence this was a cumbersome requirement. It also created opportunity for delaying tactics and political maneuvering at the legislative level.

1919 DRAFT TREATY

The Canadian-American Fisheries Conference Report of September 6, 1918, part of a joint World War I effort, included a number of proposals respecting the Pacific halibut fishery. To implement those proposals, a draft halibut treaty of October 24, 1919, was developed. The treaty provided for a joint investigation of the life history of the halibut by investigators in each country and for a winter closed season for 10 years that could be rescinded or modified after 5 years subject to the provision in the United States that any modification be made by the president upon the advice of the Senate. It also provided for the restricted sale of halibut taken incidentally to other fishing and for reciprocal port use and full tariff reciprocity. The International Fisheries Commission, provided under the convention of September 2, 1919, for the conservation of the sockeye salmon of the Fraser River, would also supervise the halibut investigations and the halibut closed season.

Despite the fact that it was a period of great expectations, created by the joint World War I effort, it is difficult at this distance to understand how anyone could have expected that such an omnibus-type agreement would have any chance of being ratified. It was never formally signed.

There had already been a long history of rejection of some of the draft provisions. In addition, there was the unrealistic relegation of halibut functions to a commission that was concerned primarily with salmon, a situation that was to arise again about 40 years later with the International North Pacific Fisheries Commission. State of Washington interests made the usual strong representation against the treaty despite the fact that it provided that changes in the closed season would only be made with the approval of the United States Senate. The latter would have permitted the senators from the State of Washington to bring their influence to bear.

There were, however, some landmark features in the 1919 document including provision for a halibut closed season and for its modification. It provided for a joint investigation of halibut and for the inadvertent taking of incidental amounts of halibut while fishing for other species.

THE 1923 HALIBUT CONVENTION

Despite the sudden demise of the 1919 draft treaty, the halibut industry in both countries persisted in advocating some international understanding regarding the fishery. The main thrust was still a winter closed season, motivated mainly by economic rather than conservation considerations. However, a 1922 Royal Commission in Canada did place primary emphasis upon the need for conservation. There was a strong case made showing the decline that had occurred in the stocks. This was evident in studies published by the Province of British Columbia (Thompson, 1916).

Preliminary exchanges between the two countries finally resulted in a draft convention that wisely excluded such sensitive provisions as reciprocal port use and tariff reciprocity. However, interests in the State of Washington again voiced strong opposition to any agreement that did not provide for legislative review of any control measures. Notwithstanding such opposition a Convention for the Preservation of the Halibut Fishery of the Northern Pacific Ocean including the Bering Sea was signed on March 2, 1923.

The convention provided a three-month winter closed season subject to modification after three years and an International Fisheries Commission of four members, two from each country. It also provided for an investigation of the life history of the halibut and for a report on results with recommendation as to regulations deemed desirable for the preservation and development of the fishery. The convention added that incidentally caught halibut during the closed season may be used for food for the crew of the vessel and that excess fish had to be surrendered to enforcement officers. Each country had to pay the expenses of its own commissioners, but joint expenses would be paid in equal moieties.

Though the provisions of the convention policy would permit an orderly attack upon the problems of the resource, implementing the convention was destined to open up a series of far-reaching constitutional consequences. The convention also became an important landmark on Canada's path toward attaining independent nationhood.

There had been a considerable exchange between Ottawa and London as to whether the Dominion of Canada or Great Britain should be designated in the title of the convention. Fortunately, that problem was overcome by the simple Solomon-like act of having no title on the convention as signed. Since that time it should be noted, however, that it has been variously titled in official sources. In registering the convention on February 2, 1925, with the League of Nations, Canada described it as the "Convention between the United States of America and Canada" and proclaimed the convention accordingly. In contrast, the president of the United States in proclaiming the convention in 1924 designated it as a convention between the United States of America and Great Britain.

Notwithstanding the problem of entitlement, the most contentious question was who should sign for His Britannic Majesty? Commencing on January 16, 1923, there were many three-way communications between the colonial secretary in London, the governor-general of Canada in Ottawa, and the British ambassador in Washington, D.C. The latter seemed so reluctant to relinquish historic imperial prerogatives that the records suggest some intentional lack of comprehension of instructions received from the Foreign Office in London.

It was not until a late hour on March 1, 1923, or early on March 2, which was the previously agreed-upon signing date of the convention, that *the British Ambassador was explicitly instructed by cablegram that Great Britain per se would not participate in the signing process; that the duly-named plenipotentiary from the Dominion of Canada would alone sign on behalf of His Britannic Majesty.* This solitary signing was to have many domestic and international consequences as well as to cause delays in ratification by the United States.

The independent action by Canada aroused a furor of parliamentary debate throughout the British Empire. There was a welter of journalistic comment in many parts of the world. Remarks ranged from a restrained level that "it impaired the unity of the Empire" to the more extreme that it was a "formal withdrawal of Canada from the British Empire." The parliamentary opposition in Canada labeled it as a "new-fangled manner" of executing a treaty and an "act of discourtesy to the British Ambassador."

Prior to this halibut convention Canada had, in some instances, signed and even negotiated a number of international agreements, conventions, or treaties on her own behalf, but *always* in conjunction with plenipotentiaries of Great Britain.

Unquestionably the signing of the 1923 Halibut Convention by Canada *alone,* unaccompanied by any British plenipotentiary, was a significant episode in the development of Canada's autonomy and of an empire to be based upon equality of nationhood.

This assertion of equality and independence and the solitary signing of the convention by Canada were of concern to the United States. The Senate Foreign Relations Committee in consenting to ratification of the Halibut Convention included an understanding that Senator Wesley Jones from Washington State had introduced "that none of the nationals and inhabitants, and vessels, and boats of any other part of Great Britain shall engage in halibut fishing contrary to any of the provisions of this treaty."

The inclusion of this understanding of Senator Wesley Jones by the Senate Committee naturally created further problems and delays. It had to be acceptable to Canada. It also brought into focus United States recognition of the independent statehood of Canada and other former Dominions of the now British Commonwealth. Earlier this had been one element in the failure of the Senate Foreign Relations Committee of the United States to ratify the Treaty of Versailles, as well as to reject the League of Nations.

Charles Evans Hughes, the United States secretary of state, noted to President Warren Gamaliel Harding at the time, June 16, 1923, that:

> The reservation of the Senate Foreign Affairs Committee had placed the Canadian Government in the embarrassing position of being obliged to go down on bended knee not only to the British Government, but to all of the British self-governing Dominions to secure their consent before Canada could be in a position to carry out legislation of purely domestic character.

In the ensuing discussions it was felt that concurrent domestic legislation by Canada and the United States might be an interim solution pending reconsideration by the United States Senate to exclude the reservation. Accordingly, the Parliament of Canada assented to such legislation on June 30, 1923. While the act did include some ambiguities, there was the further problem in that the United States could not, on constitutional grounds, execute similar concurrent legislation.

At this point of impasse a joint statement was issued October 19, 1923, by Canada and the United States noting that the closed season would not take effect in the winter of 1923-1924 pending achievement of some solution and binding agreement. The announcement at least relieved some uncertainty in the marketplace for the stocks of frozen halibut.

On May 27, 1924, Canada revised some enforcement provisions in the act of June 30, 1923, repealing Section 9, and the respective Order-in-Council was signed by the governor general on May 29, 1924. Both the original June 30, 1923, act and May 27, 1924, amendment were consolidated in 1927 into a single act. (Chapter Seventy-five, Revised Statutes of Canada.)

On May 31, 1924, the United States Senate gave its consent to ratification without reservation. Ratifications were finally exchanged on October 22, 1924, and the closed season then went into effect on November 15, 1924.

The year delay brought about by the solitary signing by Canada of the 1923 convention soon had a significant sequel. As it later materialized, the concern of the United States Senate Foreign Relations Committee as expressed in the reservation of Senator Wesley Jones was not unjustified. Also, the associated legislation possesses some implications that continue to the present.

During the House of Commons debate in 1923 on the applicability of the legislation covering the 1923 convention the minister stated that:

> As far as the extra-territorial waters are concerned no vessel from any country can carry on halibut fishing without using either United States or Canadian ports. It will be easy in the legislation which will have to be enacted by the United States Congress and by the Parliament of Canada to close Canadian and American ports to any vessels carrying on halibut fishing.

The weakness of this position was to be verified 13 years later when in 1936 the British mothership *Thorland,* with a freezer capacity of about 4 million pounds, planned to fish halibut in the Northeastern Pacific. There was also concern that Japan, already engaged in some salmon fishing in Bristol Bay, might also commence to exploit halibut with motherships in the convention waters of the Northern Pacific Ocean and Bering Sea.

Halibut fishing by the *Thorland* was only averted by formal representations made to Great Britain in 1937 by both Canada and the United States. Extended operations by Japan were undoubtedly discouraged by strong representations made to that country by the

United States through Cordell Hull, secretary of state, with respect to salmon fishing and growing clouds of war. However, neither Great Britain nor Japan renounced any rights to fish off the North American coast.

To prevent similar events such as the *Thorland* incident from occurring in the future and to discourage halibut fishing by other nations such as Japan in the convention waters, it was necessary to expand and strengthen the interdiction in legislation regarding the 1923 convention against the use of the ports of Canada and the United States by fishing vessels of a country not a party to the Halibut Convention.

In the United States Enabling Act implementing the 1937 convention, the prohibition against the landing in United States ports of halibut caught in convention waters, the convention waters by any nation not a party to the Halibut Convention, was extended to *include any type of vessel.* With "type of vessel" defined as "every description of watercraft or other contrivance used or capable of being used as a means of transportation in water," it was intended at the time that the prohibition would effectively prevent the use of United States ports not only by foreign fishing vessels but would also preclude water-borne imports of halibut caught in convention waters, namely the Northern Pacific Ocean and Bering Sea by other than the parties to the Halibut Convention, Canada and United States.

In 1928, the International Fisheries Commission made a comprehensive scientific report to the two governments with recommendations of what was required for the preservation of the resource. The winter closed season alone had been ineffective in providing any substantial protection and the stocks of halibut continued to deteriorate.

The commission recommended that authority be provided to establish areas where the catch of halibut could be reduced by a predetermined percentage annually until there was stability of yield. The commission recommended a revision of the percentage from time to time in order to prevent any increase in the amount of fishing and to close nursery grounds. It also recommended preventing the use of unduly destructive gear and the extension of the three-month closed season by two weeks. It asked for provisions for future modifications of the closed season, and the licensing of all vessels for statistical and other purposes of the convention.

While Canada was prepared to implement the commission's recommendations under the 1923 convention by Order-in-Council, constitutional requirements in the United States necessitated concluding a new convention to supplant that of 1923. A draft of the new convention was submitted on May 29, 1929, by the United States. There were amendments to the draft by Canada and protracted exchanges over various details. Again there were strong representations from Washington State interests not to expand the authority of the commission without providing for direct legislative controls over its actions. Although the convention was signed May 9, 1930, ratifications were not exchanged until May 9, 1931, a year later.

THE 1930 HALIBUT CONVENTION

The general provisions of the 1930 convention were similar to those of the 1923 convention. The essential change was the broad regulatory authority conferred upon the International Fisheries Commission as a result of its 1928 recommendations to the governments, all of which had been accepted.

The commission was empowered to divide the convention waters into areas and to limit the catch of halibut taken from each area during its fishing season, to regulate the licensing and departure of vessels for purposes of the convention, to collect statistics, and to fix the type and character of gear to be used. It could also close grounds found to be populated by small immature halibut, and conduct such investigations, as were necessary, into the life history of the halibut. Enforcement of the closed season and the regulations continued to be the responsibility of the individual governments.

Regulations enacted by the commission were subject to the approval of the governor general of Canada and the president of the United States of America. This procedure was continued under the conventions of 1937 and 1953. To expedite approval or rejection, by the United States, of actions of the International Pacific Halibut Commission taken pursuant to Article III, paragraph 2 of the 1953 convention, the presidential authority was delegated by Executive Order No. 11467 of May 1, 1969, to the secretary of state. It was also ordered that in carrying out such authority the secretary of state shall consult with the secretary of the interior.

Under prevailing conditions, as well as for the then foreseeable future, the provisions of the 1930 convention were adequate. They conferred upon the commission no more authority than was deemed necessary to accomplish the stated purposes, a desirable limitation in a democratic society. However, the sharp improvement in the stocks attracted many vessels from other fisheries, and despite some increase in the permitted catch, the season became shorter. This created some serious enforcement problems.

The "food for crew" provision in the 1923 convention became inadequate in controlling the incidentally caught halibut by vessels fishing for other species during the now longer-closed season for halibut. Also, with the fishing season now closing at the height of fleet activity, there was considerable overtime fishing that was difficult to control by surface vessel patrol. There was only limited air patrol possible in those early years.

THE 1937 HALIBUT CONVENTION

The provisions of the 1930 convention could not cope with the situation and a revision was required. The 1937 convention signed January 29, 1937, with ratifications exchanged July 28, 1937, permitted more effective control of the capture of halibut caught incidentally to fishing for other species in areas closed to halibut fishing. In addition, it provided for prohibiting the departure of vessels to any area, when it was deemed that those which had already departed would suffice to take the catch limit for the area.

The provision for the control of retention of incidentally caught halibut remained in effect for the next 28 years until 1965. The need for such a provision declined sharply in later years with changing conditions along with the lengthening of the open season, particularly in Area 2 after 1957.

As to the last date of departure, despite some benefits to observance of the law and its initial popularity with the fleet, the provision was put into effect only in 1937, 1938 and 1939. Considerable objection soon arose to the long notice that had to be given of final closure. Also, the measure did tend to reduce the freedom of individual decision or choice. Furthermore, the shortening of the fishing season posed a difficult administrative problem by having to determine and announce a last day of departure and a final closing date, practically at the outset of the season. The arrangement created more problems than it solved. While the authority to establish a last date of departure was continued in the 1953 convention, the year 1939 was the last year that it was exercised.

In succeeding years, the stocks continued to rebuild and the further shortening of the fishing season caused some stocks to be underfished and others to be overfished. The halibut fleets had since 1933 endeavored to offset the undesirable economic effect of the concentration of production by a voluntary plan designed to spread the catch over a longer season.

The commission, whose sole function under the conventions of 1930 and 1937 was conservation, could not provide legal support to the voluntary program despite repeated requests by the fleets. The commission did, however, assist the fleets in their effort for economic improvement.

In 1938, as a result of unanimous requests of the halibut fleets, the commission presented to the two governments a report commending to their consideration the fleets' requests for an extension of the commission's treaty authority to provide legal support for the voluntary program. The commission pointed out that the objectives of the fleets' requests were primarily economic, although there were some unfavorable biological effects from the shortening of the season beginning to be evident. The effects were not yet great as the seasons were still four and seven months long in Areas 2 and 3 respectively. In 1940, however, the commission found through public hearings that the previous unanimity in the fleets no longer existed, and it recommended to the governments that consideration of any treaty change be deferred.

By the end of World War II hostilities, unfavorable effects of the still shorter season, reached the point where the commission on its own initiative in 1948 recommended treaty changes to Canada and the United States that would enable it to lengthen the fishing season. In this action, it had the unanimous support of the fleets and most of the dealers in halibut.

The commission repeated the request in 1949 and it was suggested by Canada that the commission might already have had the required powers under the 1937 convention. It was subsequently advised by the United States that neither the wording nor intent of the 1937 convention would permit the exercise of the required authority. Since Canada found to the contrary, there was no other alternative than to seek a new convention.

THE 1953 HALIBUT CONVENTION

The new convention was signed March 2, 1953, the anniversary date of the signing of the first effective Halibut Convention exactly 30 years earlier.

Commemorative of the first convention, it was arranged that one of the two plenipotentiaries named by the government of Canada in 1953 would be the Honourable Hugues Lapointe, minister of Veterans Affairs, son of the Honourable Ernest Lapointe, minister of Marine and Fisheries of Canada and plenipotentiary in the signing of the first effective Halibut Convention in 1923.

The 1953 convention contained important changes. The commission was authorized to establish one or more open or closed seasons each year in any area. This permitted the commission to establish supplemental seasons on certain sections of the coast where the short single season had resulted in some underfishing. Other sections of the coast that showed indications of having been overfished during the single regular season were closed and opened at other times.

Membership in the commission was increased from four to six commissioners, three from each country. This permitted a broader representation of membership, and in the United States there was a domestic understanding that at least one United States member would be from Alaska.

The International Fisheries Commission was renamed the International Pacific Halibut Commission, a name more descriptive of its functions, having been concerned only with the halibut under the various conventions.

The new convention also specifically provided the commission with the authority to implement size limits. It also provided for the control of halibut taken incidentally to other fishing in the open season as well as in the closed season.

Also, the commission was charged with the responsibility of developing and maintaining the halibut stocks at levels which would provide the maximum sustainable yield. While this directive had been implied in earlier conventions it was not explicitly stated.

Under the 1953 convention the ultimate objective of the past 30 years was attained early in the 1960s. Studies indicated that the halibut stocks in the Northeastern Pacific, including those in southeastern Bering Sea, were then close to their maximum sustainable levels and were being fully utilized.

Since the early 1960s, expanding trawl fisheries, both foreign and domestic, and foreign setlining had been collectively placing an increasing and substantial drain upon the halibut resource. To maintain the halibut stock at the level of maximum sustainable yield, as provided in the convention, the commission was required to sharply reduce the level of catch permitted North American fishermen, to partly offset losses resulting from the operation of such concurrent fisheries.

In recounting the development of effective agreements on fisheries matters between the two countries several features stand out. First, securing the required unanimity was a long, slow process extending over many decades. This becomes understandable if one remembers the backgrounds of the parties involved and the nature of the fisheries question. The very close cultural and economic ties between Canada and the United States did more to complicate than facilitate the process. The only recourse left to retain some individuality was to unduly stress their political independence or nationalism. This tendency is accentuated when the two principals are of very different size as in the case of the United States and Canada. The smaller country develops some fear of being dominated by the larger. Also, in the case of Canada there were continuing allegiances to Great Britain, including British trade preferences. Such differences, whether real or imagined, are usually exploited out of all proportion by politicians.

The process of attaining a workable agreement on fisheries matters was also prolonged by constant shifts in United States policy in response to political turnovers. So long as any agreement included tariff, or even any trade provisions, one could be almost sure that it would be either rejected at some level of government or terminated at the next turnover of party power. Also, from the time the Washington Territory became a state, political and economic interest almost without exception, expressed strong opposition to any international fishery agreements that lacked provision for legislative review, or sanction for controls enacted thereunder. This stemmed in part from the basic supremacy question of states' rights but mainly from the fact that interests in Washington felt they could bring more influence to bear on the legislative branch of government, state or federal, than on the executive.

Another element contributing to the prolonging of both negotiations and enactment of any agreement has been the constitutional requirements for review and the more complex structure of government in the United States. In Canada, the government in power can use the Order-in-Council procedure, a direct yet seemingly democratic device. In the United States, the separation of powers with its checks and balances is pursued with all actions subject to time-consuming reviews and approvals. It is interesting to note that, although some may aspire to such power, no person in government in the United States, not even the president has the *absolute authority* that the present Canadian Fisheries Act confers upon the responsible minister. To reconcile the systems in the two countries was time consuming.

The underlying concern with nearly all the agreements was the general apprehension as to whether international controls would diminish each country's sovereign rights. The four Halibut conventions have clearly safeguarded the sovereignty of the participating countries.

Under each convention the authority of the administrative body, the commission, has been severely restricted to certain clearly defined functions. In turn the administrative bodies under the respective treaties during the past 45 years did not attempt to transcend their authority beyond that which was envisaged at the time the respective convention was negotiated, except possibly on one instance. In 1944 the commission placed a size limit on the individual halibut that might be retained. This was done on the basis that the 1937 convention then in effect granted the authority to limit the catch. Although the size limit was a popular provision in the regulations, the 1937 and preceeding conventions did not envisage nor intend such an interpretation.

Further protection against the erosion of sovereign rights is the convention requirement that all regulatory proposals must be approved by the executive branches of the two national governments. Also, the legislative branches in both countries control the funds without which the commission could not function. In addition, members of the commission are appointed by their respective governments for an indefinite period and are thus subject to removal for any cause at any time. It is also provided that, while nationals or inhabitants of each country in violation of any regulation may be apprehended by either country, the prosecution and judicial determination of offenses under the conventions remain the exclusive prerogative of the country to which the offender belongs. Lastly, the conventions can be terminated upon due notice of either party. In light of the foregoing safeguards any expressions about loss or reduction of sovereignty have an empty sound.

The concern over sovereignty probably contributed to considerable time elapsing during negotiation of each of the four Halibut conventions as shown in Table 42. The time lapses indicate, amongst other things, the critical appraisal that each country had made of the content and implications of each agreement. The time leading up to the 1953 convention was greatly extended due to conditions arising from World War II.

TABLE 42. Elapsed Time Between Initial Discussion and Effective Date of Halibut Conventions

Halibut Convention of	Discussions Prior to Signing		Additional Time to Effective Date		Total Elapsed Time	
	Years	Months	Years	Months	Years	Months
1923	1	6	1	8	3	2
1930	2	—	1	—	3	—
1937	1	—	—	6	1	6
1953	8	4	—	8	9	—

INTERNATIONAL NORTH PACIFIC FISHERIES CONVENTION

Other international fishery agreements that have affected the Pacific halibut include the International North Pacific Fisheries Convention for Canada, Japan and the United States. This convention embodies the principle that the high seas fisheries should be conducted in a manner to assure their maximum sustained productivity. It also includes the principle of abstention from fishing certain exploited stocks of fish, including halibut.

At the time of signing in 1952, Japan recognized that the stocks of halibut originating along the coast of North America fulfilled the required conditions and abstained from fishing halibut. The North Pacific Commission had to determine annually whether the stocks continued to qualify for abstention, but no recommendation could be made before the convention had been in force five years.

From 1958 to 1961 Japan continued to abstain from fishing for halibut despite nonagreement with Canada and United States as to the continued qualifications of the eastern Bering Sea stocks for abstention. In 1962, the North Pacific Commission determined that the eastern Bering Sea halibut stocks did not qualify for abstention, and Japan entered the fishery in 1963.

Thereafter the halibut in eastern Bering Sea were under the dual control of the two commissions. A satisfactory *modus operandi* was worked out to effect such control. (Bell, 1969, p 20.) In 1979 a new convention with amended protocol and no abstention provision replaced the 1952 convention. With enactment of the 200-mile legislation, the abstention provision was no longer of consequence.

BILATERAL AGREEMENTS WITH JAPAN AND THE USSR

To relieve some of the pressure that foreign trawling had placed upon the halibut, United States and Canada each negotiated and signed a number of bilateral agreements with Japan and USSR. The major effect of such agreements was to close to trawling certain areas at times of high abundance of juvenile halibut. Initial results suggest that the measures had some salutary effect.

UNILATERAL 200-MILE LIMIT LEGISLATION IN CANADA AND UNITED STATES

The enactment on March 1, 1977, of the Fishery Conservation and Management Act in the United States and of similar legislation in Canada in late 1976 introduced an entirely new dimension into the international fisheries situation. The exclusive fishery zone of 200 miles established by these unilateral acts radically changed the functions and applicability of the 1953 Halibut Convention. Also, the Halibut Commission could be radically altered.

The United States gave notice that as of April 1, 1977, it would withdraw from the Halibut Convention on April 1, 1979, as stipulated in the terms of the convention. Experience of the past 50 years has repeatedly demonstrated that unless the management of the Pacific halibut is placed under some single, unified and non-national authority, it will be only a matter of time before the resource will be relegated to the uncertain vicissitudes of a multispecies trawl fishery, domestic or foreign and probably both.

Shortly before the termination date of April 1, 1979, Canada and the United States renewed the Halibut Convention. Under the agreement, the United States fishing for halibut off British Columbia, already at a low ebb, was ended. Canadian halibut fishing off United States coasts was sharply reduced in 1979, and was to phase out completely in 1980. While for the time being, the Halibut Commission continues as the research arm of the governments with respect to halibut and continues to propose management policy, its destiny over the longer term is uncertain.

12 THE HALIBUT COMMISSION

The 1923 Convention for the Preservation of the Halibut Fishery of the North Pacific Ocean and the Bering Sea provided for a four-member commission — two members from both the United States and Canada — to carry out the mandates of the convention. This commission was established in November, 1924, and named the International Fisheries Commission. It was renamed the International Pacific Halibut Commission in 1953, and membership was increased to six people, three from each country.

Charged with providing the best management of the halibut fishery through maximizing long-range halibut yields, the commission has met annually for three- to four-day sessions to examine, with its staff, the results of investigations into the state of the fishery, the effectiveness of its present regulations, administrative and budgetary matters and to adopt regulations for the succeeding year. At its annual meetings, the commission has also conferred with industry groups regarding the management of the halibut resource.

Members from both countries have come from the private and governmental sectors. Members from the private sector received no remuneration except nominal travel costs to the annual meeting. Members from the governmental sectors, who were usually employed by fisheries agencies, received their regular remuneration.

Under the Halibut Conventions the cost of the Halibut Commission and its programs of investigation and regulations were shared equally by the two countries. Since 1925 the commission has had its headquarters in Seattle on the campus of the University of Washington. In 1965 the United States made a grant of $500,000 to build quarters on the campus in which the commission would have an established interest specified by federal law.

SUBSIDIARY BODIES

None of the Halibut Conventions provided for subsidiary bodies to formally advise the commission. However, as the need arose, the Halibut Commission did establish groups to fulfill specific purposes, more often than not to act as liaison agents between itself and fishery-related groups.

The oldest liaison body is the Halibut Conference Board, established in 1931 to secure facts about the halibut fleet and to pass along information to it. The board was originally comprised of persons, appointed by the associations of vessel owners or the fishermen's unions, who were well acquainted with the fishing banks, the fleet and its needs.

In 1955 the Industry Advisory Group was established to bring together for discussion the wholesale dealers of halibut and the fleet. Heretofore, the dealers had not felt the need to be represented to the Halibut Commission. However, by the mid-1950s the divergent activities of the dealers and the fleet were such that the commission deemed it appropriate for representatives from the two groups be brought together. The dealers had been working to increase consumer demand for halibut at a time when the fleet was affecting volunteer lay-in programs that limited fishing activity. Bringing the two groups together to discuss their problems proved successful, and in 1960

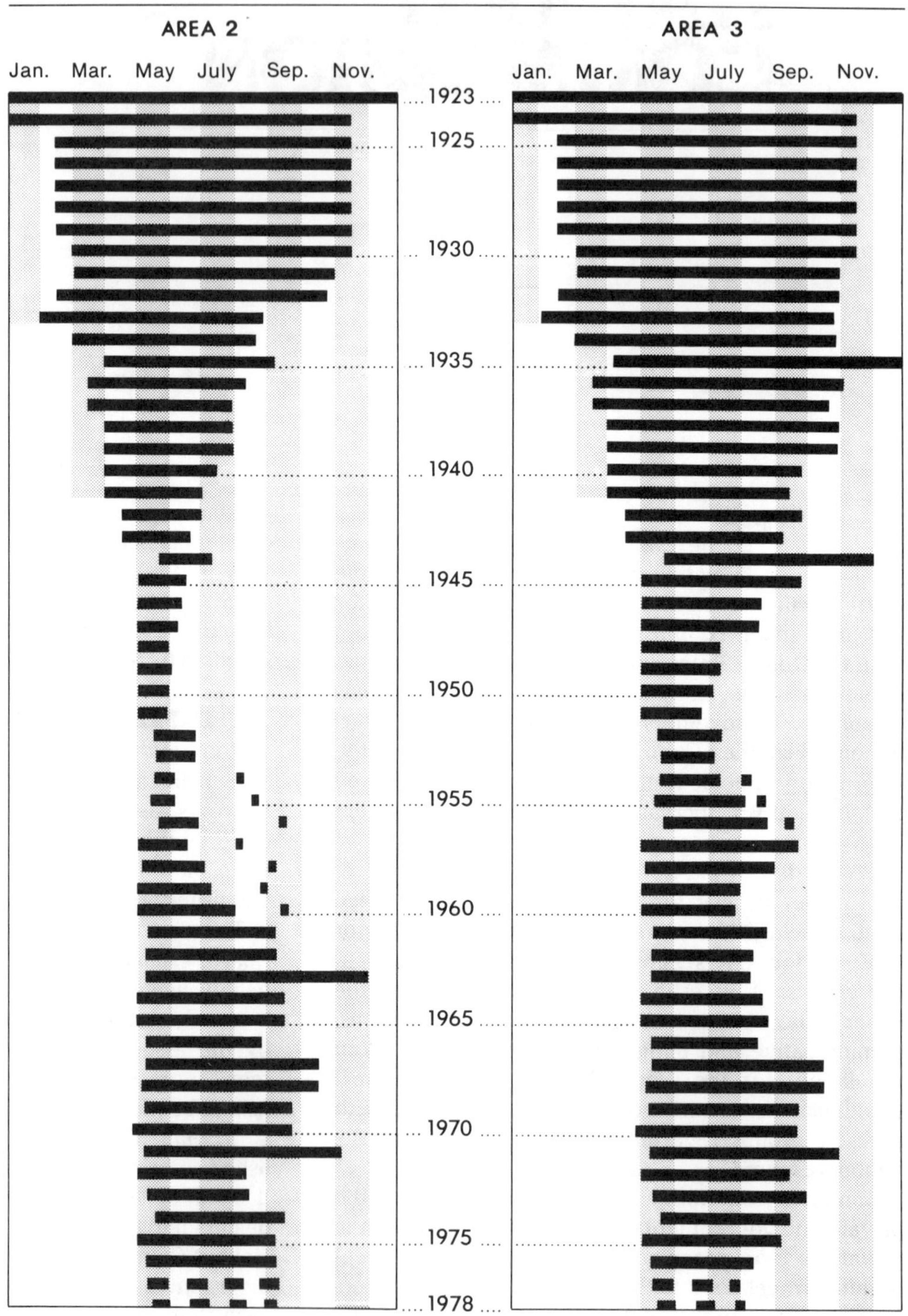

FIGURE 80. Length of fishing seasons for Areas 2 and 3 from 1923 to 1975.
F.H. Bell

the Industry Advisory Group was consolidated into the Halibut Conference Board.

The first advisory group to be formed was the Scientific Advisory Board, made up of four experts in various fields of fisheries studies who could counsel the commission and the director of investigations on programs of research. The Scientific Advisory Board was formed in 1924, but had a rather ineffective existence and was no longer recognized as an entity after 1936.

Communication with the public and the scientific community has also been accomplished through the annual meetings, publications and seminars.

REGULATIONS OF THE HALIBUT COMMISSION

The scope of the commission's power to regulate the fishery developed over time. Under the 1923 Halibut Convention, the commission was limited in regulatory power to a three-month winter closed season, which after the third such season could be modified or suspended by the commission. This measure probably slowed the decline in the resource, but it did not stop nor reverse it. After five years of investigation, the 1930 Halibut Convention empowered the commission to:

1. divide the convention-controlled waters into areas.
2. limit the catch from each area.
3. specify the type of gear to be used in taking halibut.
4. change or abolish the closed season.
5. close grounds populated by small, immature halibut.
6. license vessels and regulate their departure.
7. collect statistics
8. conduct necessary investigations upon the life history of the halibut.

The division of convention waters into areas that could be managed separately was needed for several reasons. First, the resource was not evenly distributed over the entirety of its 3,000-mile range, which meant that what were permissible levels of fishing in one portion of the range, might be deleterious fishing in another. There were also seasonal changes in the availability of the fishable stocks and differences in the size and age composition from one region to another that also required individual treatment. Furthermore, various stocks of halibut had histories of sharply differing levels of exploitation, and a balanced utilization of the fishable stocks was needed to assure attainment of the maximum yield. For all these reasons, separate conditions had to be imposed to maintain stocks in separate areas.

The power to limit the catch taken from each area was perhaps the most important authority granted the commission. It allowed adequate spawning, which in turn provided sufficient numbers of young halibut that would grow to the optimum size for maximizing the physical yields of the fish. Catch limits were set by adjusting the length of the closed fishing season, another authority of the commission after 1930. Figure 80 shows the changes in season length from Areas 2 and 3 from 1923 to 1975.

The authority to specify the types of gear used in taking halibut was used extensively by the Halibut Commission to reduce the taking of halibut below their optimum harvesting size. The subject is discussed at some length in the sections of this book dealing with trawling.

As with limiting the annual catches from each area, closing grounds populated by small halibut had the effect of limiting the mortality of the young and allowing for maximum yields of optimum-sized fish. Being able to close certain ground inhabited by small halibut circumvented the commission's inability to set a size limit for catchable fish.

The licensing of vessels and the regulating of their departure times from the ports for the grounds were used for administrative and enforcement purposes only. Licensed vessels were required to keep fishing logs of use to the commission, and any violation of the commission's other regulations meant revocation of the license.

13 INCIDENTALLY CAUGHT HALIBUT

Halibut is taken in a number of fisheries where it is not necessarily the primary target species. Controlling the retention of such fish in areas closed to halibut fishing poses a number of difficult problems. Similar difficulties are now being encountered in many areas of the world as the separate control of single species in multispecies bottom fisheries is contemplated.

The imposition of the winter closure in the 1923 convention created few complications. In the 1920s there was practically no winter activity in other demersal fisheries and the troll salmon fishery terminated in the fall. The convention permitted any incidentally caught halibut to be used as food for the crew and any excess was to be immediately turned over to the authorities.

Substantially, the same provision was included in the 1930 convention, but by that time and increasingly thereafter the closed season was longer by reason of earlier attainments of the catch limits. It no longer encompassed the main period of activity in other fisheries such as the setline fishery for blackcod *(Anoplopoma fimbria)* in which significant amounts of halibut were caught incidentally.

SETLINE BLACKCOD FISHERY

Under these new conditions the provision of the 1930 treaty governing incidentally caught halibut became increasingly less satisfactory to the enforcement agencies and to the halibut fleet while fishing blackcod. From the enforcement standpoint, the detection of sales of illegal fish was made almost impossible by the fact that the boats, even when found unloading halibut at the dock, could claim that they were going to turn it over immediately to the authorities. Also, when contacted at sea, it could be claimed that all halibut would be surrendered in excess of what was consumed as food for the crew. This was conducive to some vessels selling catches illegally instead of surrendering them.

Since a certain amount of halibut was caught by setline boats fishing for other species during the closed season and in view of the difficulty of controlling the landing of such halibut under the convention of 1930, the regulatory provision relating to the incidental capture of halibut was revised in the 1937 convention. The revision stated that the commission could permit, limit, regulate and prohibit in any area or at any time when fishing for halibut is prohibited, the taking, retention and landing of halibut caught incidentally to fishing for other species of fish, and the possession during such fishing of halibut of any origin.

The provision was designed to avoid unnecessary wastage in the setline fishery for blackcod, lingcod *(Ophiodon elongatus)* and mixed rockfish *(Sebastes* sp.). In the late thirties there was no significant bottom trawl net fishery and the incidental capture of halibut by the troll fishery was not a problem as the open halibut season still covered most of the active salmon trolling season.

The 1953 convention continued the substance of the above provision but broadened the authority so that the limited capture of incidentally caught halibut could be permitted in areas or in portions of areas both open or closed to halibut fishing instead of only in closed areas. The latter modification provided flexibility in the event it became desirable at some time to permit some type of gear to retain limited amounts of incidentally caught halibut during the open season.

The proportion permitted to be retained by setliners was based upon what was in fact taken incidentally when fishing with setline gear for other species on the typical fishing ground. Only users of setline gear could secure permits for such closed season retention despite repeated petitions by users of other types of gear. The setline fishery was the only fishery in which the possession of incidentally caught halibut in an area closed to halibut fishing could be effectively administered and controlled. Other claimants often made unfounded recriminations of the commission favoring the setline halibut fleet.

On a number of occasions the commission was requested by the setline fleet to raise the proportion of halibut permitted to be retained. The proportion had been one to seven. Such requests were rejected. The convention clearly stated that the halibut retained under permit was that which was *in fact caught incidentally*, and the proportion allowed by the commission was determined in as objective a manner as possible. Because of the changing abundance of halibut and of blackcod, the main target species of the permit fishery, the validity of the ratio was tested from time to time.

On several occasions during the 29 years from 1937 to 1966, due to infractions of the provision by the fleets, the commission took under serious consideration the advisability of terminating the program. But the governments' strengthening surveillance and enforcement of warnings to the fleets were effective, and it was possible to justify the program's continuation. The benefits continued to outweigh disadvantages.

Landings under the permit provision increased from 274,000 pounds in 1937 to 1,180,000 pounds in 1946 as the length of the closed season increased. The permit season extended from the closing of an area until the beginning of the statutory winter closed season or sometimes earlier. In the next decade to 1955, permit landings averaged about 600,000 pounds annually with as many as 180 vessels participating in some years.

By the mid-1950s the halibut season began to lengthen which in turn shortened the post-season period when permits were in effect. This resulted in a parallel reduction in the activity in the permit fishery with a steady decline in the amount of halibut taken incidentally while fishing for other species in areas closed to halibut fishing. By the mid-1960s, fleet participation in the permit fishery had reached a point where the benefits were outweighed by surveillance and other costs. Permit landings were about 136,000 pounds with only about 59 vessels participating in 1965. The provision was terminated in 1966 despite some complaints by a few of the fishermen affected.

TROLL FISHERY FOR SALMON

Halibut has been a variable, and at times a very significant item in the United States and Canadian hook-and-line troll fishery for salmon. During the halibut fishing season, the retention of troll-caught halibut has been permitted without any restriction, but after the close of the halibut fishing season, the retention of such halibut has been prohibited.

The proportion of the total catch taken by trollers while fishing for salmon varies widely from place to place and from time to time. Trollers can adapt their methods of fishing depending upon the relative availability and market value of the two species, salmon and halibut. In the 1950s and 1960s trollers caught up to two million pounds of halibut in some years.

Despite comprehensive and repeated studies by the commission, it was not administratively feasible to extend the permit provision to the troll salmon fishery. Any system of control would have created more problems than it would solve. At that time the provision was not withheld for any conservation reason, but solely because of the impossible administrative and enforcement problems it would have posed.

In recent years conservation considerations have arisen with respect to the catch of halibut by troll salmon gear. The halibut taken by troll gear in some grounds off the British Columbia coast have consisted chiefly of small fish. In Dixon Entrance in 1971 over 81% of the individuals were below 26 inches, the legal size at that time, while in Queen Charlotte Sound only 2% were in that category. Since a large proportion of the troll-caught halibut can be released alive, probably over 95%, the commission's raise of the legal size limit in 1973 to ten pounds or 30 inches was expected to effectively reduce the catch of such small fish by salmon trollers without undue wastage of released individuals.

SETNET FISHING FOR HALIBUT

The taking of halibut by bottom set nets had not developed on this coast up to 1938, although a few such nets had been purchased in Norway by a Pacific

Coast fisherman and several had gone to Norway to purchase them. In that year the commission, anticipating the introduction of such gear to the fishery, prohibited its use and has continued the prohibition to date. The basis for the action came from indications of very high selectivity of the gear for large spawning-sized fish (Devold, 1938).

TRAWLNET CAPTURE OF HALIBUT

Prior to 1944 there was no restriction upon the taking of halibut by trawl-net gear. However, since the late 1930s the quantity of trawl-caught halibut landed increased from 80,000 pounds in 1941 to 528,000 pounds in 1943 in Oregon and Washington alone. In addition, another 100,000 pounds was recorded for California and British Columbia.

In 1944 the Halibut Commission prohibited the retention of halibut taken by bottom trawlnet gear both on an incidental basis and as a target species. While this action forestalled the development of a domestic trawl fishery with halibut as a primary target species, the inadvertent taking of halibut by an expanding domestic trawl fishery posed many problems amongst which were losses that accrue to the stocks of halibut through wastage. Also, the enormous expansion of foreign fishing in the late 1950s, chiefly trawling in the eastern Pacific Ocean including Bering Sea, would have had a disastrous impact upon the stocks of halibut, both of adults and juveniles.

What the future holds for the effects of trawling upon the Pacific halibut stocks under the new 200-mile limits in Canada and United States contains many imponderables. A great deal will depend on how each of the two countries exercises its new authority and manages its multispecies demersal fisheries. Will Canada and United States accede to the longstanding demands of their trawl fleets to retain the incidentally caught halibut—a position having considerable support in the research and administrative departments of the respective countries?

Magnitudes of Halibut Catches by Trawler by Sections of the Pacific Coast

While the Halibut Commission may prohibit Cana-

FIGURE 81. The deck of a foreign trawler fishing off Kodiak Island. Note the white-sided halibut apparent on the deck from their size and shape.
National Marine Fisheries Service

dian and United States vessels from retaining halibut taken by net gear, whether inadvertently or with intent, the demersal fish community is such that both adult and juvenile halibut are present in varying amounts over the entire continental shelf. Consequently it is inevitable that they will be caught in various degrees and proportions when trawling for other bottom species. The amount and proportions taken vary widely, and seasonally, from ground to ground, as well as with the type of net gear used and with what the target species of the trawling might be.

The increase in trawling effort both domestic and foreign has not been uniform on all sections of the coast. The intensity of the domestic fishery off the British Columbia coast has probably at least tripled during the 1960s. Although the recorded hours fished by domestic trawlers appeared to have only doubled (Hoag, 1971) during that decade, such a measure of effort does not make adequate allowance for the sharply increased efficiency of the gear and of the vessels. Trawling during the same decade by Japanese and USSR vessels off British Columbia became an operation of large proportions which continued into the 1970s.

Due to the migratory nature of the adult halibut ranging from distant St. Matthew Island in the northern Bering Sea southward over 2,000 miles to the coasts of California, Oregon, Washington, and British Columbia and due to the interchange over long distances of the eggs and larvae and of the developing young or juvenile halibut, management of the resource must be implemented on a coastwise basis. Heavy removals of halibut at any stage of its existence on one section of the coast can have major repercussions anywhere over the entire range of the species.

It has been difficult to estimate how much halibut was caught by the foreign fisheries. Events have shown that the estimates made by this writer in the 1960s were too low. By the early 1970s, the Halibut Commission was able to develop estimates that were more supportable and were also higher. They were still regarded as conservative according to Hoag and French, 1978, Halibut Commission Report 60.

Those authors stated that no allowances were made for trawlers observed to have been targeting on halibut; that such targeting can be directed to adult-sized halibut; that halibut credited by Japan to have been caught in the western Bering Sea where Japan could retain halibut under the Tripartite Convention might well have been taken elsewhere and that in one important area only 7 metric tons had been estimated taken when, in fact, 1,564 tons had been caught.

Until the late 1970s the conditions for observing on Japanese vessels were dictated by Japan. Some initial "observing" had to be conducted from the bridge of large motherships. Occasionally deck observations during some period of around-the-clock fishing were permitted, but seldom was the aggressive observer allowed to "trespass" below decks at night. Interestingly enough halibut "incidence" usually increased sharply from each new vantage point. Also, in the North Pacific only 3,000 hauls were observed over a period of seven years and covering a catch that totaled many billion pounds in that period. Unfortunately the expanded observer program in the late 1970s contributed little to past estimates due to the low level to which the stocks of halibut had fallen.

South of Cape Flattery

This section of the coast off California, Oregon and part of Washington includes the southern extremity of the commercial range of the species.

The halibut were found in localized and discontinuous concentrations, but they had very little capacity to sustain any substantial yields from year to year. Despite a strong demand and accessible markets, production remained at a very low level, and in later years all of the recorded catch has been taken incidentally to fishing for other species. On this section of the coast losses in yield to the setline fishery stem from two sources: those caught by domestic trawlers and either destroyed at sea or sold illegally and those taken by foreign trawlers.

Prior to 1944 halibut taken by the domestic trawler fleet could be sold, and between 1930 and 1944 such net-caught fish accounted for about 20% of the total halibut catch on this section of the coast. After 1943 and coinciding with the Halibut Commission prohibiting the retention of net-caught halibut, there was also a great wartime increase in domestic trawling in the region. Conservative estimates based on very reliable industry information and by direct and indirect observations indicated that a large proportion of the recorded catches of halibut in the region consisted of those landed illegally by domestic trawlers. In addition, unrecorded landings of halibut by trawlers were also considerable.

The normal proportion of halibut in trawl catches on this section of the coast has been of a magnitude of 1 to 2% of the total weight of the demersal food fish catch. In California recorded landings of trawl-caught food fish from 1930 to 1943 included 2.3% halibut (Scofield MSS). Sampling the catches of commercial bottom trawlers at sea off California in 1952 and 1953 showed 1.1% were composed of halibut. Observations at sea off California on commercial trawlers in the 1960s by the Halibut Commission showed 1.1% halibut. Sampling at sea on commercial trawlers off Oregon on 41 trips between 1950 and 1960 by Oregon officials showed an average incidence of 0.5% of halibut in a total catch of food fish of slightly over 1 million pounds. Though not retained as food fish, most of the 41 trips sampled had a hake content that ranged from about 5 to 35% of the total catch. When the large Soviet hake fleets in later years turned to using midwater trawls, the incidence of halibut in their catches was low. More recent observations of 74 bottom trawl hauls made by the Halibut Commission in 1971 off Oregon and Washington showed a trawl catch of 96 halibut, which was about 1.1% by weight of the total fish catch.

With the present level of domestic bottom trawling in this region and the continuing difficulties of surveillance on this long coastline, 100,000 pounds of halibut annually would be a reasonable estimate of unreported removals by domestic trawlers taking into consideration the present very low stock levels of halibut in the region.

Beginning in the 1960s large quantities of bottom fish have been taken in this section of the coast by foreign trawlers, chiefly the USSR. The only accurate statement that can be made about the USSR catch data is that they involved extremely large amounts of fish. There has been an obvious reluctance upon the part of USSR to provide the United States acceptable catch statistics except vague figures largely unsegregated as to species and areas of origin that were reported to the Federal Accounting Office annually. While few halibut were observed in the foreign trawl catches by United States observers in the mid-1970s, such was not necessarily the case in the 1960s. Most of the halibut have since probably been caught during the passing years.

Nevertheless, with the large magnitude of the foreign effort and the observed incidence of halibut in the catches of domestic trawlers as recounted above, it is certain that quantities of halibut were taken notwithstanding the very limited halibut stocks on these grounds and despite the usually very low incidence of halibut found with hake, which was the chief target species of the foreign trawlers. The total loss of halibut was probably in the general vicinity of 250,000 to 500,000 pounds annually. This annual drain was levied not only on any small indigenous population of halibut still remaining in the area in the 1970s, but as shown by tagging, it also drew upon the stocks throughout the range of the species even to the central Bering Sea. In addition, tagging showed that adults and juveniles from grounds off British Columbia were caught.

The total amount of halibut estimated above as representing the probable losses to the foreign fleets on this section of the coast may be regarded by many as small. However, when such amounts are multiplied many times over the entire range of the species, they cannot be ignored. Indeed, collectively they have been disastrous for the resource and for the special interest that Canada and United States have in the halibut fishery. Most of the published estimates that have been made of the losses that have accrued to the resource from foreign fishing and domestic trawling have been on the conservative side.

Cape Flattery to Cape Scott, off the west coast of Vancouver Island

This section of the fishing grounds lies off the west coast of Vancouver Island. Though not one of the more important halibut producing areas the long term trend of the catch up to 1960 suggests that this section might yield about a million pounds of halibut annually. Since the early 1960s the catches by Canadian and United States vessels have declined sharply and the condition of the stocks as shown by the CPUE have deteriorated badly.

In the early 1960s, the region was the scene of heavy concentrations of foreign trawling on such halibut grounds as Big Bank, Kyuquot, Esperanza and other locations with rockfish and other demersal species the main targets. Based on the incidence of halibut in the trawl catches of domestic trawlers (Hoag, 1971) and the magnitude of the foreign operations, catches of halibut by foreign fleets on this section of the British Columbia coast probably range up to 2 million pounds in those years of more intensive fishing. Even the average annual catch of halibut by foreign trawlers has probably been as much as 1 million pounds. Such large removals along with the losses that accrue from domestic trawling in the same region, which in some years may have reached about 500,000 to 1 million pounds, provide sufficient reason for the very sharp decline that has occurred in the availability of halibut on this section of the coast. Canadian and United States vessels, both setliners and trollers, probably accounted for no more than 30% of the total annual removals from this area during the decade of the 1960s.

Cape Scott to Dixon Entrance

This section of the coast is contiguous to British Columbia and can be considered the heartland of the halibut fishery. It had the greatest primeval productivity of halibut of any section of the Pacific Coast. With a controlled fishery, the region appeared to have been able to sustain an average annual removal of about 16 million pounds until the early 1960s as indicated by the catch trends and yield models. Since 1960, the removals permitted the Canadian and United States vessels have been drastically reduced. By 1970 the reduction had amounted to over 40%. Although the permitted removals by the United States and Canada were down to the 10-million-pound level by 1970, there was no improvement in the stocks as shown by the CPUE. Even more drastic reductions in the permitted longline catches by Canada and the U.S. since 1970 have shown little response in the CPUE. Obviously there have been major losses from other sources. To credit such loss in productivity to a natural decline in recruitment or a fall-off in growth is not indicated by the available facts. While no one can rule out the possibility of the latter considerations, there are more indications that domestic and foreign trawling must be assigned some but not necessarily the sole responsibility.

Observations on the domestic trawl fishery conducted by the Halibut Commission at the time the writer was director of investigations involved 3,031 hauls during 120 trips on 32 different Canadian and United States trawlers. It has been shown (Hoag, 1971) that the calculated incidental catch of halibut by the Canadian and United States trawl fleets off the British Columbia coast alone increased from an estimated 2,726,000 pounds in 1962 to an estimated 4,228,000 pounds in 1969. Although the foregoing estimated loss of fish to the setline fishery is due to domestic trawling, it probably represents but a part of the total halibut that may have been removed by other than the United States and Canadian setline fleets. They do not include the illegal landings of halibut of domestic trawlers who from time to time did catch substantial quantities of halibut, either inadvertently or by targeting them. By the late 1960s this practice declined and probably would account for no more than about 500,000 pounds annually by that time. In the late 1950s and early 1960s the total could have been four to five times that amount as almost pure trips of halibut were illegally landed by some domestic trawlers in Puget Sound and British Columbia.

In addition to the catches of halibut by domestic trawlers off British Columbia there have been those taken by foreign trawlers and Japanese setliners, the latter fishing for blackcod. In 1965 United States authorities calculated that Japan caught about 400,000 pounds of halibut over 65 centimeters long off the British Columbia coast (INPFC reports). In addition, it was estimated that about 2,500 small halibut were also taken the same year. It is unlikely that any of the halibut, large or small, were released. Since the estimates were probably based upon the extrapolation of observed proportions of halibut taken with catches of ocean perch in the region, they would not provide for the fact that a few large Japanese stern trawlers made a practice of moving from the deeper water perch grounds under the cover of darkness onto the shallower Goose Islands and other grounds in Hecate Strait where they could not possibly avoid making large catches of halibut. One Japanese trawler in particular was observed engaging in the practice on the Goose Islands grounds *two years in succession,* 1968 and 1969. Since single hauls of halibut weighing 10,000 pounds and up can be taken with the small domestic trawlers from such areas as the Goose Islands grounds, the above estimated annual catches of 400,000 pounds for the large Japanese stern trawlers off British Columbia could be *conservatively quadrupled.*

With the USSR also mounting trawl fisheries during the past decade off the British Columbia and Southeastern Alaska coasts, there were probably as large catches of halibut made by its fleet. That country was not obligated by treaty to abstain from retaining halibut as in the case of Japan. They admit to taking halibut in their trawl fishery off the coasts of British Columbia and Southeastern Alaska as a "by-catch" or "bonus catch."

Dixon Entrance to Cape Spencer

This section of Area 2 includes the grounds off Southeastern Alaska. No significant amount of domestic trawling has been conducted on this section of the coast to date. As stated elsewhere, large Japanese setliners have been observed by the halibut fleet fishing on these grounds ostensibly for blackcod as early as 1962. They were first observed by United States patrol planes in April 1964. This was many years before the Japan Fishery Agency had licensed such vessels to fish in the Gulf of Alaska. Also about 1961, North American halibut vessels began reporting many hundreds of Japanese hooks in halibut on well-known blackcod grounds off Cape Ommaney, Biorka Island, Coronation Island and other locations. On such grounds Canadian and United States vessels would, when fishing blackcod, take up to 9% halibut, except when they were restricted to the 1 to 7 ratio. Also, within two to three minutes run from such grounds the catch might consist almost exclusively of halibut.

Circumstantial evidence, which often is as good as direct evidence on some fishery matters, leads to the conclusion that the catches of some of the Japanese longliners in this area have consisted of a large proportion of the more valuable halibut, probably

FIGURE 82. A common modification of a standard Japanese longliner fishing for blackcod is the addition of a temporary shelter deck over the main working deck, as shown in the photograph. Butchering occurs out of sight under the shelter deck rendering any air surveillance of procedures useless.
National Marine Fisheries Service

involving even millions of pounds. The ineffective surveillance of the Japanese blackcod fishery in the 1960s was markedly altered in the 1970s and undoubtedly reduced the impact of such fishing on the halibut stocks.

South of Cape Spencer

The annual losses of halibut from all grounds south of Cape Spencer due to domestic and foreign trawling were probably at the very least 8 to 11 million pounds annually and conceivably even much greater in some years. In addition to losses due to trawling, there were those that stem from the fleet of large Japanese setliners fishing primarily for blackcod south of Cape Spencer, particularly off Southeastern Alaska. Their catch of halibut could have amounted to 2 to 3 million pounds annually, as estimated from occurrence of Japanese hooks in halibut taken by the Canadian and United States setline fleets. Eastern Bering Sea fishing by the Japanese and Russians has also greatly reduced the migration of halibut to the stocks in the Gulf of Alaska and as far south as the grounds of British Columbia.

The above unrecorded losses from all sources both foreign and domestic with a probable minimum magnitude of 11 to 14 million pounds annually during the 1960s from grounds south of Cape Spencer required a drastic reduction in the catches permitted the Canadian and United States setline fleet. The Halibut Commission during the 1960s reduced the permitted level of removals from a test level of about 31 million pounds at the beginning of the decade to 20 million by 1970. Despite this 30% reduction, there was no significant response in the condition of the stocks. Such lack of response is entirely contrary to what was shown by the catch and stock reactions of the previous 50 years. There could be no better indication that large quantities of halibut were being removed from the grounds south of Cape Spencer by other than those permitted the Canadian and United States setline fleets.

In the 1960s it would have been naive to have reduced the North American setline catches at any greater rate or in larger amounts in face of the removals by the foreign fleets, unless of course, it was to benefit the latter. With the decline in the foreign depredations in the 1970s, and provided they are further reduced, some benefits may ultimately accrue to the Canadian and United States setline fleets by holding catch limits at fairly low levels.

West of Cape Spencer

In the Gulf of Alaska and westward to beyond Unimak Pass, there has been a large foreign trawl fishery since the mid-1960s. Japanese setliners have also fished this region since the mid-1960s.

It was once believed that trawling in the gulf tended to catch only young or juvenile halibut. On the contrary, there are many documented instances that show that at the appropriate location, season and depth of water, large trawl catches of halibut averaging 20 pounds or more may be taken. These would be easy targets for trawlers wishing to capitalize on the more valuable halibut. Also, in the late 1960s, foreign trawlers extended their operations into the spawning season of halibut with the attendant dangers to the spawning populations

It is generally claimed that the stocks in Area 3, west of Cape Spencer, commenced to decline before foreign fishing had extended to the Gulf of Alaska and that removals by the United States and Canadian setline fleets initiated the decline. Be that as it may, there are many factors that can enter into such a discussion. It soon became clear by 1964, and continuing into the 1970s, that losses to foreign fishing were undeniably very high and would nullify any further reductions in the catch limits allowed the domestic setline fleets.

The Bering Sea

In the Bering Sea the stocks of halibut underwent a process of rapid destruction, even more rapid than in Areas 2 and 3. The large fleets of foreign trawlers, chiefly Japanese, have been exacting a toll upon the halibut, both young and adults, since the late 1950s.

The interests of Canada and the United States in the halibut of the Bering Sea were claimed to have been bartered away in 1962 in a so-called subtle attempt to sustain the abstention principle with respect to Pacific salmon, particularly those of Bristol Bay. While admittedly, the salmon resource is much more valuable than halibut in the eastern Bering Sea, more thought could have been given to the fact that the halibut in the eastern Bering Sea probably possessed the key to the survival of the entire North Pacific halibut fishery. Persons in authority in the Canadian and United States governments gave repeated voice to the idea that the jurisdiction of the Halibut Commission over the halibut in the Bering Sea should be terminated, and the exclusive control over the resource turned over to the International North Pacific Fisheries Commission.

Prior to the onset of the multi-billion-pound bottom fish fishery by Japan and the USSR, the Bering Sea was populated by an extremely large group of young halibut. In the very late 1950s there were probably as many as 60 million one- to three-year-old juvenile halibut in the southeastern Bering Sea alone, based on early king crab surveys of the United States Bureau of Fisheries. The size of the population seemed much larger than could have been derived from the numbers of spawning halibut estimated to be in eastern Bering Sea at the time. Also the life history of the species and the water transport patterns of the North Pacific Ocean indicated that a large proportion of the young halibut in the Bering Sea were the progeny of and in turn the recruits for, the adult populations spawning on the grounds of the Gulf of Alaska and British Columbia.

In the Bering Sea it was determined (Halibut Comission Report 60) that the losses in potential yield of halibut have been high. In the late 1950s and early 1960s when the foreign fleets were decimating the yellowfin sole stocks, the losses may have been greater than the aforesaid report estimated. In 1961, based on calculations credited to K.S. Ketchen in Report 60 and combined with catch data of a research vessel (Best, 1970), a probable total of 23 million pounds of small halibut might have been taken in the eastern Bering Sea by the foreign fleets.

On different sections of those grounds, the halibut ranged from an average of under 1 pound to a 59-pound average. In 44% of the sections, the halibut averaged over 11 pounds, a size readily available to setline gear.

To what extent the above early losses in the Bering Sea would have reduced the emigrants to Area 3 and lowered recruitment to the fishable stock in that area is problematical. In addition, the Canadian and United States setline fleet from 1958 to 1961 had taken 15.9 million pounds of halibut out of eastern Bering Sea, some of which fish would have otherwise emigrated to Area 3 or even Area 2.

Until the late 1970s, the conditions for observing Japanese vessels were dictated by Japan. Some initial "observing" had to be conducted from the bridge of large motherships. Occasionally deck observations during some period of around-the-clock fishing were permitted, but seldom was the aggressive observer allowed to "trespass" below decks at night. Interestingly enough halibut "incidence" usually increased sharply from each new vantage point. Also, in the North Pacific only 3,000 hauls were observed over a period of seven years and covering a catch that totaled many billion pounds in that period, a very small and probably biased sample!

To arrive at an estimate of what the total removals have been from the Pacific halibut stocks, the commission in its 1975 annual report presented the data in Figure 83. However, this otherwise useful presentation possessed the defect in that, in arriving at such a total, dressed heads-off weights were used for the domestic setline catch and were added to round live weight figures for the catch in other fisheries. This materially distorts the figure as well as requires corrections to the accompanying texts of that annual report. Also, the source or sources of the totals shown for fish removed or destroyed by other fisheries were not given in the report.

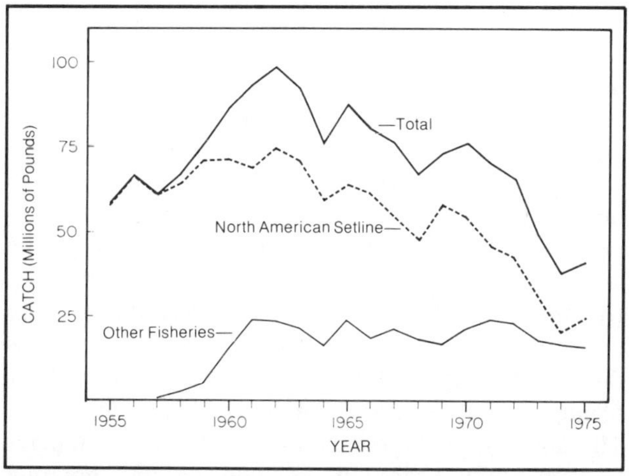

FIGURE 83. Total catch of halibut (round weight) in the North Pacific Ocean and the Bering Sea, from 1955 to 1975.
F.H. Bell

ENFORCEMENT AND VIOLATIONS

One of the more important considerations in managing a fishery is assuring a reasonable adherence to required control measures. To regulate the taking of halibut, which is conducted both in territorial and high seas waters by several nations, poses many difficult problems of enforcement. In addition to those created by the domestic fleets, the massive development of Japanese and Russian fishing, with both bottom net and setline gear in the North Pacific and the Bering Sea since the late 1950s, greatly compounded the difficulties. Without reasonable observance and effective enforcement all efforts to manage the resource would be futile.

In the past a preponderance of the participants in the domestic fishery both vessels and dealers did exercise a satisfactory degree of compliance with the rather stringent regulations. It is estimated that only about 10% of those in the industry accounted for about 90% of all detected violations. Close contact with both dealers and fishermen and an intimate familiarity with the fishery itself also made it possible to detect any significant noncompliance. Furthermore, procedures were developed that enabled allowances and corrections to be made for the effects of violations upon the catch statistics and the management of the resource.

The problems of regulating the domestic fishery are reviewed in detail for several reasons. It will show the degree to which the original participants, the fleets and dealers of the two countries, Canada and United States, complied with the restrictions placed upon them. It also brings into closer focus some of the problems that were encountered in making acceptable annual estimates of the unreported removals from the stocks by both foreign and domestic vessels. It will show that contrary to what has been said (IPHC Technical Report No. 14) the volume involved in some years was significant and that the amounts were even more precisely determined than what has been reported as legal landings. More important, however, a recital of past events in the domestic scene may place the situation regarding the foreign fisheries in sharper perspective. Accordingly, with an understanding of what occurred in the relatively simple structure of the domestic fishery in the 1930s and 1940s one should not be surprised at what happened in the large complex foreign fisheries conducted off the Pacific coasts of Canada and United States. Also, an understanding of the some of the potentialities of the situation might permit realistic estimates to be made of the unreported removals of halibut by the foreign fleets.

It should be noted that with any government or with any industry there is a reluctance to admit to the possibility that infractions of fishery regulations are as extensive as they often are. Unfortunately, in the meantime the fish stocks in question may be undergoing drastic losses of undetermined magnitudes.

Prior to the initial Halibut Convention of 1923, the halibut fishery was relatively free from any restraints. There were no restrictions upon the time, the location, or manner of fishing halibut or upon how much might be taken each year. There were only the international navigation rules, the restrictions against vessels fishing within the territorial limits of another country (a basic three miles in the two countries), and the usual and accepted principles of human and "property" rights that are involved in peaceful and innocent pursuits on the high seas. Whether to fish or not to fish depended upon the vicissitudes of the market for halibut, the profit motive and the personal inclination to undergo the effort and dangers involved.

Suddenly the Northeastern Pacific Halibut fishery was catapulted into an entirely new dimension. Overnight, as it were, in 1932 it became and continued to be for the next 40 years, the most rigidly controlled marine fishery in the world. The abruptness of the change combined with the depressed economic conditions of the 1930s along with other situations resulted in a considerable resistance to restraint leading to many violations of the commission's control measures. These were exclusively domestic as there were no vessels other than those of the Canadian and United States flags engaged in bottom fishing in the Northeast Pacific until the late 1950s. The magnitude of the violations fluctuated from place to place and at times involved considerable tonnages of halibut.

Types of Violations

Group I. Violations at sea at the time of fishing were

- (a) fishing for halibut in an area during the closed season in that area.
- (b) fishing for halibut in closed halibut nursery grounds.
- (c) fishing for halibut in an open area with a halibut license valid for some other open area.

Group II. Violations on shore at the time of unloading were

- (a) falsification of weight of halibut landed and sold.
- (b) false declaration as to the origin of the catch.
- (c) landing incidentally caught halibut in excess of that permitted by the halibut regulations.
- (d) retaining and landing halibut caught with illegal gear.

There are many other types of violations, but despite their visibility, they had little direct effect upon the management of a resource involving the use of catch limits except that collectively they could create a climate of lawbreaking. They included such infractions as commencing to fish a few hours or days before the opening time, fishing overtime after the closing hours, and failing to secure a proper validation of the halibut license.

Violations at Sea At the Time of Fishing

Fishing for halibut in an area during the closed season in that area

This activity became a major problem soon after regulation began in 1932. Most of such illegal fishing occurred in Area 2 south of Cape Spencer as this area was closed to halibut fishing much before Area 3, the grounds west of Cape Spencer. The vessels that fished Area 2 were smaller and at that time predominantly Canadian. They were largely precluded by their limited seaworthiness from diverting to Area 3 after the earlier closure of Area 2. Illegal fishing or poaching in Area 2 increased as the difference in closing dates of the two areas widened.

Conservative estimates of the halibut taken in the closed Area 2 rose from almost negligible amounts in 1932 to 2 million pounds or more annually by the end of that decade. In the early 1940s the estimates reached 3 million pounds in some years. These amounts were for the regular halibut vessels. In addition, the rapidly growing fleet of small boats, including salmon trollers, was estimated in some years to have retained and sold from 300,000 to 500,000 pounds, caught while fishing for other species, after Area 2 had closed.

In an attempt to avoid apprehension, the fleets engaged in a number of strategems in addition to flagrantly fishing in a closed area with a license validated for an open area.

One fleet, in particular, developed the strategy of "doubling up," or operating in pairs. Vessel A with a halibut license validated for Area 3, still open to halibut fishing, would anchor in one of the innumerable secluded places in Area 2 which was closed to halibut fishing. Vessel B, with its license validated to retain halibut caught incidentally to fishing for other species in closed Area 2 in the amount of one pound of halibut to seven pounds of other species, as permitted by the regulations under the 1937 convention, would proceed to fish for halibut. Every few days, Vessel B would use Vessel A as a receiving station.

As soon as Vessel A had accumulated a reasonably sized load, both vessels would proceed to port, usually one or two days apart to allay suspicion. Vessel A would unload a full fare of halibut ostensibly from Area 3, and Vessel B would unload a blackcod trip from Area 2 with the permitted ratio of halibut. For obvious reasons, the latter trips were usually small. The vessel had been preoccupied with fishing for halibut rather than blackcod. This practice assumed fairly large proportions in the late 1930s when air surveillance was very limited and difficult with the equipment then available.

Fortunately, there were a number of means whereby it was possible to make satisfactory corrections in the catch statistics. The distinctive size composition of the illegal catches often revealed the probable true area of origin of the catches. Also, an intimate knowledge of the fishing patterns of every vessel in the fleet and the timing of their movements permitted a tentative "pairing up" of the vessels and confidential sources in the fleet were invaluable in identifying and confirming suspicions. A few vessels, with their licenses validated for Area 3, were apprehended by fishery patrol vessels in the sequestered harbors in closed Area 2. However, the crews would claim that they were in transit from the open Area 3 and had been forced to anchor using bad weather or engine breakdowns or some other transparent subterfuge as an alibi.

There are indications that a modern version of the above procedure has been practiced by Japanese vessels. Halibut taken by both trawlers and setliners fishing in areas other than the Bering Sea and where Japan had agreed to abstain from retaining any halibut, may readily be transferred to motherships servicing fishing vessels outside the abstention area. Such illegal halibut thus lose identity as to origin and can be landed with impunity in Japan. There is no way of estimating what the magnitude of such a practice might be, but it could be very considerable under prevailing conditions.

Hoag and French (1976) also suspected such a situation. They found that the reported catch in some years from some areas was much higher than what their estimates indicated. According to those authors, during an 11-year period over 28 million pounds were credited to an area for which only 1.75 million pounds had been estimated.

In addition to doubling up or just flagrantly fishing in closed areas, some Canadian and United States vessels engaged in other dissembling tactics during the 1930s and early 1940s. They would clear customs from their home port for another port in closer proximity to the open regulatory area. They would then clear for the open area for which their halibut license had been validated. Instead of proceeding to the open area, they would return toward their home port and fish on a ground in the closed area sometimes as much as 500 miles south of the open area. On a number of occasions, it was possible to anticipate such movements and advise the sea patrol of the respective country of the firm probability of a specific vessel's presence on a particular closed ground at a certain time. The size composition of landed catches of such suspect vessels frequently confirmed the true grounds where they had actually fished. Anomalous release-recovery patterns of tags reported also provided very useful clues. However, some vessels withheld reporting recovered tags when they suspected that the tags might indicate illegal retention of halibut. It is quite likely that the dearth of tags recovered from foreign vessels may proceed from the same acts of caution. Yet there have been Japanese vessels reporting recovered tags with completely anomalous release-recovery patterns, which is strong circumstantial evidence of illegal fishing.

There is no exact Japanese counterpart to the above attempt of vessels to elude suspicion by "backtracking" and then end by fishing in some closed area close to their home port with a license for an open area. However, it is known that very many Japanese vessels do not fish in the area for which they are licensed by the Japan Fisheries Agency. If they do report their catches, they tend to be credited to the area for which they were licensed, and not where they fished. A fully verified example of this is that no setline-caught blackcod from the Gulf of Alaska was reported by Japan until many years *after* Japanese setliners were first observed blackcod fishing on a large scale in that region.

Fishing for halibut in closed halibut nursery grounds

The convention permits the commission to close to all halibut fishing on those grounds found to be populated by small, immature halibut and to designate them as nursery grounds.

During the 1930s there was at times considerable poaching in the two areas closed as nurseries by the commission, off Massett, B.C., and the Timbered Islet in Southeastern Alaska. While the fish were small at that time, they were relatively abundant and found a ready market either as whole fish or as fillets. From time to time, occasional whole fares of Masset "fleas" were landed in Prince Rupert. However, the poaching did not reach a magnitude to be any threat to the resource. The imposition of a size limit by the commission in 1944 became a further deterrent to fishing in closed nurseries.

In 1967 the Halibut Commission designated a large section of the southeastern Bering Sea as a nursery area closed to all halibut fishing.

An intensive trawl survey initiated by the commission in 1962 and earlier by the Bureau of Commercial Fisheries had demonstrated that the 25,000-square-mile southeastern flats of the Bering Sea had at least a very large population of young halibut. Japan, as a domestic measure, had designated a very small portion of this nursery area closed to trawling. This was to prevent conflicts between Japanese crab fishing and trawling interests and not for halibut protection. Aerial surveillance by United States authorities had shown that the Japanese prohibition against trawling was being flagrantly violated by both Japanese stern trawlers and Danish seiners fishing at will in their so-called closed area. One can only imagine the slaughter that must have taken place of the formerly dense concentrations of young halibut. During the decade of the 1960s the population of young halibut on the Bering Sea flats declined over 75%.

Actual observations made by the Halibut Commission research vessel in March, 1970, alongside Japanese Danish seiners, have shown hundreds of juvenile halibut being taken per seine haul in the southeastern Bering Sea. Tagging showed that these young halibut would provide recruits to the adult stocks as far away as the British Columbia coast. Thus the survival of the total North American setline fishery was being impacted and endangered by such destruction.

The retention of such small halibut, whether taken anywhere in the eastern or western Bering Sea, was prohibited not only to North American vessels but to those of Japan as well, under the International North Pacific Fisheries Convention. Notwithstanding, Japan, in 1972, exported probably millions of pounds of halibut fillets taken from fish under the legal size limit, most of which came to the United States where demand and price for imports of halibut had increased sharply due to the severe decline in the halibut stocks available to the domestic fleets in their own waters.

To forestall such importations as these, the Northern Pacific Halibut Act of 1937, had specifically, and with stated intent, expanded and strengthened the interdiction in Canadian and United States legislation regarding the 1923 convention against the use of the ports of Canada or the United States by *fishing vessels* of a country not party to the International Pacific Halibut Convention.

In the United States act for 1937, implementing the 1937 convention, the prohibition was extended to include *any type of vessel.* There was already the expressed concern in 1937 that Japan would soon begin a program to exploit halibut, as it had already done with respect to the salmon of Bristol Bay. With the type of vessel now carefully defined, and with expressed intent as "every description of watercraft or other contrivance used or capable of being used as a means of transportation in water," it was believed by many that the use of United States ports by foreign fishing vessels, motherships, and common carriers, whether transshipping halibut at sea or in a Japanese port, would no longer be permitted. Nevertheless, unrestricted waterborne imports of halibut into the United States from Japan have been permitted, seemingly upon some sufferance or favored nation policy or possibly upon some other interpretation of the 1937 act — an interpretation entirely contrary to the stated intent of that law. Failure to control Japanese importations into the United States, and then later condoning them, has made a mockery of the whole concept of conservation and international fishery agreements.

Fishing for halibut in one area with a license validated for another open area

Initially, under the halibut regulations vessels could have their licenses validated for more than one open regulated area. This allowed vessels to declare catches made in Area 2 as originating from Area 3. By so doing, they gained from the larger trip limits that the fleet's voluntary curtailment program assigned to vessels fishing in Area 3. At the same time such violators believed they were extending the fishing season in Area 2 at the expense of that in Area 3 and the fleet therein. Such false declarations were conservatively estimated to have amounted to at least as much as 2.25 million pounds in some years.

In 1940, the commission limited the validation of licenses to a single area per trip. This did reduce the practice to some degree as vessels fishing in Area 2, with licenses validated for Area 3 in order to qualify for the larger self-imposed trip limit, would at least be subject to detection at sea.

Violations On Shore At the Time of Landings

Violations of this type usually attract less attention than those occurring at sea, but in the Pacific halibut fishery the tonnage involved in such violations and the number of violations greatly exceeded the latter. Most of the violations involve a third party, namely the buyers, in addition to the fisherman and the authorities.

False declaration of the weight of halibut landed and sold

Probably the major reason for the false reportings of landings in the 1930s was the fishing fleet's voluntary program to reduce the rate of landing of the catch limits set by the Halibut Commission. The provision of the program that limited the amount of halibut each vessel might land per trip, per crew member, was fraught with much temptation. While there were penalties for any evasions enforced by the fishermen's unions and in Canada, the program at times had the force of law under the British Columbia or the Canada Marketing Acts, in neither instance were the deterrents very effective. A system of union and vessel owner's inspectors or an exchange of delegates between vessels landing the same day in a port did in some areas maintain a semblance of compliance.

One of the relatively simple techniques used by some offending vessels, in collusion with the buyers, was the "double invoice" or "ticket." The buyer would issue an invoice for the poundage permitted the vessel under the fleet's program, which the captain would then present to the fleet's curtailment board. The balance of the catch, recorded on a second invoice, was used for reporting the complete catch to the commission as required by the halibut regulations that had the force of law. This practice did protect the accuracy of the commissions catch data although the fleet made constant demands that the commission turn over the information to their curtailment boards. Needless to say, these annual demands were never met. Although the commission was always in full sympathy with the fleets efforts to achieve an orderly marketing of the permitted catch, the commission had to risk some alienation of support by the fleets in order to protect the integrity of the statistics.

Much of the halibut caught and retained contrary to the regulations either of the commission or of the curtailment board was usually sold by the fishermen below the current market rate. This provided such buyers a considerable trading advantage in the market. They could sell slightly below other dealers yet still maintain the necessary profit margin. Competition would force other buyers into similar retaliatory practices, a situation which became acute at times particularly between companies that maintained receiving facilities at places where little or no effective union surveillance existed. Occasionally, such companies desirous of cooperating with the commission would jointly agree to terminate the practice (at least for a time).

During the 1930s and to the mid-1940s covering a 15-year period almost every type of strategem was used by buyers and fishermen to subvert the halibut regulations in one way or another. The economic exigencies of the depressed 1930s and later the high wartime demand for fish created an economic and political climate conducive to infractions of the regulations. This was aggravated by the fact that there was no adequate force of fishery enforcement agents in either country equipped to cope with the complexities of the problems. The uniformed enforcement agents in both countries could provide but limited assistance in the situation due to the devious methods that some vessels and buyers employed to hide their manipulations. For a brief period in the United States a special agent was assigned to work undercover—unknown to any local enforcement agency—reporting only to Washington, D.C. Since this writer was in charge of the statistical work of the commission, at the time the agent maintained informal contact with this writer.

In turn the Halibut Commission used all means at its command to maintain the validity of the data collected. While it lacked any enforcement authority under the convention, it was able to use persuasion of one type or another to reduce the violations or at least acquire knowledge of their magnitudes.

The support and confidence of several individuals in the halibut industry also provided the commission with excellent information as to poundage being handled in contravention of the regulations. In the United States this situation applied in Seattle, as well as at such trouble points as Eureka, California, Warrenton, Oregon, and Port Angeles, Bellingham, Tacoma, Marysville, Hat Island, Blaine, and occasionally Kingston and La Conner, Washington. In British Columbia, good information was also available upon what was transpiring at many points on the west coast of Vancouver Island and at several salmon centers between Vancouver and Prince Rupert that possessed unloading and freezing facilities for halibut. In Southeastern Alaska there was information regarding what was occurring in that region with respect to the landing of illegally caught halibut or not reporting landings.

The foregoing methods could only be effective in a more or less closed organization such as the Halibut Commission. There was no large bureaucracy and very few staff changes. The confidential sources and the data were known and identifiable only by the three directors and one staff member over a 40-year period. Furthermore, the commission specifically authorized the director "to take whatever measures deemed necessary to maintain the validity and integrity of the commission's statistics." While one or two commissioners from the private sector wished that such confidential information secured by the staff be identified to the commission, even they finally realized that to do so would eventually foreclose the sources. This would have been particularly true as the two commissioners in question were very closely connected with the halibut industry and the conflict of interest would have been pronounced.

In using sources within the industry, it was necessary to check one against the other to reduce the possibility of inadvertent under-or-over-estimations. In some instances, individuals other than the established confidential sources would volunteer information. These would have to be verified in case of bias, intended or otherwise. For example, it was once claimed by a very responsible group of Canadian dealers that "several millions" of pounds of halibut had been illegally landed one fall in a particular British Columbia port. The commission about the same time had uncovered the practice and by reliable methods had been able to calculate that the total was about 800,000 to 900,000 pounds—a very substantial quantity but not "several millions."

In addition to the information received from confidential sources, the commission used other means of ensuring the reasonable integrity of the catch and fishing statistics collected.

The demand for halibut livers and their increasing value resulted in all of them being retained. This enabled the commission to establish statically valid ratios of livers and halibut sold for the various grounds and seasons of the year. These ratios when checked by other direct observations proved to be excellent detectors of poundage sold, but unreported. When confronted privately with the fact that the commission possessed close approximations of what they actually received, some dealers decided to make available complete data of their purchases. At the outset, they believed we were securing the complete record from the captains of the vessels involved, which, in some cases, was true. After three years, when some of the few less-cooperative members of the industry suspected use of liver ratios, they held out some livers and shipped them to firms in other ports for sale to pharmaceutical houses. However, this also was detected through the records of the pharmaceutical companies showing receipts of halibut liver from fish companies which had bought no halibut.

In the case of some companies, it was found that they were distributing more halibut than their purchases indicated, even with due allowance being made for stock carry-overs, glaze on frozen sales, and the normal difference between buying and selling weights. With many vessels providing the commission with the correct amounts of their landings, it was usually not difficult to eventually persuade most dealers to provide complete records. Sometimes such a process of persuasion took several years. Some firms used various code signals in their books to camouflage halibut, or to identify altered amounts, purchased in excess of that permitted the vessel or bought from trawlers. Usually illegal or unreported halibut would be purchased at below market prices, so signaled in the company records by a code mark such as an asterisk for stock and cost purposes. There was almost an endless variety of ingenious ways used in United States and Canada to contravene the halibut regulations. There is no reason to believe that fishermen and fish dealers today in other fisheries or in other countries are not equally ingenious.

Landing incidentally caught halibut in closed areas in excess of that permitted by the halibut regulations

The retention of incidentally caught halibut when fishing for species, particularly blackcod, in areas closed to halibut fishing created many problems. Lately they have become particularly acute with the development of a major Japanese setline fishery for blackcod since 1964 in the Gulf of Alaska chiefly off Southeastern Alaska and British Columbia. In this region Japan agreed to abstain from retaining any halibut. Previously the problem was restricted to the North American setline fishery that fished for blackcod, a relatively modest operation compared to that pursued by Japan.

The 1923 convention provided that incidentally caught halibut during the closed season may be used for food for the crew of the vessel. Any remaining were to be landed and immediately turned over to enforcement officers. This provision was continued almost without any change in the 1930 convention. At that time there was very little fishing for other species in the closed season which then extended from November 15 to February 15. Consequently the question of the retention of incidentally caught halibut in the closed season arose very infrequently.

This food-for-the-crew provision possessed a sentimentality not in keeping with the modern halibut fishery in which relatively little fish reaches the galley table. The sooner the provision was abolished, the better the fishery would be.

As the length of the fishing seasons declined with the improved stocks and larger fleets, the open season no longer encompassed the blackcod fishing season in which considerable quantities of incidentally caught halibut were taken. From an enforcement standpoint the detection of sales of illegal halibut was made almost impossible because fishermen, even when found unloading halibut at the dock, could claim that they were going to turn it over immediately to the authorities as required by the 1923 and 1930 conventions. When confronted at sea they claimed that all halibut would be surrendered in excess of what would be consumed as food for the crew. Some crews would claim that they would freeze the "food" for later consumption at home. These conditions were conducive to many vessels illegally disposing of catches instead of surrendering them. In order to eliminate the food for the crew device, an amendment of the 1930 convention was needed.

The 1937 convention included a provision under which the commission could permit the retention of incidentally caught halibut in a closed area. Regulations controlling and limiting such retention remained in effect for the next 28 years to 1965. By that time the need for such a requirement had declined with the changing conditions in the fishery, including the lengthening of the halibut fishing season, particularly in Area 2, south of Cape Spencer.

During those 28 years, 1937 to 1965, retention of

incidentally caught halibut in a closed area was restricted, for enforcement or biological reasons or both, to vessels fishing with setline gear. The permitted retention, one pound of halibut for each seven pounds of other species, resulted in many violations and on several occasions the commission came close to abolishing the provision due to the lack of compliance by the setline fleet with the allowed ratio. Also otter trawlers, not permitted to retain the incidentally caught halibut both in the open and closed seasons, went to great lengths to sell such illegally retained fish. Salmon trollers were in a somewhat similar situation although they were allowed to retain incidentally caught halibut in the open halibut season with no restrictions as to amounts.

Most violations with respect to the incidentally caught halibut were carried out in collusion with the buyers. However, there were infractions that did not require buyer collusion. Sometimes an otter trawler, in the open season would unload its halibut to a setline vessel in an isolated harbor. An unusually high proportion of net-marked halibut or "chalky" individuals—those with the opaque flesh of a cooked fish—would make a vessel suspect. It was necessary to carry out somewhat the same counter measures that were used to counteract the falsification of landing records by those attempting to evade the fleet's curtailment plan in the 1930s and early 1940s. These measures permitted estimates to be made of the amount of incidentally caught halibut that was retained and sold in excess of that permitted by the regulations. While it is not possible to secure the same hard evidence about the present Japanese blackcod fishery as was ascertained by the writer in the North American blackcod fishery many years ago, some good estimates may be made. In 1969, about one third of the North American halibut fleet was questioned on the incidence of Japanese blackcod hooks in their catches. A total of 295 hooks were reported by 73 vessels on 111 trips. Had the total Canadian and United States fleet been questioned each trip during the year, the total would have been at least 1,000 hooks. This would be also a minimal estimate because many of the hooks would have been shed before or unobserved after recapture. Translating such a number into one based upon the chances of their recovery, determined from controlled tagging experiments, the total could represent a minimum of 10,000 halibut so marked with Japanese hooks. An accurate figure could well be double that number.

It was noted also that the halibut recaptured with Japanese hooks involved a high proportion of the larger-sized fish. Such sizes would tend to break or part the fairly fragile but effective Japanese blackcod "hachi." The hauling rate, which is fairly rapid, could cause the loss of the larger-sized halibut.

From such circumstantial evidence, it was imagined that blackcod vessels were retaining large quantites of halibut of the smaller sizes, particularly during night operations when they could move undetected one or two miles off blackcod grounds and set on productive halibut grounds.

It may be significant to note that the sharp increase in activity of these land-based blackcod longliners in the Gulf of Alaska coincided with the reported abrupt increase in landings of halibut by the land-based vessels from the western Bering Sea where Japan unilaterally continues to catch and retain halibut with relatively no restrictions as to the size of the fish. It is significant that the reported land-based fish production in the western Bering Sea, caught by dragnets, trawls or seine gear, consistently showed a much higher proportion of halibut (8%) than Japan reported being taken inadvertently when trawling in the Bering Sea.

Large Japanese setliners up to 300 tons with 25 to 30 crew members were first observed, by United States patrol planes, fishing blackcod in the Gulf of Alaska in April, 1964. The fleet increased to as many as 23 vessels by 1967. Blackcod catch increased from 4 million to an estimated 20 million pounds by 1968 (Branson, 1968). It is an interesting commentary that setliners were first licensed by the Japan Fisheries Agency to fish in the Gulf of Alaska in September, 1967. It has been said that some were penalized for fishing without a license (Branson, JH 1968) although the same source states that no Japanese patrol vessel was ever seen on the blackcod grounds prior to 1968. It is strange that penalties could be invoked without any evidence. The above report stated that the setliners had been boarded and inspected whenever possible by United States personnel, and only one ship was found to have halibut aboard and "that there was reason to believe those fish were taken in the Bering Sea." It is stated that halibut were observed on Japanese gear on only two occasions between 1964 and the date of the article, 1968.

It is not stated how many vessels had been boarded nor how complete the inspections of the vessels were, nor for what length of time the hauling of the gear was observed, nor whether the observations were conducted at night. These are crucial considerations in light of what occurred in years past with the North American fleet. The Japanese vessels for the most part had covered decks after late 1966 and possessed a crew size that would permit day and night fishing. Operations extended throughout the Gulf of Alaska but most were conducted off Southeastern Alaska.

Fishing was conducted from 150 to 550 fathoms, mostly about 200, and often within a few miles of very productive halibut grounds. The Japanese reported no incidentally caught halibut, which is indeed a strange statement in light of the hook evidence.

Accordingly, there is strong circumstantial evidence indicating that they did retain many hundreds of thousands if not millions of pounds of halibut annually during those blackcod operations in the Gulf of Alaska. There is also no evidence to indicate that they were less capable or resourceful than were the North American fishermen who during the mid-1940s landed relatively large quantities of incidentally caught halibut in excess of what was permitted, sea surveillance and shore inspections notwithstanding.

North American vessels fished primarily for blackcod in much the same depths off Southeastern Alaska as did the Japanese. Before there were any restrictions on the ratio of halibut and blackcod, catches contained about an average of five to seven percent halibut. The North American vessels also found that a very small shift in location, of even the length of a few units of gear, would frequently sharply increase the incidence of halibut in the catch. The Japanese setline gear, although less robust than the North American gear used, was quite capable of retaining a large proportion of the fishable stock of halibut.

Retaining and landing halibut caught with illegal gear

The prohibition of the retention of halibut caught with net gear created considerable enforcement and violation problems, first with respect to the domestic trawl fishery and subsequently with the foreign trawl fishery.

Landing of illegally retained halibut by domestic otter trawlers has been a continuing problem since 1945, although it has fluctuated in magnitude from time to time and place to place. Northern California, Oregon, Puget Sound and British Columbia ports have at times been scenes of unacceptably large landings of illegally retained trawl-caught halibut. Such landings reached a high level in the mid-1940s and the mid-1950s. During the 1960s they were at a tolerable level, except that an occasional trawler would land a relatively large catch of halibut in the 20,000- to 30,000-pound category. However, in recent years as far as the domestic fleets are concerned, it is probably not the surreptitious landing of such halibut that is the problem. It is the mortality caused on the grounds by the inadvertent taking of halibut.

Like the setliners, the North American trawlers used a wide variety of strategems to avoid detection when landing illegally caught halibut. Many unloaded the halibut at some sequestered point en route to the normal port of unloading. In other instances they landed at their regular place of landing and sometimes after regular working hours, depending upon whether an enforcement officer might be expected. For example, one plant in Puget Sound bought about 530,000 pounds of trawl-caught halibut in one season before being detected. For many years there was practically unrestricted landing of trawl-caught halibut on the northern California and Oregon coasts. One trick used by a few Puget Sound vessels was to fillet the halibut at sea and bring them in large five-gallon cans once used for containing fish livers. These cans of fillets were unloaded with impunity along with those filled with livers.

It is extremely difficult to make a complete search of even a small fishing vessel if it has any fish or ice on board. Trawlers have been boarded along the docks by officers, inspected, and after the departure of the boarding officer, have proceeded to unload substantial quantities of halibut from some previously unopened portion of the fish hold. This situation has occurred many times down through the years. At sea it is practically impossible to conduct an adequate search. Should such a vessel be a large foreign setliner, stern trawler, or mothership, even boarding it at sea with the expectation of finding any contraband halibut would be the utmost in self delusion, unless of course, the vessel had no other cargo, then every freezer, cold room, and hold space could be closely searched and checked against the construction plans of the vessel. Boardings can only act as a deterrent if penalties are high.

Under the enforcement conditions that prevailed from 1932 to about the mid-1940s the commission had to carry out its own informal surveillance of both the setline and trawl fisheries. It was necessary at times to conduct many round-the-clock telescopic surveillances of a waterfront from some strategic point. Trawlers not permitted to retain halibut, or setliners with halibut in excess of what they were allowed to possess when fishing blackcod, were on occasion observed unloading large quantities of halibut in the dark of the night or early morning after the uniformed shore patrol went off night duty. Tedious two- to three-day vigils at minor outports brought to light some of the suspected sources of illegally caught halibut both by setliners and trawlers. Despite the utmost willingness and diligence of the few officers available, most were not temperamentally equipped to carry out nonroutine patrols.

They were frequently provided with clues requiring immediate action, but they would not be able to pursue them because of other demands upon their time. The opportunity for seizure would be lost. Notwithstanding, there was a large number of apprehensions and convictions secured in the 1930s and 1940s by federal and state officers, and this certainly exercised some deterrent effect.

It should also be remembered that in the 1930s and later years, there was no radar nor effective air patrol. Patrol vessels were few in number, sometimes slow and were usually diverted in the summer months to the patrol of the more valuable salmon fishery. Bad weather made halibut patrol a difficult assignment. It was a tedious task at the best. At times it was difficult to secure convictions in local courts if the infraction did not involve an emotional consideration such as catching *baby* halibut in a *nursery* area or fishing on a *spawning* ground. Long days at sea for the patrol ship often ended with a discouraging acquittal, a suspended sentence, or a minimal fine for the offender. Frequently the fish involved, worth several thousands of dollars, would not even be confiscated. In some jurisdictions fines levied would be remitted at a later date, depending upon the economical and political climate.

During the years a program of patient persuasion was pursued by the commission to reduce the level of violations generally. Eventually, even the least cooperative of the dealers attempted to make available reasonably good records of their total purchases, including that of illegal fish. The conditions that prevailed in the 1930s and 1940s did not continue in the North American fishery. This is not to imply that there were no violations after that time, but they were of a much lesser magnitude. They were soon greatly overshadowed in the late 1950s by the unaccounted losses resulting from foreign fishing. Minimal estimates of these are reported in Halibut Commission Reports Nos. 60 and 61.

14 HALIBUT RESEARCH AND THE COMMISSION

Without an understanding of the life cycle of the species and of the condition of the stocks, efforts to manage the halibut resource or to evaluate the effects of such management would have been of little avail. This was recognized in the very early discussions on regulation of the fishery. The aborted 1919 Draft Treaty required that a "thorough joint investigation be made into the life history of the Pacific Halibut," and research has been a primary requirement of each subsequent Halibut convention.

Under the 1923 convention the function of the commission was primarily investigational. It was directed to

> . . . make a thorough *investigation* into the life history Pacific halibut . . . and such *investigation* shall be undertaken as soon as possible . . . [the commission shall] report the results of its *investigations* to the two governments. . . .*

In addition the commission was to make recommendations as to the regulation of the fishery and changes in the closed season.

The 1930 convention further emphasized investigation. Unlike that of 1923, it stipulated that changes in the closed season could be made by the commission only "when it finds *after investigation* such changes are necessary. . . ." It continued the general requirement that the commission "*make such investigations* as are necessary into the life history of the halibut in the convention waters. . . ."*

The 1937 convention had the same investigational requirements as that of 1930.

In the 1953 convention the research requirements were stated even more specifically than theretofore. While it was implied in earlier conventions, the 1953 convention clearly stated that the stocks were to be developed to levels which would permit the maximum sustainable yield. It goes without saying that the determination of such yields and stock levels requires both extensive investigations and basic research. The 1953 convention also continued the long-standing directive that the commission "shall make such *investigations* as are necessary in the life history of the halibut in convention waters,"* but included the further requirement that it should publish a report not only of its activities but also "of its investigations from time to time." It is to be noted, however, that although the requirement for the publication of investigational reports was not specifically stated in conventions prior to that of 1953 many scientific reports were nevertheless published prior to that time.

The 1953 convention, unlike its predecessors, also stated that only "after investigation has indicated such action to be necessary" could the commission exercise its regulatory functions such as dividing the convention waters into areas and setting catch or size limits. This was to guard against any capricious or ill-considered act on the part of the commission. The commission was thus made accountable.

While the foregoing directives provided the commission with its terms of reference, the need for extensive investigations was of course obvious from the outset of the work in 1924. However, on several occasions the presence of those stated directives in the conventions were of value. Very early there were persons in government who believed that the commission's function would terminate on the presentation in 1928 of its first recommendations to the governments as to the regulation of the fishery. Some felt that the fishery could then be managed under

*Author's italics.

annual regulations enacted concurrently by the two parties.

It was pointed out that the 1923 convention placed no time limit upon the completion of a "thorough investigation," nor did it state that the commission should make only *one* report on its studies or only *one* set of recommendations as to regulation. Once it was understood that the "preservation and development" of the resource, the prime objective of the convention, would have to be a continuing responsibility for some international body, the idea that the commission had completed its task with its first report to the two parties ceased being an official view. This does not mean that in the next 45 years the question of terminating the Halibut Commission or its scientific investigations was not brought up by individuals both in and out of governments.

While the conventions discouraged capricious actions on part of the commission, their restrictions on a few occasions might have unduly hindered needed action. There were occasions where the commission by necessity had to act on the basis of the evidence at hand, incomplete though it might have been, to avert some trend that could have been irreversible.

From the above it is apparent why a large proportion of staff time and about 85% of its funds were expended over the 50 years of its existence since 1924 on applied research. Fiscal necessity and the directives of the various conventions kept the commission from engaging in the pure research which more properly belongs in universities.

Despite the importance of the subject, only the main features of the life cycle of the Pacific halibut can be described here. However, a few references will also be made to some of the findings of workers other than of the Halibut Commission, both recent and early, including some comparisons with Atlantic halibut. More detailed and extensive information on the Pacific halibut are available in the scientific reports of the Halibut Commission.

REPRODUCTION AND DEVELOPMENT

In the Northeastern Pacific spawning takes place over a fairly protracted period of the year. From November to March mature males and females about to spawn are found concentrated at places along the continental shelf at depths from 125 to 250 fathoms (228 to 456 meters) deep. Spawning also occurs to a lesser degree at many other locations.

In the Bering Sea, according to Novikof (1964), spawning occurs from October to March, and on the Asiatic coast spawning was observed from December to January. McCracken (1958) concluded that available observations suggest that spawning occurs off Nova Scotia in the winter and early spring. In the Northeast Atlantic according to Taning (1936), it extends as late as May. Devold (1938) reported halibut spawning from November to April off Norway. Moiseev (1953) noted that "halibuts begin spawning during late autumn or in the winter (August to December)" in the Northwestern (Asiatic) Pacific. Rae (1959) indicated that the spawning time in the North Sea to Iceland area was from April to May, and at east Greenland it was from the last days of May into the first week of June.

FIGURE 84. The spawning grounds of the North Pacific Ocean: (A) Cape St. James; (B) Whaleback; (C) Cape Omanney Spit; (D) Eastern Spit; (E) Yakutat Spit; (F) the "W"; (G) Seward Gully; (H) Trinity Islands; and (I) Chirikof Island.
F.H. Bell

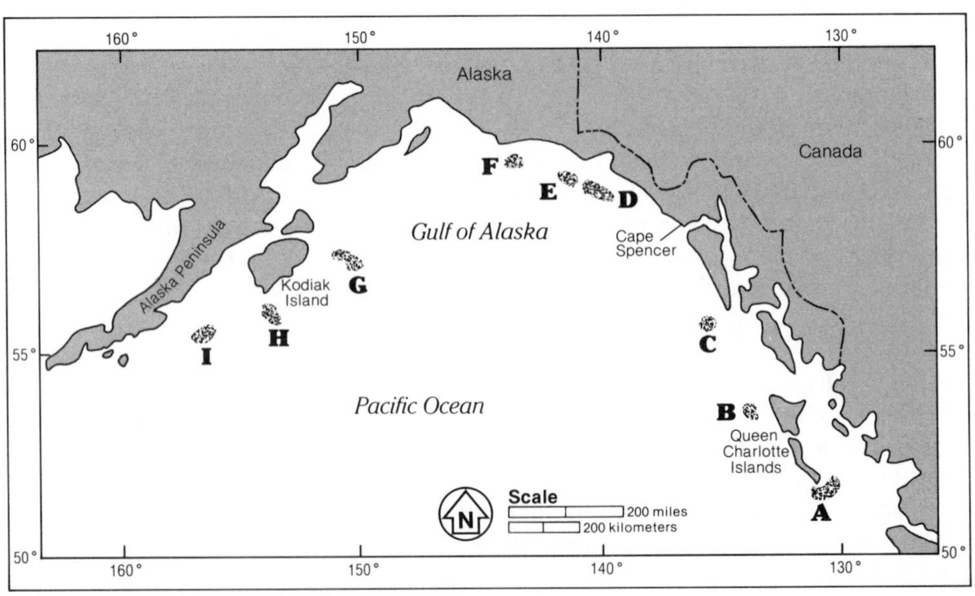

It is evident from the foregoing that, generally, halibut spawn in the winter. Due to the nature of the observations one cannot state with precision when the height of the activity occurs in the various regions nor exactly the length of the period over which the process occurs in each area. Off the British Columbia coast exhaustive studies from 1935 to 1946 of the availability of first and second stage ova off Cape St. James at the south end of Moresby Island indicated that spawning in that area took place between December 1 and March 10 with the height of the activity occurring from January 1 to February 10 (Figure 85; Van Cleve, R. and Seymour, A.H., 1953).

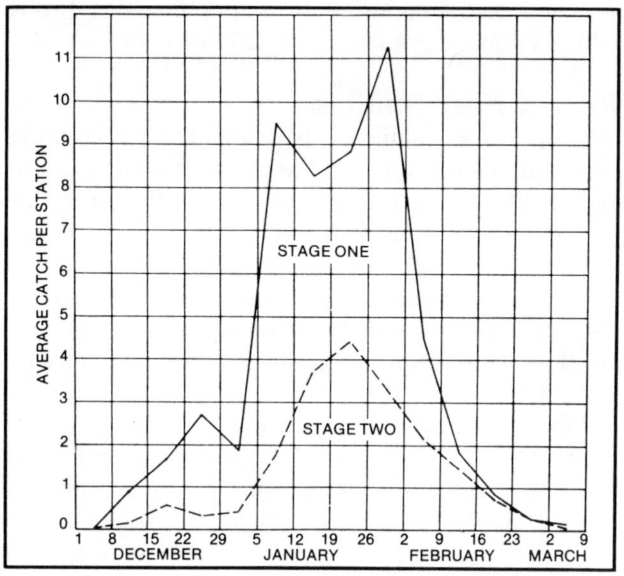

FIGURE 85. The average corrected catch of Stage 1 and Stage 2 eggs per station per week for all years of sampling on the Cape St. James grounds off British Columbia.
International Pacific Halibut Commission

There seems to be little confirmation for summer spawning from June to September such as reported in the Northwest Atlantic by Goode (1887) on the basis of random statements of fishermen. They reported halibut spawning in late December, January, April, May and June on Le Have and Banquereau banks and had often been seen in August and well into September on Sable Island grounds off Nova Scotia with running milt and extrudable ova. The spawn of the halibut was reported "to have been a favorite food of southern New England though never eaten by those of Cape Ann." It is quite possible that with some individual fish the ova may develop very early in the fall and be classed as a spawning fish, i.e., with eggs ready to be extruded.

The Halibut Commission (IPHC Report 66) generally confirmed Thompson's finding (1916) that the average age of first maturity for female halibut is about 12 years with some maturing as young as 8 years of age and some as old as 16 years. Males on the other hand, mature much younger with an average age of first maturity of about 7 to 8 years of age. Any attempt to refine or change the foregoing estimates is rather difficult, if not futile, because sampling errors, growth rate changes and difficulty of determining maturity at some seasons of the year.

Novikov (1964) reported that in the Bering Sea most males mature at an age of 7 to 13 years and most females at 9 to 15 years. He reports isolated cases of males maturing at 4 years and females at 6 years of age.

McCracken (1958) observed on the basis of limited data that Atlantic females off Nova Scotia matured between 10 to 12 years and males slightly younger, from 8 to 11 years. This confirmed the findings of Jespersen (1917) in the Northeast Atlantic.

In light of the above it is probable that with adequate samples of material from broad areas in each of the regions mentioned above and the use of some objective means of determining maturity rather than the visual one would probably find no statistically valid difference between regions in the average age of first maturity.

Spawning appears to occur annually, although Novikof (1964) was of the view that they spawned once in two years or even more seldom. Vernidub (1936) was also of the opinion that it was every two or even three years. These views may have arisen from the fact that at the end of the spawning season considerable numbers of fish have ovaries with ova at the early stages of development but did not presumably spawn that year. That they are not merely regenerating from earlier spawning in the season is based on the assumption that the regenerative process is prolonged, hence the two or more years interval between spawning.

When one considers that some individual halibut might not spawn for the first time until 16 years or even older and that probably only about 50% of the females are mature at about age 12, it is no small wonder that a large number of individuals will not have spawned at the end of the spawning season and whose ova will be in the early stages of development. The assumption that the process of regeneration after spawning is a prolonged one is not borne out by the fact that in the summer one cannot distinguish with any assurance size for size, between females that had spawned the previous winter and those that have never spawned.

No other observer seems to have the same views as do Novikof and Vernidub on the frequency of spawning, although the evidence in these matters is always rather vague and at the best, circumstantial. It should be noted, however, that on the assumption that his determinations were valid, Devold (1938) found that the annual rings in halibut otoliths undergo a change in character at a certain age. Like others have done with cod, plaice, and other species, Devold attributed such changes to the onset of spawning. Such spawning "marks" occur each year after their first appearance, not every two or three years as one would expect if halibut spawned other than annually as suggested by Novikof and Vernidub.

Sex Ratio

The proportion of the sexes in the commercial catches varies from ground to ground and by the season of the year. Report No. 66 of the Halibut Commission attempted to show that there had been an 8% increase in the percentage of females on the grounds west of Cape Spencer between the 1950s and early 1970s. This would have been a radical change in such a basic population parameter.

However, in the 1960s a large proportion of the samples were from catches off Chirikof Island made during bait studies in which a high proportion of octopus bait was used. Vessels using a high proportion of octopus bait had a large proportion of large fish (over 60 pounds) in their catches from the same grounds resulting in a high proportion of females. After properly weighing the data from the grounds west of Cape Spencer, no statistically significant change in the sex ratio was observable. In fact the data showed a slight decline from 69.5% in the 1950s to 68% in the early 1970s, not the 8% increase presented in IPHC Report No. 66.

What the true proportion of sexes may be in the total population cannot be determined from the commercial catches or from the landings. If a true sample of the entire population could be studied the probability would be that the sexes are equally divided.

Also, despite the errors inherent in determining an average age of first maturity, the Halibut Commission Annual Report 1978 claims that there has been a decrease in the age of maturity of females in Area 3 from 12 to 11 years with a lesser change in Area 2. The reality of any such change rests upon rather tenuous grounds.

Egg Production

Egg production or fecundity of the halibut is very high. This might be expected of a species that spawns in the open ocean, whose eggs and larvae drift long distances and one in which the young receive no parental care, all such features being conducive to large losses. Egg counts are proportionate to the weight of the female and do not seem to be age-dependent. A 50-pound halibut may have about 750,000 eggs, a 100-pounder from 1.5 to 2 million. The largest count made on Pacific halibut was 4,001,350 for a 199-centimeter individual (IPHC Report 66). Large counts have been reported by various workers. G. Browne Goode (1887) reported a calculated count of 2,182,773 for a 190- to 200-pound halibut caught September 13 on Banquereau Bank off New England. Malm (1877) reported a 135-pound Baltic halibut with 3.5 million eggs in ovaries that weighed 35.5 pounds. Fulton (1904) quotes Brook with a 91-pound female with 1,327,000 eggs.

Recruitment Studies

Despite the pitfalls of oversimplifying a complex problem, management of the halibut resource can be reduced to two basic objectives: first, to assure that the optimum numbers of recruits are produced each year; and second, that the optimum use is made of such recruits when they reach a fishable size. From the beginning most of the investigations of the commission have been directed to these two primary objectives.

One of the first tasks was to make a comprehensive study of the early life of the species, from the egg stage to the time the larvae change or metamorphose into an "adult" form. Since very little was then known regarding that aspect of the halibut an adequate description of the eggs and larval stages and where they might be found were first prerequisites.

Halibut eggs had been described from the Northwest Atlantic by several early workers including Brook (1886) and Schmidt (1904), the latter having reported upon both the eggs and some of the larval stages. The Halibut Commission, using precautions against distortion arising from preservation, found that a large and representative sample of eggs from a spawning female had an average diameter of 3.17 millimeters (about ⅛ inch) with a range from 2.05 to 3.85 millimeters. For fish, such a size of ova was relatively large, but at least three other species of fish in the Northeastern Pacific have eggs as large or even larger. The halibut eggs or ova are bathypelagic, being extruded and fertilized in proximity to the bottom but subsequently to rise and drift in the middle-to-upper water areas as shown in Figure 86.

At this fairly typical sampling section in the Gulf of Alaska (off Cape Cleare) eggs were found as deep as 680 meters. None were taken in the upper 100 meters of less saline and lighter water. Ninety-eight percent

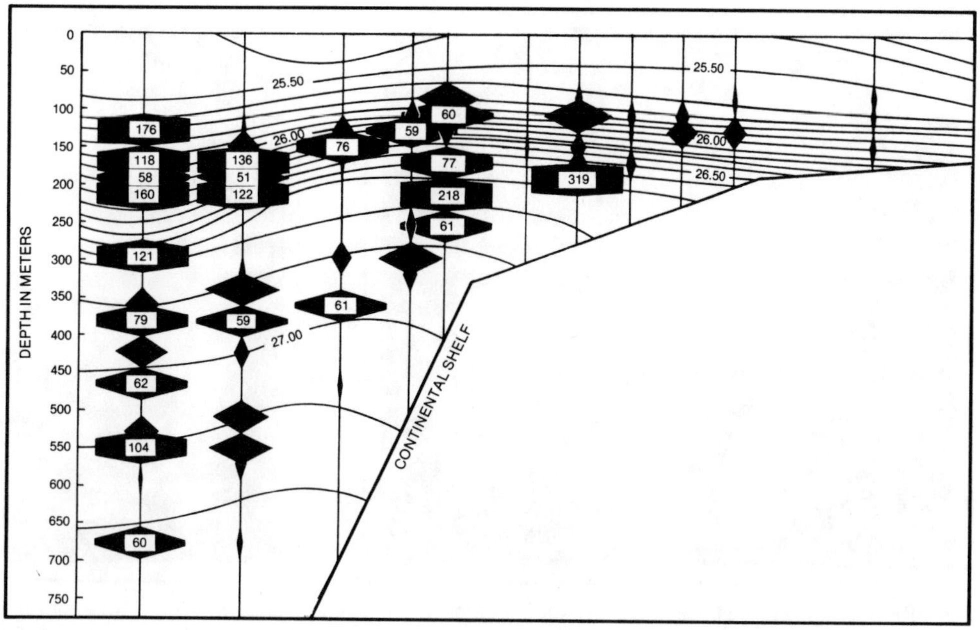

FIGURE 86. Depth distribution of halibut eggs for a typical sampling section (Cape Cleare) across the continental shelf showing water density contours.
International Pacific Halibut Commission

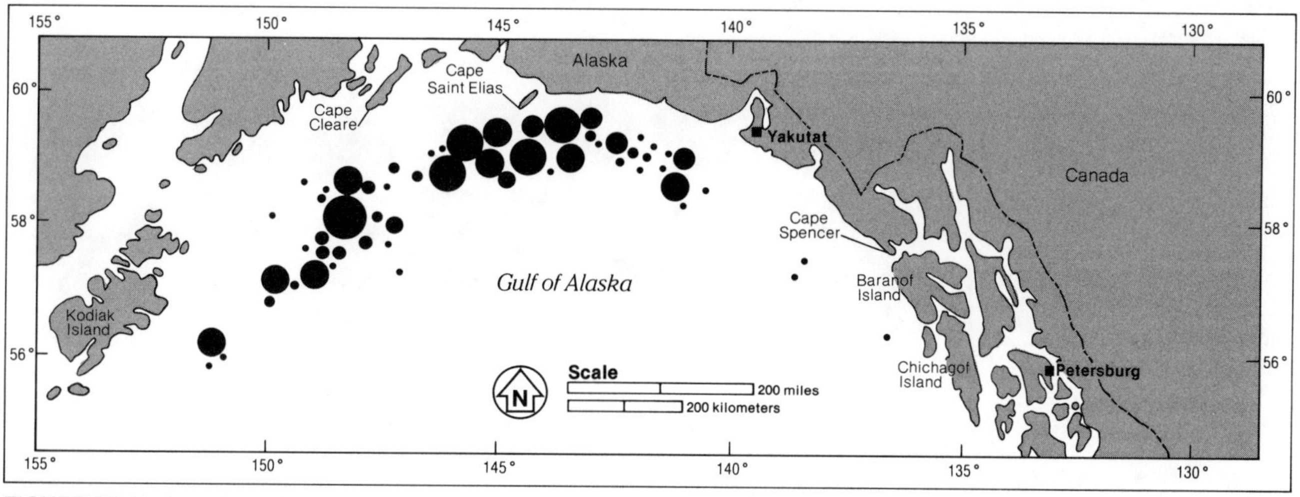

FIGURE 87. Horizontal distribution of all halibut ova. The area of each circle is proportional to the total number of eggs taken at that locality in all net hauls. A • equals one egg recovered.
International Pacific Halibut Commission

of the eggs were taken over depths greater than 275 meters and 67% over depths more than 400 meters.

In the Gulf of Alaska the eggs are transported by the westward flowing ocean currents, and their displacement off the spawning grounds is fairly rapid (Figure 87).

Development

The *germinal disc* of the egg goes through the normal processes of cell division (Figure 88) to form the embryo (Figure 89) that lives off the yolk that makes up the main mass of the egg.

After about 15 days they hatch, still living off the yolk sac (Figure 90).

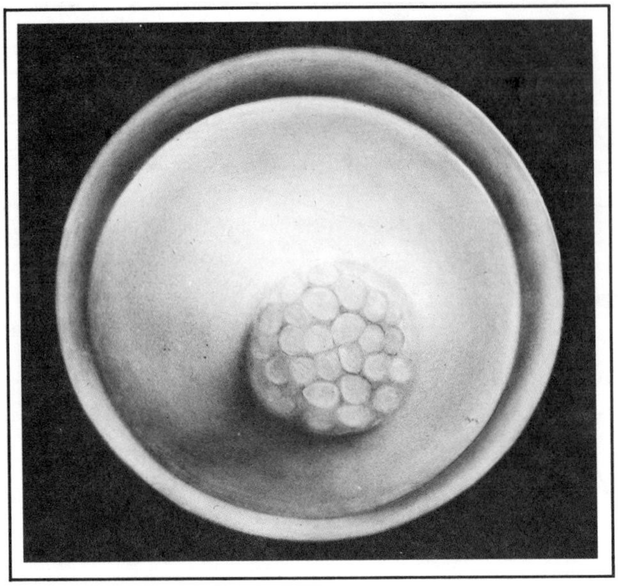

FIGURE 88. Early stage of halibut ovum showing germinal disc. Size: 3.3 millimeters (0.125 inch).
International Pacific Halibut Commission

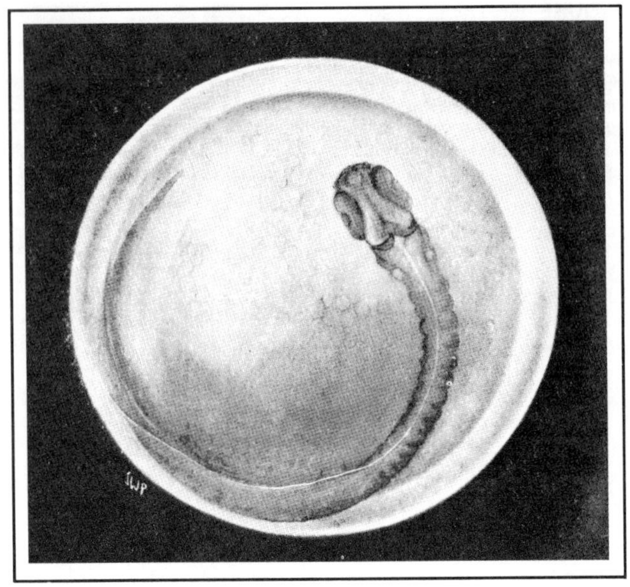

FIGURE 89. Halibut ovum showing the embryo shortly before hatching, shown 21 times natural size.
International Pacific Halibut Commission

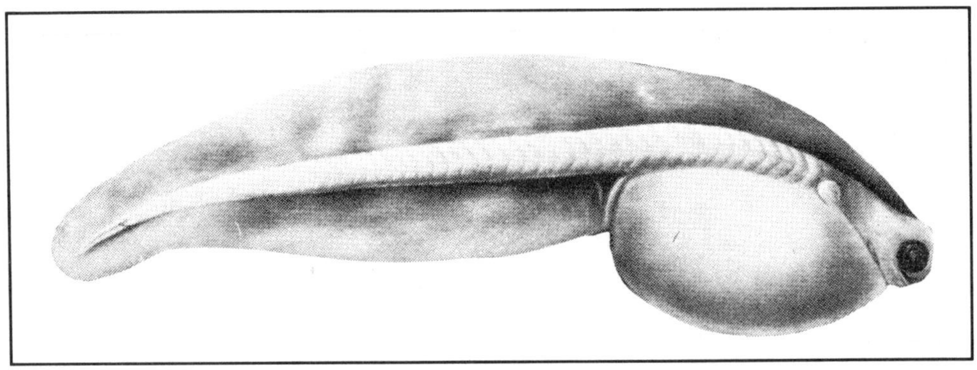

FIGURE 90. Newly-hatched halibut larva showing yolk sac. Size: 1 centimeter (0.38 inch).
International Pacific Halibut Commission

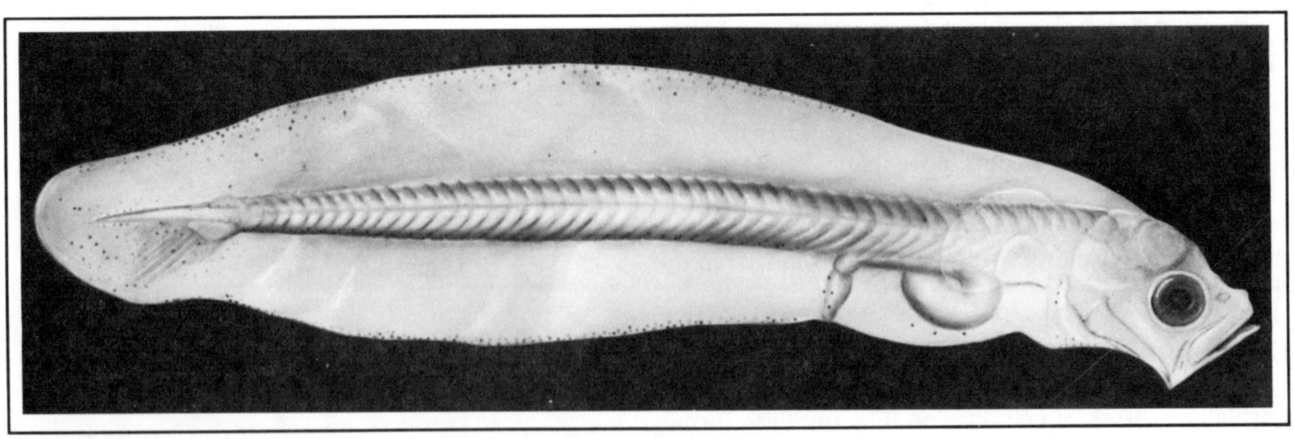

FIGURE 91. Halibut postlarva. Yolk has been absorbed. Size: 1.7 centimeters (0.67 inch).
International Pacific Halibut Commission

After absorption of the yolk the postlarva (Figure 91) must depend upon the external environment for its food.

Like the eggs the larvae and postlarvae continue to be free floating and are also transported many hundreds if not thousands of miles by the westward-moving ocean currents. The velocity of some parts of this westward-moving Alaska stream may reach as high as 10 knots per hour.

The center of gravity of the abundance of the early larval stages is still well over deep water outside the continental shelf (Figure 92).

While the larvae show some movement toward the upper water layers, 71% were over the abyssal depths beyond the slope. They are also displaced still farther west (Figure 93).

During the free floating stage covering about six months the postlarvae go through a number of developmental stages, two of which are indicated in Figures 94 and 95.

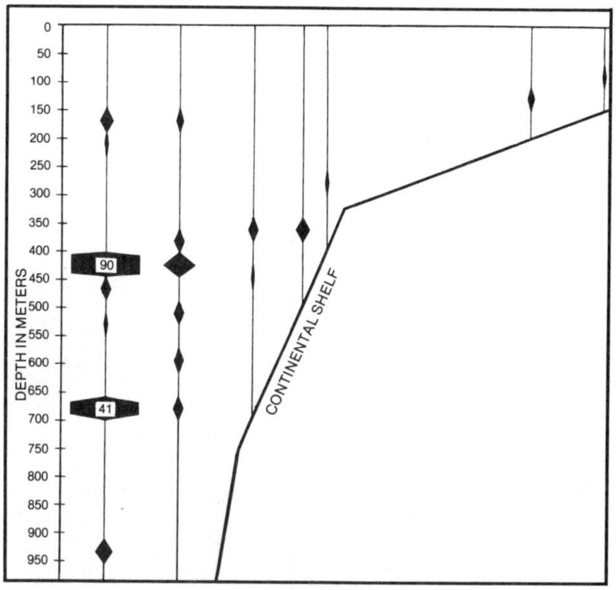

FIGURE 92. Depth distribution of early larval stages.
International Pacific Halibut Commission

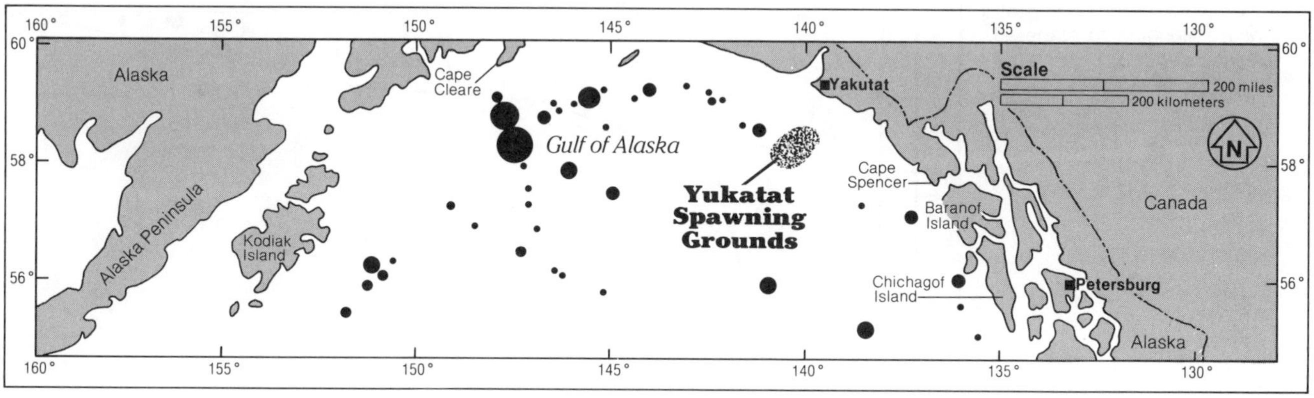

FIGURE 93. Horizontal distribution of all Stage 2 larvae. The area of each circle is proportional to the total number of larvae that have been taken in that locality. A • equals one larva recovered.
International Pacific Halibut Commission

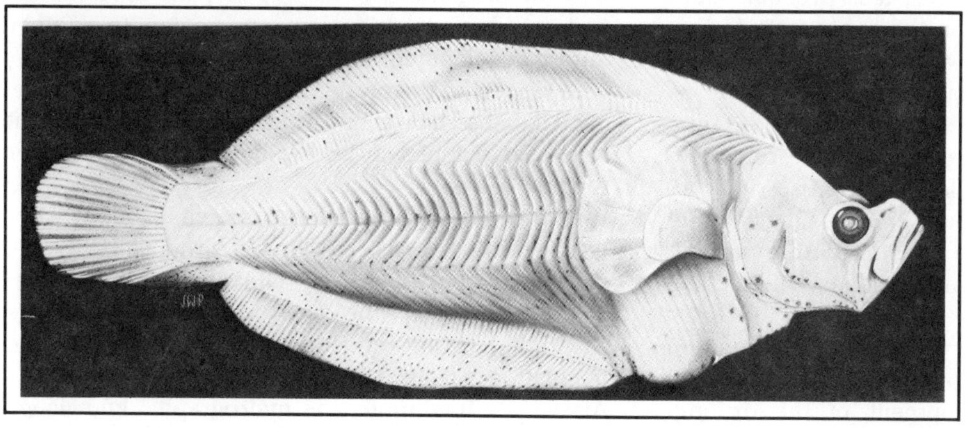

FIGURE 94. Halibut mid-postlarva. Size: 2.2 centimeters (0.87 inch).
International Pacific Halibut Commission

FIGURE 95. Halibut late postlarva, showing the beginning of eye migration. Size: 2.5 centimeters (0.98 inch).
International Pacific Halibut Commission

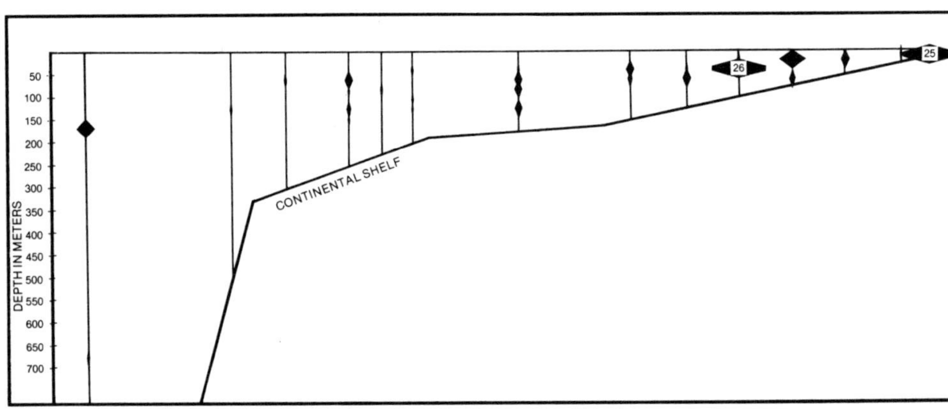

FIGURE 96. Distribution according to depth of all postlarvae belonging to stages 5 to 9 inclusive.
International Pacific Halibut Commission

Having risen into the surface water layers (Figure 96) the later stages tend to be propelled by the prevailing winds toward the shallower sections of the continental shelf. About 93% were taken in water averaging 12 meters. However, it is probable that young halibut settle down on the banks outside the bays and perhaps in deeper water than is indicated by their distribution in the net hauls.

The larvae undergo metamorphosis and commence their bottom existence distant from the spawning grounds as juvenile halibut possessing the characteristic adult form (Figure 97). Thus the floating eggs, the developing larvae and the postlarvae are dispersed far westward from the points where they were produced.

The whole early life history of the halibut from the time of spawning on the continental slope through fertilization and the development of the drifting stages of the eggs and larvae and their rise to the surface layers is closely adapted to the general ecology and movements of the water masses of the North Pacific Ocean and the Bering Sea.

From a practical standpoint, the knowledge regarding the size and locale of the spawning population and the number of its progeny, and the direction and speed of its drift from one region to another during development possesses important implications with respect to the proper husbandry of the resource. Such data had a direct bearing upon the management strategies pursued by the Halibut Commission over the past 45 years.

One of the considerations was whether the westward drifting eggs or larvae produced by the spawning populations in the Gulf of Alaska would circulate and enter the eastward moving offshore subarctic current and thereby contribute to the population off British Columbia and Southeastern Alaska. By integrating the hydrographic knowledge of those times with some studies of its own along with estimates of the rate of development under different

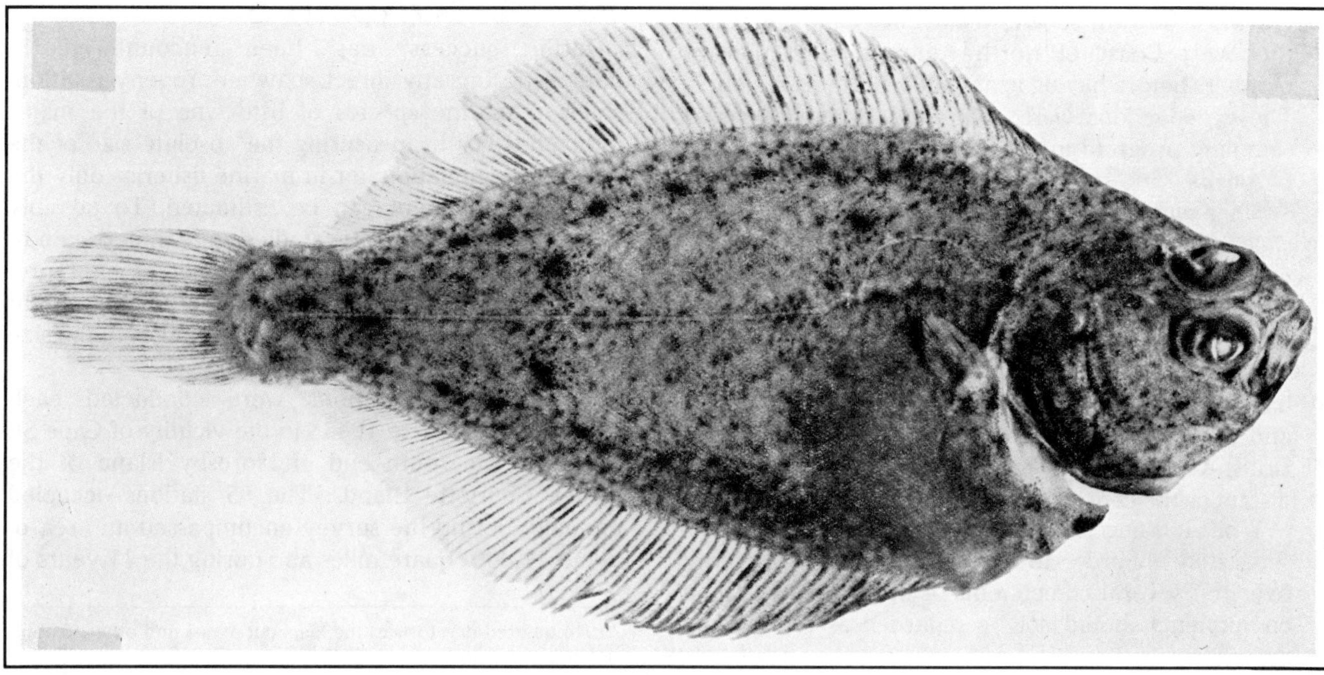

FIGURE 97. Young halibut (six times natural size). Size: 2.88 centimeters (1.13 inches).
International Pacific Halibut Commission

FIGURE 98. An adult halibut
International Pacific Halibut Commission

temperature conditions (Tagesgrade), the commission deduced that any larvae that might enter the eastward flowing subarctic current could not reach the West Coast of North America in significant numbers before having to assume a bottom existence. These deductions were confirmed by an extensive sampling program in the offshore waters of the Gulf of Alaska.

Since eggs and larvae produced on the edge of the shelf off British Columbia and Southeastern Alaska could enter the relatively fast moving subarctic current as it turns westward as the Alaska stream there must be some counterbalancing movement within the species. Tagging has shown that the balance is maintained by the migration of both adults and juveniles to the coasts of Southeastern Alaska, British Columbia and points south from places as far distant as the central Bering Sea.

While it is necessary to regulate the different stocks separately in order to spread the removals evenly over the several components of the population such components should not be regarded as biologically separate.

SPAWNER-PROGENY RELATIONSHIPS

Little success has been encountered in demonstrating any direct spawner-progeny relationships in marine species of fish. One of the major difficulties lies in measuring the absolute size of the spawning population, for in marine fisheries only the fishable component can be estimated. To possibly circumvent the difficulty, studies were undertaken to measure the variation in the number of eggs produced on the principal spawning grounds that might in turn give some indication of the changes in the size of spawning population in the region.

Intensive investigations were conducted each winter from 1937 to 1945* in the vicinity of Cape St. James at the south end of Moresby Island of the Queen Charlotte Islands. The 65 stations occupied annually during the survey encompassed an area of about 11,000 square miles and during the 11 years of

*1944 omitted due to wartime blackout at sea and other national defense restrictions.

TABLE 43. Number of Stations on the Cape St. James Grounds Sampled by Three-Day Periods from 1934-1946.

Date (3-day period ending)	1934 -35	1935 -36	1936 -37	1937 -38	1938 -39	1939 -40	1940 -41	1941 -42	1942 -43	1944 -45	1945 -46	Total
Dec. 7	—	—	—	—	7	8	—	—	—	—	—	15
10	—	—	2	—	11	9	—	4	—	—	—	26
13	—	—	4	—	16	3	—	3	—	—	—	26
16	—	—	2	—	14	4	—	10	—	—	—	30
19	6	—	4	—	19	10	—	6	—	—	—	45
22	10	—	0	—	0	11	—	5	—	—	—	26
25	1	—	0	—	0	0	—	0	—	—	—	1
28	0	—	6	—	0	0	—	0	—	—	—	6
31	7	—	4	—	0	1	8	0	—	—	—	20
Jan. 3	18	—	13	9	0	7	8	11	—	—	—	66
6	5	—	3	11	10	15	2	15	—	—	—	61
9	9	—	6	7	4	8	0	0	—	—	—	34
12	—	8	4	6	9	7	8	0	3	—	4	49
15	—	19	6	3	10	10	12	1	3	4	12	80
18	—	0	4	5	5	14	3	2	0	14	12	59
21	—	0	9	7	6	0	0	0	0	7	0	29
24	—	12	3	0	0	0	4	0	4	0	0	23
27	—	13	6	0	0	0	7	0	2	9	0	37
30	—	4	5	3	10	17	7	10	1	15	4	76
Feb. 2	—	—	3	10	4	0	1	10	0	2	11	41
5	13	—	—	6	5	11	6	7	0	0	5	53
8	12	—	—	4	4	—	4	8	6	5	9	52
11	3	—	—	16	9	—	10	5	5	1	—	49
14	8	—	—	1	—	—	0	8	2	0	—	19
17	3	—	—	0	—	—	10	12	2	0	—	27
20	—	—	—	8	—	—	20	12	6	5	—	51
23	—	—	—	11	—	—	3	2	4	4	—	24
26	—	—	—	1	—	—	14	—	—	1	—	16
Mar. 1	—	—	—	8	—	—	19	—	—	—	—	27
4	—	—	—	—	—	—	12	—	—	—	—	12
Total	95	56	84	116	143	135	158	131	38	67	57	1,080

sampling a total of 1,080 stations were sampled, each involving three net hauls. Their seasonal distribution each year is shown in Table 43.

As indicated by the abundance of eggs at their earliest stage of development, the spawning season was from December to March inclusive.

The abundance of eggs from year to year was calculated by several methods of which the density-contour method of Buchannan-Wollaston (1926) was apparently the most satisfactory. The calculated ova counts are shown in Table 44.

TABLE 44. Calculated Ova Counts from Cape St. James

Year	Count	Year	Count
1935	882	1941	719
1936	997	1942	431*
1937	2,774	1943	1,294
1938	1,508	1944	*
1939	1,149	1945	1,520
1940	1,420	1946	1,294

*The coverage in 1942 was not entirely satisfactory as operations had to be foreshortened due to hazards arising from hostilities with Japan. In 1944 operations were finally abandoned due to blackout of ships at sea. Without day and night operations, a single vessel could not adequately sample the area, and fiscal restraints precluded chartering and manning two vessels.

The most noteworthy feature of the above ova counts was the high value in 1937. This apparently persisted through the juvenile stage and produced a strong year-class that was observable in the commercial fishery in 1943 and 1944. However, despite the comprehensive character of the investigations, it was not possible to obtain any statistically significant relationships between egg production and the abundance of spawning adults in the region.

The variations in egg abundance did, however, show some general relationships to the changing levels of the halibut stocks. The relative scarcity of eggs in the samples might well have contributed to the failure to secure statistically significant correlations with the size of the resulting year-classes.

There was, however, a negative correlation between egg abundance and water temperatures which was significant at the 5% level. It should not be construed that high temperatures might be lethal. The correlation is more likely due to the greater rate of development at higher temperatures reducing the time that the egg would be at any one stage of development which would cause the number of such eggs on the grounds at any one time to decline.

Although the primary objective of the 11-year study was thus not attained, many useful findings resulted from the prolonged study. It was

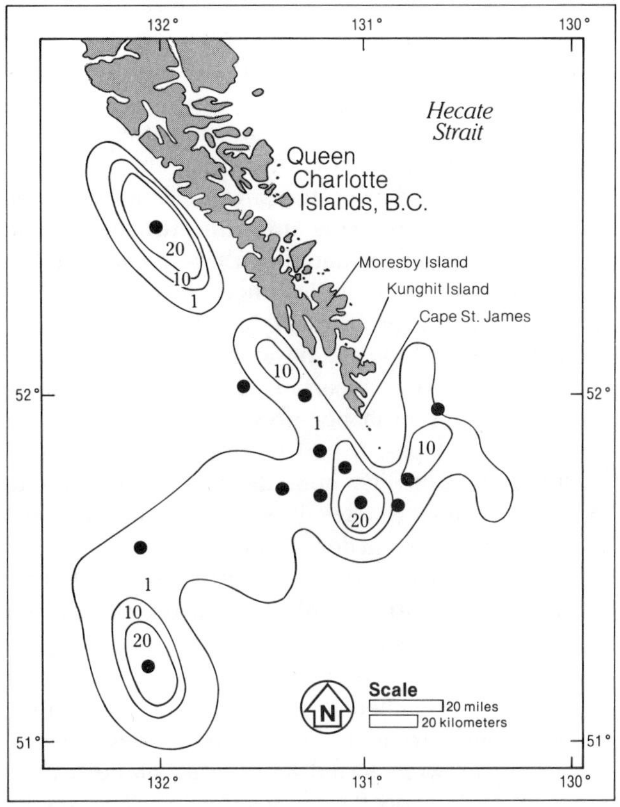

FIGURE 99. The centers of gravity (●) of halibut egg abundance off Cape St. James from 1937 to 1945. The contour lines show the density of distribution in a typical year.
International Pacific Halibut Commission

demonstrated that plankton nets and hydrographic equipment could be developed that could withstand the rigorous North Pacific weather conditions that prevailed during the 11 winters of operation. It was also shown that there was a remarkable constancy in the location of the center of gravity of the egg abundance from year to year (Figure 99).

Such constancy suggested that the sampling design and procedures as well as the method of analyzing the data were probably valid. Also, should the area be monitored in the future, the above findings would provide a data base for comparisons.

RECRUITMENT OF YOUNG AND JUVENILE HALIBUT

As a matter of definition the term juvenile is used to describe those halibut between the time of change in form or metamorphosis from when the larval stage is complete and the age at which halibut enter the North American setline halibut fishery, which is usually from four to seven years of age. The age of entry depends, amongst other factors, upon the prevailing growth rate and the condition of the stocks. In the early 1970s, the growth rate was high and the stocks of halibut were low, both factors being conducive to an entry at an early age into the setline fishery.

In view of the high cost and the rather indeterminate results of studies attempting to relate egg abundance to the size of the spawning stock or to the forthcoming recruitment into the fishable stock, investigations were initiated in 1955 to evaluate the size of the stock of the younger juvenile halibut. Approaching the problem of recruitment at a later stage than that of the egg or larva had the advantage of by-passing the uncertain and highly variable mortalities that probably prevail at the earlier or intermediate stages. It had the further value that it would throw some light upon the losses that accrue to the young halibut as a result of their capture by trawl gear, both foreign and domestic.

The investigations were begun on a modest scale commensurate with available funds first in 1955 off the coast of British Columbia. It was necessary to ascertain where the young could be taken from year to year and in numbers large enough for quantitative studies of the strengths of the year-classes. Several types of gear were tested and the small-mesh otter trawl proved to be the most effective. The reconnaissance survey was extended to Southeastern Alaska in 1956, Kodiak Island in 1957 and the Shumagin Islands in 1958. In the following years most of the sampling was concentrated at those locations that appeared to possess adequate and consistent concentrations of juvenile for quantitative baseline studies of year-class strengths and growth. Some reconnaissance work was continued.

The development and increasing extent of the program is evident in Table 45.

In the early 1960s the abundance and destiny of the young or juvenile halibut in the Gulf of Alaska and in the Bering Sea came to possess an added significance. Russian and Japanese bottom fishing in the Bering Sea since the late 1950s was beginning to expand into the Gulf of Alaska. The governments of Canada and the United States were concerned as to what the potentialities of the Gulf of Alaska were for bottom trawling and to what effect any such fishing might have upon the supply of adult and juvenile halibut in that region.

At the request of the two governments, the Halibut Commission began on May 1, 1961, and continuing until 1963, a comprehensive year-round investigation of the demersal fish stocks on the continental shelf from Cape Spencer to Unimak Pass a distance of 1,030 miles along the coast and involving about 65,000 square miles of shelf area in the Gulf of Alaska.

In each region a round of sampling was conducted in the spring, summer, and winter to ascertain the seasonal distribution and availability of halibut and other demersal species to a trawling operation. The two-year study required the continuous charter of three otter trawlers with crews and extra field personnel. Approximately 1,560 stations were occupied in the two years. The intensive sampling pattern of the year-round operation and the broad area covered qualified the survey as the most comprehensive of its type ever conducted anywhere (Figures 100 and 101). The two countries provided the Halibut Commission $635,805.75 equally divided to cover operational costs. The commission supplied the supervisory staff, the direction and administration of the entire program and reports on the results to the governments.

In the Gulf of Alaska young or juvenile halibut of ages two to four were found distributed generally throughout the entire sampling area out to a depth of

TABLE 45. Summary of Field Work Conducted on the Juvenile Populations

Year	No. hauls	No. halibut caught	Region
1955	54	133	British Columbia Coast
1956	54	156	British Columbia & SE Alaska
1957	58	1,123	British Columbia Coast to Kodiak Island
1958	50	4,663	British Columbia Coast to Shumagin Islands
1959	79	5,604	British Columbia Coast to Kodiak Island
1960	214	11,462	British Columbia Coast to Trinity Islands
1961	165	9,250	British Columbia Coast to Chirikof Island
1962	124	9,495	British Columbia Coast to Chirikof Island
1963	176	6,164	British Columbia Coast to Shumagin Islands
1964	NA	8,679	SE Alaska to Chirikof Island
1965	349	7,857	British Columbia Coast to Bering Sea
1966	369	12,576	SE Alaska to Bering Sea
1967	389	26,335	SE Alaska to Bering Sea
1968	251	10,377	SE Alaska to Bering Sea
1969	251	11,149	SE Alaska to Bering Sea
1970	227	7,992	SE Alaska to Bering Sea

Not available after 1970

100 fathoms (182 meters) and occasionally deeper. There was a pronounced tendency for their abundance to be greater at depths 60 fathoms (110 meters) or less and to decrease in deeper water. They were taken at about 75 to 80% of all trawlable stations occupied (Figure 100). The halibut during the summer phase of the program represented 6% of the total weight of fish caught. In the program there was no targeting on halibut such as would occur in a commercial fishery where halibut is by far the most valuable species in the catch. A 6% by weight representation would be considerably higher on a value basis notwithstanding the fact that the average size of the halibut was low. The average size of juveniles taken by trawlers produces halibut fillets of an acceptable market size.

The proportion of halibut in the catches in the eastern portion of the Gulf of Alaska, from Icy Bay to Cape Spencer, was 3% lower than farther west where in some sections of the shelf it reached 8%. The incidence of halibut under 65 centimeters in length shown by the catch per standard haul was much lower east of Kodiak Island than from Kodiak Island west to Unimak Pass.

In 1963 the Halibut Commission as part of its regular investigational program conducted a trawl survey in the southeastern Bering Sea in the early summer when the juveniles first become available on the shallow flats. The stocks of juveniles both in the southeastern Bering Sea and the Gulf of Alaska have been monitored annually since.

In the Gulf of Alaska, although the abundance of juvenile halibut as a group showed a lesser decline than in the Bering Sea, the larger individuals in that category declined sharply in the gulf. These are the sizes that are taken in great numbers by the foreign trawl fishery and by the small hooks used in the Japanese setline fishery for blackcod. Such destruc-

TABLE 46. Number of Hauls, Catches of Fish and of Halibut in the Trawl Survey During the Summers of 1961 and 1962

Section of Coast	No. of Hauls	Catch in 1,000's pounds		Catch Per Haul (lbs.)		Percentage of Halibut in Catch	Halibut under 65 cm		Halibut over 65 cm	
		All Fish	Halibut	All Fish	Halibut		Number	Average No. per Haul	Number	Average No. per Haul
Unimak Pass to Shumagin Is.	92	52	4	567	41	7%	1,428	16	83	1
Shumagin Is. to Chirikof Is.	119	97	5	810	41	5%	1,031	9	124	1
Chirikof Is. to E. Kodiak Is.	198	164	13	830	64	8%	3,407	17	459	2
E. Kodiak Is. to Cape Hinchinbrook	119	77	4	645	32	5%	300	2	128	1
Cape Hinchinbrook to Icy Bay	54	39	2	727	41	6%	119	2	92	1
Icy Bay to Cape Spencer	33	39	1	1,180	35	3%	148	11	40	1
Total	615	468	29	760	47	6%	6,433	10	926	1

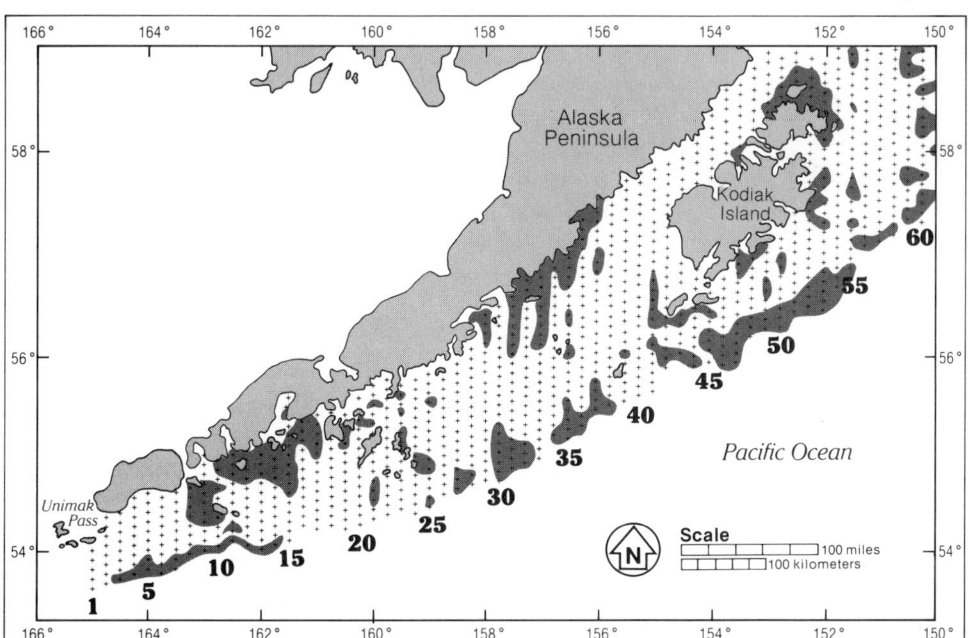

FIGURE 100. Patterns of stations sampled between the eastern end of Kodiak Island and Cape Spencer, Alaska, during the 1962-1963 survey. Nontrawlable areas are shown with dark shading. Lines of stations shown by plus (+) signs are numbered from left to right with every fifth line identified.
International Pacific Halibut Commission

tion of recruits that would have later been available to the North American setline halibut fishery has been a disaster.

In the southeastern Bering Sea the juveniles are found in relatively dense congregations that shift with the season. In the Gulf of Alaska they are fairly widely dispersed over the entire continental shelf. The concentrated density of the juveniles in the Bering Sea proved to be their undoing. They probably numbered about 60 million or more in the early 1960s. Since halibut and yellowfin sole are closely associated on the grounds, the removal of 1.3 billion pounds of yellowfin sole in 1961 alone by the Japanese and USSR trawl fleets also decimated the large population of young halibut. This was disastrous for the future recruits to the North American halibut fishery not only in the Bering Sea but in Areas 2 and 3 as well.

Side-by-side fishing with foreign trawlers and Danish seiners fishing "on-bottom" gear as well as direct observation of their catches leaves little doubt

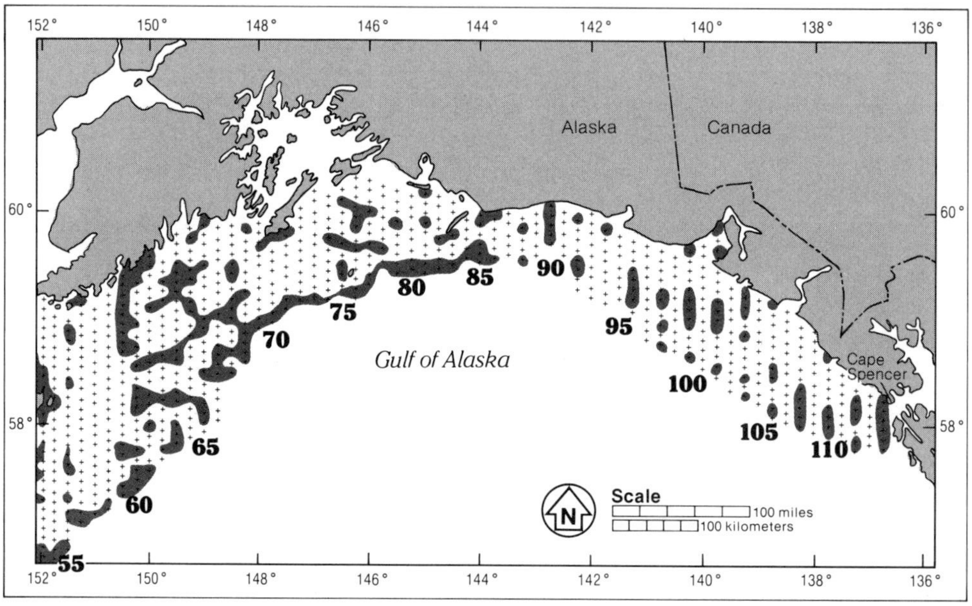

FIGURE 101. Pattern of stations sampled between Unimak Pass and the eastern end of Kodiak Island, Alaska, during the 1961-1962 survey. Nontrawlable areas are shown with dark shading. Lines of stations shown by plus (+) signs are numbered from left to right with every fifth line identified.
International Pacific Halibut Commission

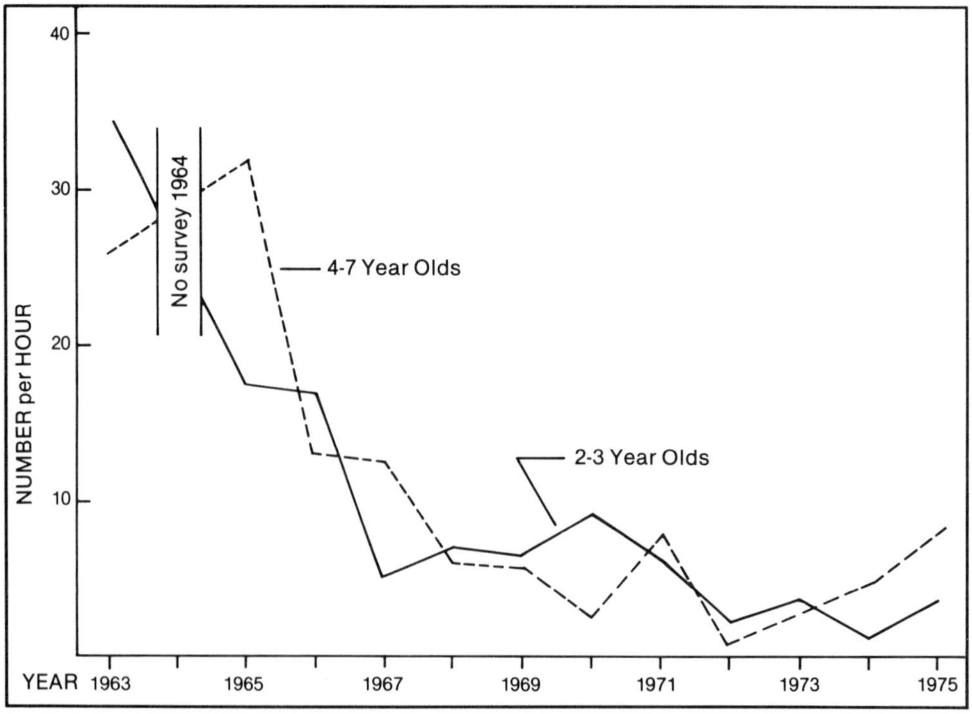

FIGURE 102. Catch per unit effort of juvenile halibut in IPHC surveys in the southeastern Bering Sea from 1963 to 1975.
F.H. Bell

as to the source and magnitude of the destruction. Such destruction is evident in the decline in the annual catch per standard trawl haul by research vessels operated each year in the region by the Halibut Commission (Figure 103).

It was assumed in Halibut Commission Report 61 that the two- to three-year olds were not vulnerable to trawl gear, and it was claimed that unfavorable environmental conditions or a reduction in the spawning stock caused by the North American setline fishery or trawling or some combination thereof caused the decline in the strengths of the year-classes.

Using an orthodox determination of year-class strengths based on the same data as used in Report No. 61 (Best, 1977) indicates that the production of young in Bering Sea did not decline as claimed but showed a slightly rising trend (Figure 104).

Year-class strengths must be determined by estimating the cumulative contribution of each year-class covering as wide a range of ages as possible. This tends to offset year-to-year variability in the availability of a year-class to the sampling gear.

The use of the correct method for estimating year-class strengths is particularly important in the southeastern Bering Sea as the population of young halibut is segregated according to size and age groups, and the concentrations shift their positions both in time and space. Consequently, with a sampling pattern that has become relatively fixed both as to time and space, it is very possible that individual year-classes will not be uniformily represented in the sample catches each year.

The crucial question is whether the remaining numbers of young in the Gulf of Alaska can alone provide sufficient recruits to the population of adults fished by the United States and Canadian setline fleets. Or will the reduction in recruits from the Bering Sea reinforce the deficiency of recruits prevailing in the Gulf of Alaska, particularly in those years of average or below-average year-class size?

Halibut Commission Report 65, Hoag and McNaughton, 1978, used catch and age composition to show that there had been a decline in year-class strengths in Areas 2 and 3 since the mid-1930s. While the method used avoided the use of CPUE with its attendant problems, the age data used also leaves much to be desired. It does, however, give the values in numbers rather than by weight.

For quantitative purposes the samples of age data are very meager and particularly so for a species as nonhomogeneous in its distribution as halibut. Also, the representativeness of the data can be distorted by the many long term changes that have occurred in the fishery, in bait and in the gear used, each of which affects the size of fish caught. Price trends in the marketplace for various sizes of halibut, stock levels and other changing conditions can also affect the sizes of fish sought by the fleets and also what may be

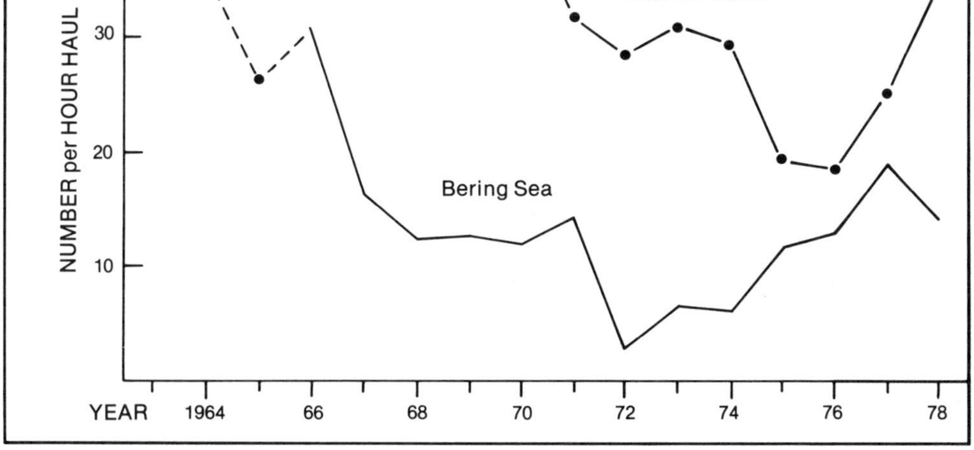

FIGURE 103. Catch of juvenile halibut per haul of research trawlers in the Bering Sea and the Gulf of Alaska.
International Pacific Halibut Commission

FIGURE 104. Strengths of year classes in the Bering Sea from 1964 to 1974.
F.H. Bell

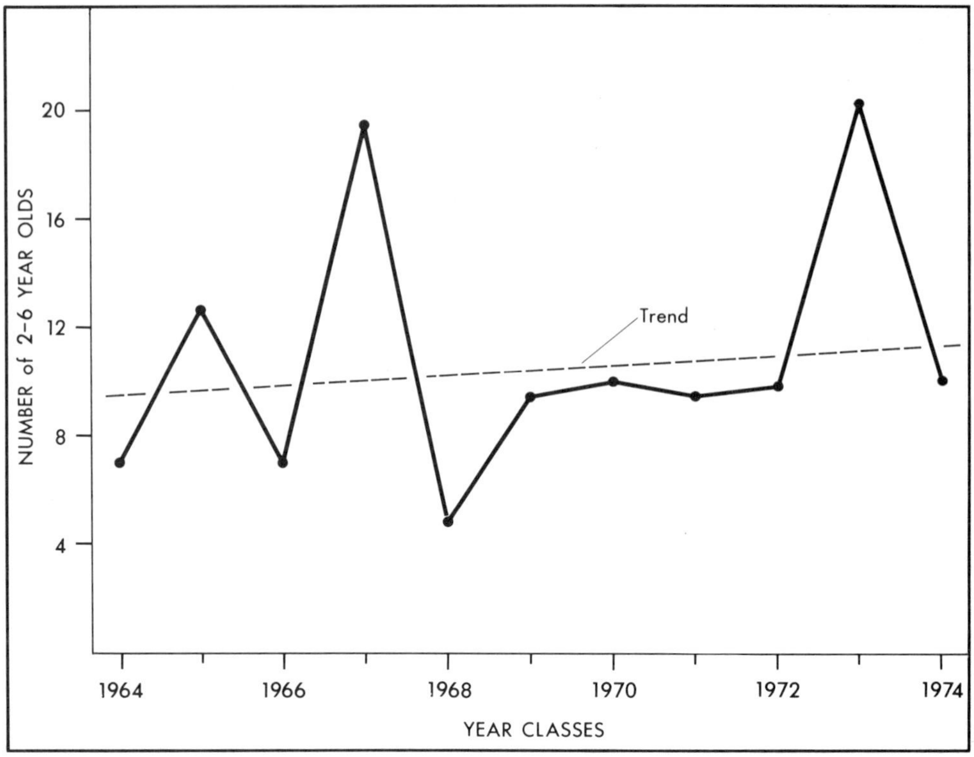

landed, whence comes the materials for age studies. Sections of the grounds fished over the years, each with its typical size of fish, have changed.

Under the foregoing circumstances, the age composition data is useful mainly for qualitative purpose, a purpose for which it was intended. As a base for determining numerical strengths of year-classes, it can only be indicative at least until it can be shown that the indications of such changes in year-class strengths can be related to some exogenous condition changing in the physical or biological environment.

In Area 2 the long term trend of the CPUE of small halibut, 5 to 10 pounds (chickens), remained relatively level from 1930 to 1950, although a slight upward trend is discernible (Figure 105). Since the growth rate in Area 2 also remained relatively unchanged during the same period, the numbers or strengths of the year-classes can be assumed to have been relatively stable, at least on the average.

The subsequent apparent decline in CPUE of chickens in Area 2 into the 1960s occurred when the 1954 to 1959 year-classes entered the fishery as fast-growing four-year-olds. They consequently passed through the chicken class rapidly, accounting for the downward trend in CPUE of that category in Figure 106 giving an illusion of a succession of weak year-classes.

In Area 2 from about 1930 to 1945 there was a succession of above average year-classes that passed through the fishery. They were particularly prominent on Goose Islands grounds, but though still present, their visibility in the catches declined as one went north in Hecate Strait.

In Area 3 conditions were different than in Area 2. In the former area there is usually a heavy rejection of chickens "at the rail roller" depending upon a number of variable economic and technological considerations both short and long term. Small fish spoil more readily on long or rough trips or with substandard ice, and they command a much lower market price. Also, when the fleet was operating a voluntary program of limiting the catch-per-man-per-trip the rejection of chickens in favor of the more valuable medium-sized fish was very high.

As the stocks rebuilt rejection of chickens at sea tended to increase as it was possible to secure full loads of the more valuable sizes. In Area 3, if fishing was good at the outset of a trip, chickens would either be "shacked" (shook) off the roller or released with a twist of the gaff hook. (The latter caused less damage to the fish.) On the last days of a trip chickens might be saved depending upon where the vessel planned to land and the expected price differential between chickens and the more valuable medium- and large-sized fish.

In Area 2 culling at sea was not the problem that it was in Area 3. The fleet in Area 2 fished on grounds that were usually closer to port than in Area 3 and the

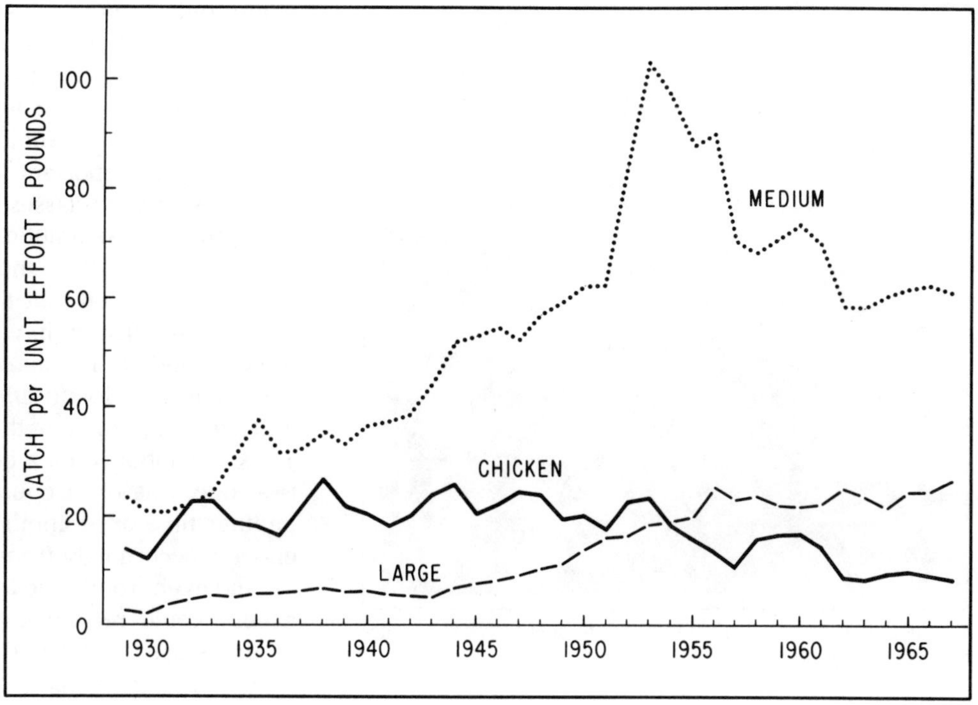

FIGURE 105. CPUE of halibut of the medium weight category (10-60 pounds), chickens (under 10 pounds) and large category (over 60 pounds) for all grounds south of Cape Spencer, Area 2.
F.H. Bell

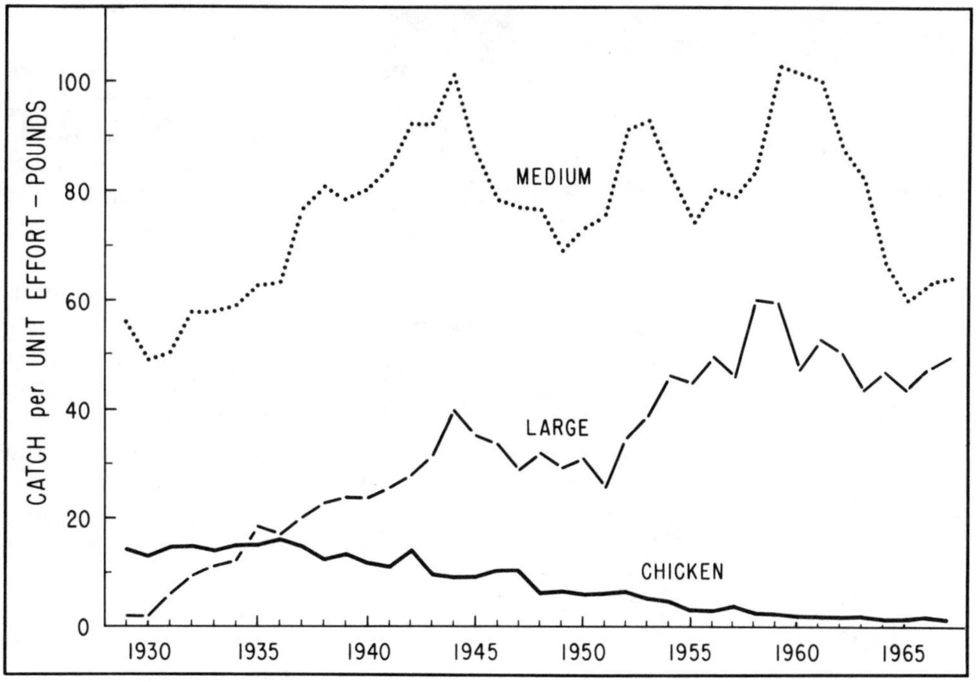

FIGURE 106. CPUE of halibut of the medium weight category (10-60 pounds), the large category (over 60 pounds) and chickens chickens (under 10 pounds) for all grounds west of Cape Spencer, including the Bering Sea.
F.H. Bell

small fish maintained their quality better. More important, however, when fishing in Area 2, one could move to other grounds or to other parts of the same ground and thus avoid the higher concentrations of small fish. Such shifting of grounds by a matter of a few miles can make a radical difference in the size composition of the catch.

In Area 3 the upward trend in the growth rate from earliest times, except for some indications of a leveling between 1935 and 1945 is, as in Area 2, inverse to the downward trend in the CPUE of chickens for the same reasons discussed above with respect to Area 2 and not due to a long-term decline in year-class strength, Figure 106.

Whether there was a succession of above-average year-classes in Area 3 in the early 1930s as in Area 2

FIGURE 107. A large halibut (with white side showing) — 629 pounds — from Greenland.
F.H. Bell

FIGURE 108. A large halibut — 455 pounds — caught with a handline near Chatham cannery in Southeastern Alaska. The man beside the fish was a cook on a nearby tender. His 6′2″ height gives an idea of the size of the fish.
Arve W. Pande II, courtesy of the International Pacific Halibut Commission

where rejection of such small fish is usually minimal, cannot be determined from this data. If they had been present they would probably have been heavily culled at the roller and would not have shown in the CPUE or samples of landings.

While the important question of trends of the strengths of year-classes cannot be resolved at this point, the vital assumptions made in the recent IPHC Report No. 65, 1978, should be rigorously tested. It should be stressed that the data used in that report came from a fishery in which selection of the fish by the gear and of the fishing locations by the fishermen can have a profound effect upon the trends of size and age composition of the landed catch. The fishable stock of halibut is not a homogeneous entity. What it is at one time or at one place is not the same as at another time or at another place, and the fisherman chooses accordingly from day to day and from set to set and even from skate to skate. Such selection in the fishery could readily impair the elegant analytical methods used in IPHC Report No. 65. Certainly the argument for a long-term decline in year-class strengths rests on some tenuous evidence concerning age composition.

SIZE AND GROWTH OF HALIBUT

Halibut are the largest of all the flounders, and its recorded maximum weight is probably greater than most if not of all the bony fishes. (Some ganoids such as sturgeon, some of the cartillaginous groups including sharks and rays, and some tunas and marlins do attain weights much in excess of the largest halibut.) The North American commercial catch of Pacific halibut, caught mostly by setline, consists of individuals that range from 5 to about 200 pounds, averaging about 25 to 30 pounds. Many larger individuals have been reported, and those 200 to 350 pounds would not be regarded as being uncommon but usually attain some recognition as trophy fish. The occurrence of large individuals is also reported in the North Atlantic, where they seem to average somewhat larger than in the North Pacific halibut.

The presence of so important a food-fish as the Halibut in America did not long escape the observations of the early English explorers. Capt. John Smith, in his "History of Virginia," wrote:

> There is a large sized fish called Halibut, or Turbot; some are taken so bigg that two men have much a doe to hall them into the boate; but there is such plenty, that the fisher men onely eate the heads & finnes, and throw away the bodies; such in

Paris would yeeld 5. or 6. crownes a peece: and this is no discommodity. (From G. Brown Goode, 1884, p. 309.)

Ehrenbaum (1936) described with some reservation a report from European waters of a 743-pound (337-kilogram) halibut quoted by Nilsson as having been taken in the Baltic Sea.

In December, 1966, a 609-pound (276.2-kilogram) eviscerated halibut was landed in Grimsby, England, by the trawler *Coldstream* from the White Sea grounds. This weight would be equivalent to over 700 pounds live weight. Another about 7 pounds less in live weight had been landed in Grimsby from the north coast of Iceland 31 years earlier on July 13, 1935 (*Fishing News,* July, 1935). The fish was 11 feet 11 inches long and 17.5 inches thick.

In 1963 an individual 9 feet 5 inches (287 centimeters) long with a 629-pound (286-kilogram) live weight was taken by a Greenland fisherman handlining alone in a 20-foot boat. The fact that the head later weighed out at 75 pounds confirms the 629-pound total live weight, for the proportion (12%) is normal for the species (*Vestkysten,* Esbjerg Denmark, September 3, 1963).

Off Cape Fairweather in the Gulf of Alaska in 1916 Captain T.P. Quinn of the *Knickerbocker* caught a large halibut which with its head on but eviscerated weighed 472 pounds when landed in Ketchikan. Live weight was estimated at 550 pounds.

Martin Holst of Juneau reported catching a large halibut at the head of Lynn Canal in 1899. The fish was dressed and headed and taken to Juneau and cut into four pieces according to what he told the author in 1945. The four pieces had a total weight of 410 pounds. The fins had also been removed. Live weight estimated at 540 pounds.

A 507-pound halibut was reported by Russian scientists to have been taken in the Bay of Datt on Sakhalin Island, USSR.

Another large individual was landed in 1974 in Seattle from the Bering Sea. Its dressed, head-off weight was 375 pounds. It might have been about 500 pounds live weight (IPHC Technical Report No. 16). A large halibut was caught off the dock in Petersburg, Alaska, early in the twentieth century with a live weight of 495 pounds.

A 363-pound, 8 feet 5 inches long, dressed, head-on halibut, which would have been about 430 pounds live weight, was weighed and measured by the writer in Seattle, March 15, 1934. The fish was taken by Captain Charles Ness of the M.S. *Venture* on Portlock Bank, off Kodiak Island.

A 455-pound individual was caught on a handline near Chatham cannery in Sitkoh, Chichagof Island, Alaska (Figure 108).

In October, 1968, the M.S. *Norma* of Petersburg, Alaska, under Captain Aril Mathiesen, unloaded from grounds off Sukoi Islets a 427-pound live-weight halibut. The same vessel two years earlier in the spring of 1966, had caught a 366 pounder.

While no complete survey has been made of the literature for all published accounts of the capture of large Pacific halibut, it can be said that while reports of halibut over 400 pounds live weight are infrequent, only halibut over 500 pounds can be considered rare.

Growth Rates

Growth studies have been a major phase of the investigations by the Halibut Commission. An understanding of growth or rate of weight increase was essential for the scientific management of the resource. The productivity of the species is derived from two main sources, of which one is growth in weight and the other increase in numbers or recruitment.

Early studies by the commission indicated that while the growth rate was generally slow it was quite fast in young halibut, and the balance between growth and the mortality from natural causes would favor the deferment of capture and thus produce a net gain in yield. This was the rationale for the early establishment of closed nursery areas and the later application of minimum size limits (Myhre, 1974). It also is the basis for restricting the taking of halibut to types of gear that tend to catch the fish at their optimum harvesting size.

The rate of growth varies between regions and from time to time, and the males grow much more slowly than females (Figure 109).

Early research showed that halibut west of Cape Spencer grew more slowly than those south of Cape Spencer. The existence of the difference supported the findings from tagging and other studies that the stocks in the two regions did not freely intermingle. Growth rates have also changed radically with the passage of the years. By the late 1940s the rate of growth west of Cape Spencer (Shumagin Islands) had increased sharply over that prevailing in the 1920s (Figure 110).

Growth investigations were augmented considerably by the mid-1950s to provide more precise estimates of the rates of weight increase for use in models for estimating the maximum sustainable yields of the various stocks. It was first shown that there was a consistent relationship between the radius of the otolith and the body length of the fish (Figure 111).

By measuring the width of the annual rings on many thousands of otoliths it was possible to calculate the lengths of the fish at each earlier age. This was carried out for a number of representative regions. Also, by using otoliths collected in 1914 (Thompson, 1915) it was possible to extend the series as far back as 1902 for some areas, a period of 65 years (Figure 112).

It was found that in all areas the average weight at each age had increased since the mid-1940s, but the trends were not the same in all areas. On grounds south of Cape Spencer the growth rate had remained relatively unchanged for nearly two decades from 1930 to 1950. It was not until the 1954 to 1959 year-classes entered the fishery as fast growing four-year-olds did the period of stability end. Since that time the growth continued to accelerate (Figure 112).

West of Cape Spencer on Portlock Bank the growth rate has shown an upward trend from earliest years, although there were some indications of a temporary

FIGURE 109. Average rate of growth difference between male and female halibut.
F.H. Bell

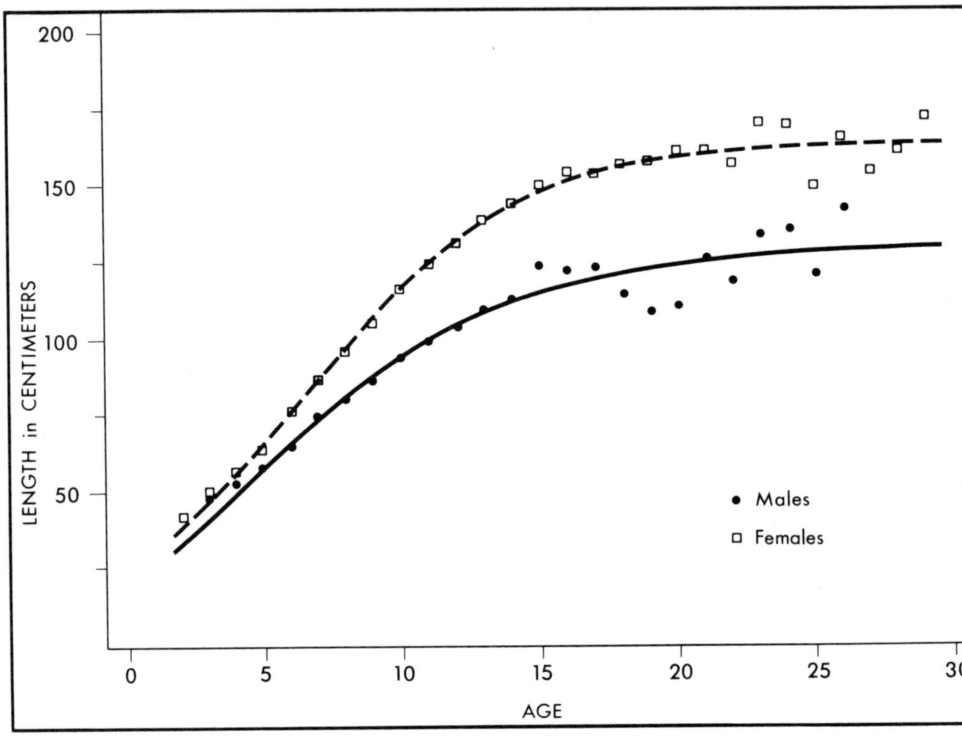

FIGURE 110. Growth rate of female halibut for, *left*, the same grounds in different years, and, *right*, different grounds within the same year.
International Pacific Halibut Commission

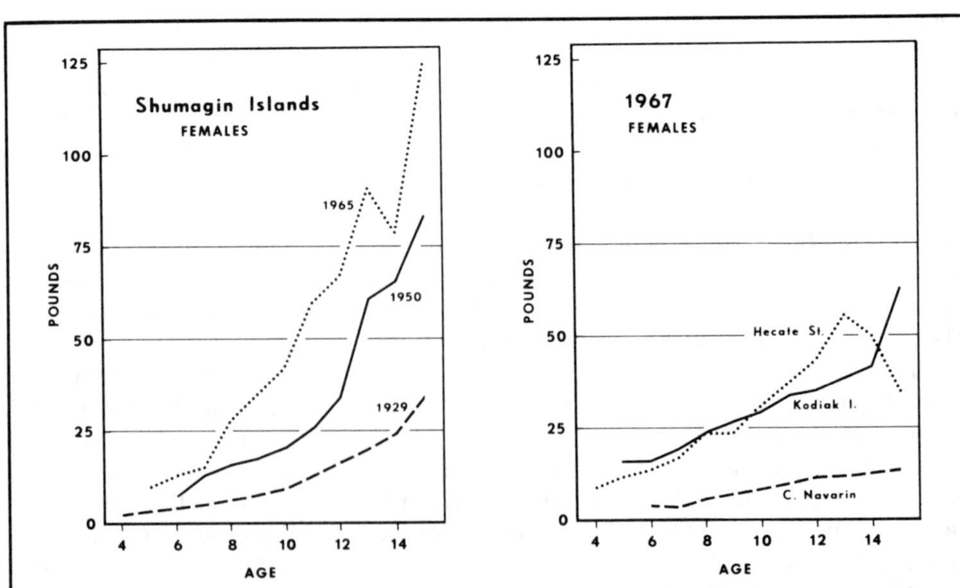

leveling between the mid-1930s and mid-1940s (Figure 113).

There are indications that some year-classes had grown at a greater and others at a lesser rate than the average. Furthermore, in some instances this can be related to the relative abundance of small halibut in the commercial catch. Assuming that such small halibut average about five years of age, there were four short periods centered about 1926, 1932, 1938 and 1944 when the catch per unit effort of small halibut in the catch increased sharply. These years coincide with years in which five-year-old halibut were at or above-average size. Also, the below-average size of five-year-olds in 1936 and 1941 coincide with low catch per unit effort of small halibut in the fishery.

Growth rate has a marked effect upon the length of time that the halibut remains in the various size categories used in the industry. With the recent high growth rates, a year-class will enter the fishery at an early age and pass through the chicken, or small-size trade category very rapidly. The present low abundance of chicken halibut in the catch may be due to this, although changes in recruitment or selectivity of the fishery should not be entirely ruled out.

There are some indications that growth rate changes may in part be density-dependent. Com-

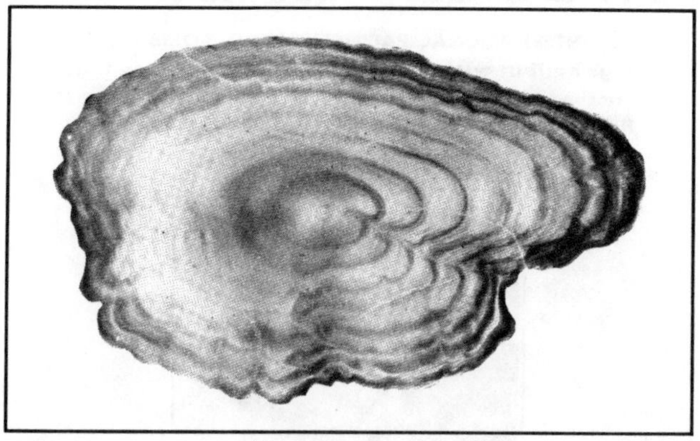

FIGURE 111. Otolith from a halibut in its ninth year.
International Pacific Halibut Commission

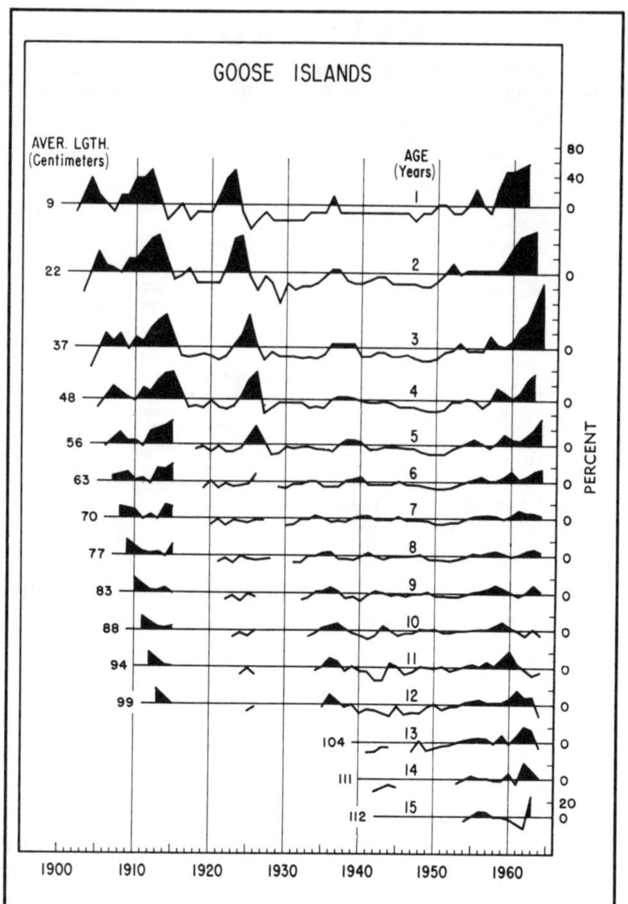

FIGURE 112. Length-frequency distributions of halibut taken on Goose Islands grounds expressed as percent deviations from a long term average length.
International Pacific Halibut Commission

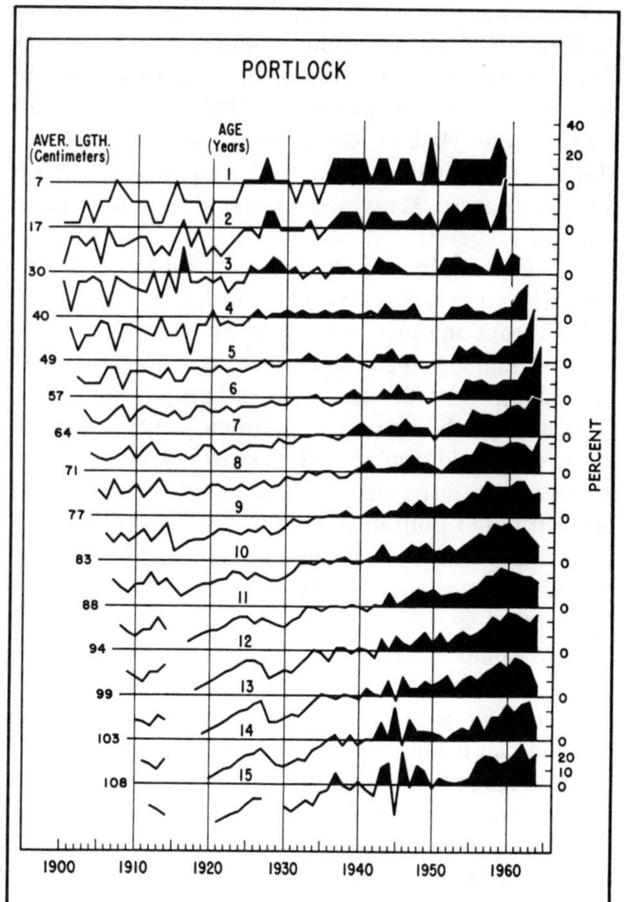

FIGURE 113. Length-frequency distributions of halibut taken on Portlock grounds expressed as percent deviations from a long term average length.
International Pacific Halibut Commission

mencing in the early 1950s the average weight of halibut at all ages increased on all grounds, and at the same time removals from most areas were at a high level from 1950 to 1960. In Area 2 south of Cape Spencer they were increased to test the upper limits of the maximum sustainable yield. West of Cape Spencer the increased removals after 1957 were also followed by an increase in weight of the recent age classes entering the fishery.

From the growth investigations to date it appears that changes in growth rates are probably due to a combination of density dependence and changes in the environment. Management of the resource currently requires that growth and any changes therein be carefully monitored. The long term increase in growth, particularly west of Cape Spencer, was an important factor in the decision to permit increased removals over the years from 1932 to 1960.

TAGGING MIGRATIONS AND MORTALITY RATES

When Tag No. 2* was placed upon a 12-pound, 34.5-inch halibut caught on a handline in Qlawdzeet (Squaderee) Cove, June 14, 1925, in northern Hecate Strait few would have expected that in the next half century the Halibut Commission would tag over 200,000 adult halibut for a total weight of about 3.5 million pounds. Many thousands of juveniles have also been tagged.

The commission conducted marking experiments primarily to ascertain the relationships between the halibut from one section of the coast to another. Also, the percentages recovered were used to determine the fishing mortality rate, i.e., the proportion of the fishable stock being removed annually by the fishery as well to determine their seasonal availability. Recovery patterns also indicated how fishing should be distributed both in time and place to assure that the utilization of the various sections of the population would be proportional to their respective productive capacities. Because of time lag, intermingling regions could not assure that each region would be fished at its optimal level.

Except for a few thousand halibut released from commercial fishing vessels by commission personnel while engaged in other field studies, practically all tagging was conducted on chartered halibut vessels or chartered otter trawlers.

A prime requirement in charter operations was the experience and productivity of the captain who was usually an owner. The owner tendered, in competi-

*Tag No. 1 was kept as a sample.

TAGGED HALIBUT

The INTERNATIONAL PACIFIC HALIBUT COMMISSION tags halibut with plastic tags and metal strap tags attached to the cheek on the dark side of the fish. Some fish have two tags. Retain all tagged halibut regardless of size or gear used.

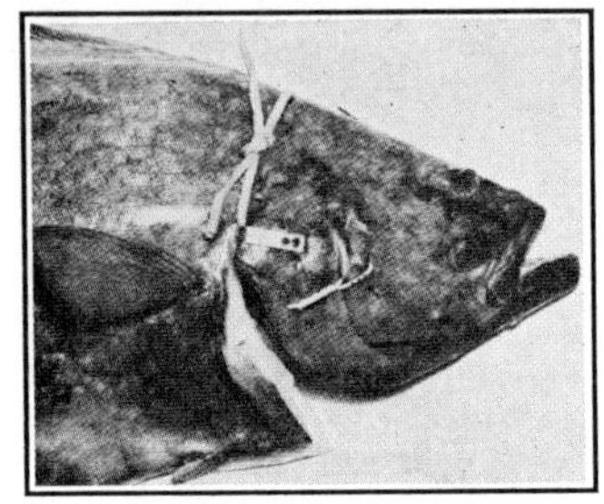

REWARD

$2.00 WILL BE PAID FOR THE RETURN OF THE TAGS AND RECOVERY INFORMATION FROM EACH FISH. $100.00 WILL BE PAID FOR SPECIAL PRESELECTED TAGS.

WHEN YOU CATCH A TAGGED HALIBUT:

1. Record Tag Numbers, Date, Location and Depth in your log book.
2. Leave Tags on the fish.
3. Mark the fish with a gangion.

WHEN YOU LAND A TAGGED HALIBUT:

1. Report fish to a Commission Representative or Government Officer

or

2. Forward tags to address below and enclose recovery information (see above), your name, address, boat name, gear, overall length of fish and, if possible, earstones from the fish.

FINDER WILL BE ADVISED OF MIGRATION AND GROWTH OF THE FISH.

International Pacific Halibut Commission
P. O. Box 9
University Station
Seattle, Washington 98105

Tag Reward Poster.

FIGURE 114. Tag reward poster
International Pacific Halibut Commission

tion, a daily rate for a fully equipped vessel. The commission then negotiated a compensation arrangement with the crew and captain of the vessel chosen for charter. There was usually a nominal fixed salary, but most of the crew remuneration consisted of a bonus per pound of halibut caught. Under some circumstances the fishermen were also guaranteed a minimum. Earnings of crews on commission chartered vessels were usually in the uppermost decile of the fleets earning range. Such compensation assured having good crews capable of fulfilling a fisherman's workday of 15 to 16 hours with the resultant lower unit costs of tagging.

From time to time it was suggested that it would be

TABLE 47. Release and Recovery Location of Tagged Adult Halibut, 1925-1976

		Recoveries by Region									
			Area 3								
Release Region	Number Released	Bering Sea	Shumagin Is.	Chirikof Is.	Kodiak Is.	Yakutat Is.	Southeastern Alaska	British Columbia	Total	Unknown
Bering Sea	20,435	756	21	69	125	116	83	53	1,223	40
Shumagin Is.	5,992	0	202	104	35	20	24	11	396	10
Chirikof Is.	9,193	0	37	473	91	20	17	10	648	31
Kodiak Is.	16,501	0	17	119	1,294	40	36	25	1,531	31
Yakutat	11,431	0	31	122	428	1,078	62	52	1,773	5
Southeastern Alaska	9,729	0	0	0	1	4	1,945	85	2,035	46
British Columbia	59,361	0	1	0	7	39	194	17,288	17,529	254
Total	132,642	756	309	887	1,981	1,317	2,361	17,524	25,135	417

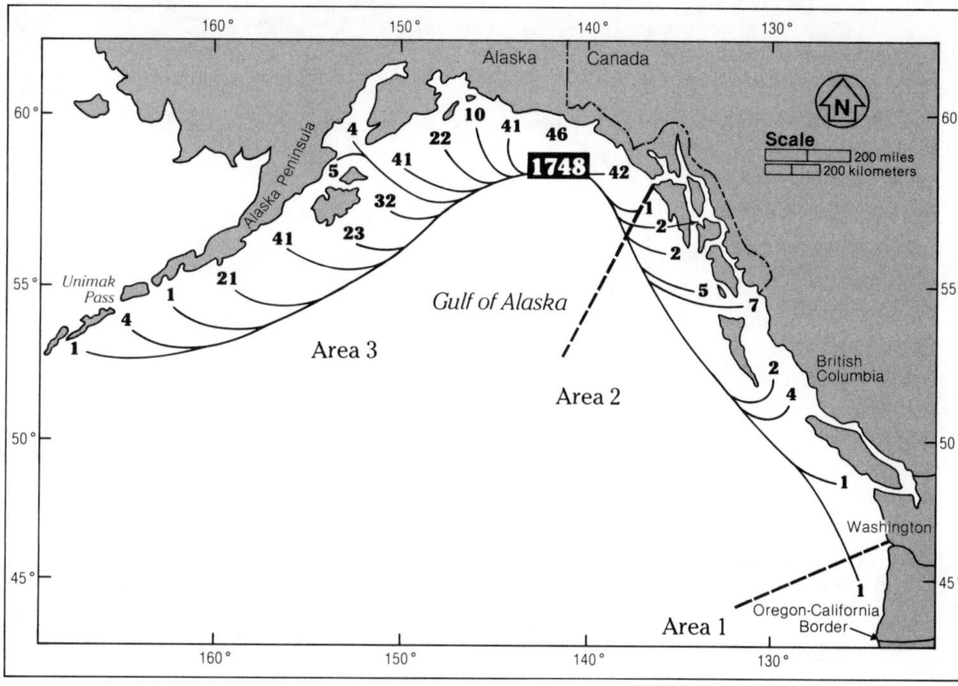

FIGURE 115. Winter tagging on Yakutat spawning grounds. International Pacific Halibut Commission

less costly to buy the fish on board a commercial fishing vessel or to contract the tagging to some other organization or even to own a vessel. Repeated cost analyses have shown that any such alternatives would have doubled or even quintupled the cost per halibut tagged.

The extent of the tagging of adult halibut on different sections of the coast and recoveries from each region are shown in Table 47. They do not include the many thousands of juvenile halibut that have been tagged in the Bering Sea and the Gulf of Alaska since the early 1960s.

A few typical tagging experiments will be briefly described. One of the earliest involved 1,748 predominantly mature individuals released on the Yakutat spawning grounds in the Gulf of Alaska in the winter of 1926. Recoveries ranged as far west as Unimak Pass and would have probably extended farther had there been a fishery beyond the pass at that time. Recoveries west of Cape Spencer were generally proportional to the amount of fishing on each section of that coast (Figure 115).

Although only 5% of the recoveries from that winter tagging west of Cape Spencer were taken on the very heavily fished grounds south of Cape Spencer, they were distributed throughout that region as far south as northern California, the southern limit of the commercial abundance of halibut. Also, somewhat similar to what prevailed west of Cape Spencer, the few recovered on each section of Area 2 were also generally proportional to the amount of fishing on such sections.

Other early, as well as more recent, winter tagging experiments on spawning concentrations in the Gulf of Alaska west of Cape Spencer gave substantially the same results as the 1926 winter Yakutat experiment.

Recoveries from summer tagging in the Yakutat region showed a different pattern from the winter experiments. Instead of a predominantly westward movement shown by the winter releases, the movement by summer releases was greater to the east (Figure 116). However, the movement to grounds south of Cape Spencer was at about the same 5% level as with the winter experiments.

Tagging south of Cape Spencer in Area 2 found recoveries being made over a much more restricted range than for releases made west of Cape Spencer (Figure 117). Also, of the many thousands of

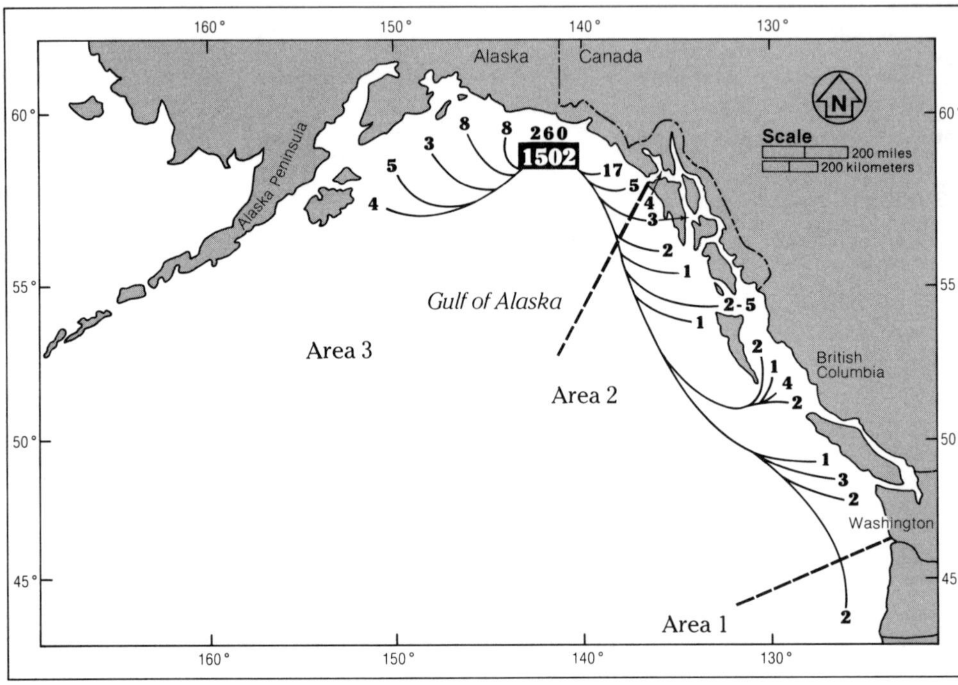

FIGURE 116. Summer tagging on Yakutat spawning grounds.
International Pacific Halibut Commission

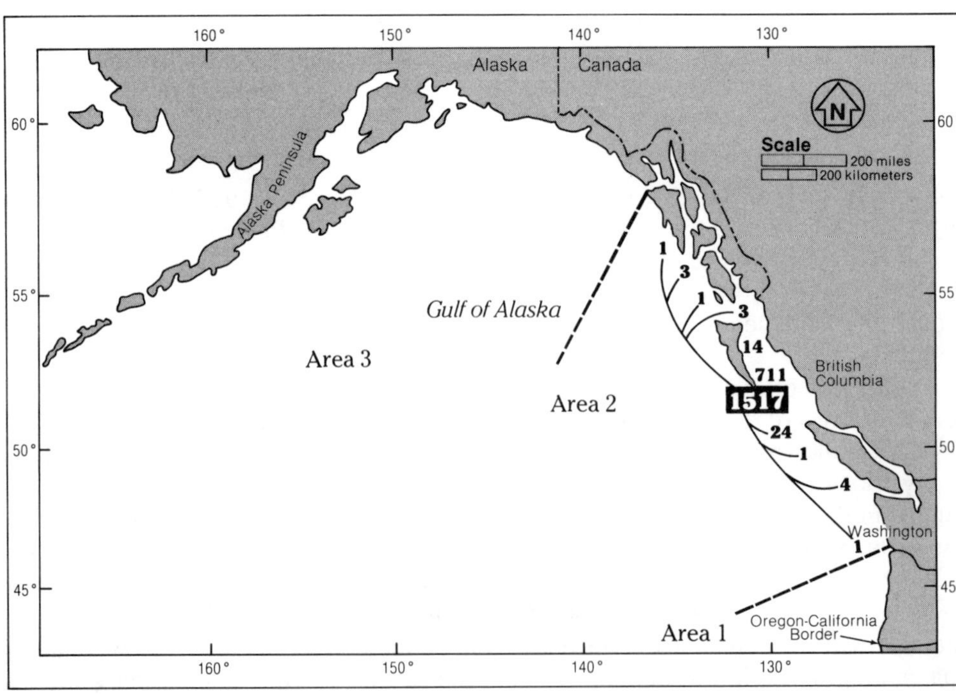

FIGURE 117. Summer tagging in Area 2.
International Pacific Halibut Commission

recoveries from Area 2 releases only a few have been reported from Area 3, west of Cape Spencer. Even the recoveries from winter tagging on mature fish off Cape St. James in Area 2 were not made over as wide an area as what prevailed in Area 3. They were all taken in Area 2, off British Columbia, Southeastern Alaska and as far south as Eureka, California.

With the winter tagging in Area 3, more of the marked individuals would proceed beyond the range of the current fishery and thus contribute to the lower rate of recovery. The same seemed to prevail with the winter tagging experiments south of Cape Spencer.

In the eastern Bering Sea, lying at the northwest extremity of the species' range, over 20,000 adult halibut were tagged as late as 1976 in many different experiments. The most significant feature of the recoveries was the very high proportion of tagged adults later recovered outside the region. About 25% of these adults emigrated to grounds in the Gulf of Alaska, off British Columbia, Washington, and occasionally as far south as northern California (Figure 118).

The geographical range of the migrants from the Bering Sea reached a maximum in the first year after release, but the numbers emigrating reached a maximum usually in the third and occasionally in the fourth year after tagging. The consistency of such timing and the distribution of the recoveries from experiment to experiment in the Bering Sea indicates that the Bering Sea stocks are a part of the 3,000-mile stock continuum that extends north from California. As for any counter movement of adults into the Bering Sea from the Gulf of Alaska, there is no authentic evidence of a single such instance. Tens of thousands of halibut have been tagged in the gulf at a time when there was an active setline fishery in Bering Sea that would have assured some recoveries.

Obviously the halibut in eastern Bering Sea are an integral part of the eastern Pacific population, a fact that was not only disregarded but was seriously questioned by some when the halibut of the eastern Bering Sea were removed in 1962 from abstention by Japan by the Tripartite Fisheries Commission.

Tagging experiments were also of value in demonstrating that the halibut even at the southern extremity of the species' commercial range, off the coasts of California, Oregon and Washington, intermingled with the main population off British Columbia and northward as far as the Bering Sea (Figure 119).

This interchange with the halibut off British Columbia was also corroborated by the similarities in age composition, growth rates and age-weight relationships of both setline and trawl-caught halibut between the two regions, both south and north of Cape Flattery. The movements of young or juvenile halibut are consistent with what appears to be the fundamental unity of the halibut population. Since 1955 the commission has tagged many thousands of juveniles. While recoveries are relatively few due to natural mortality before reaching a size to be caught by setline gear and because of the failure of foreign trawl

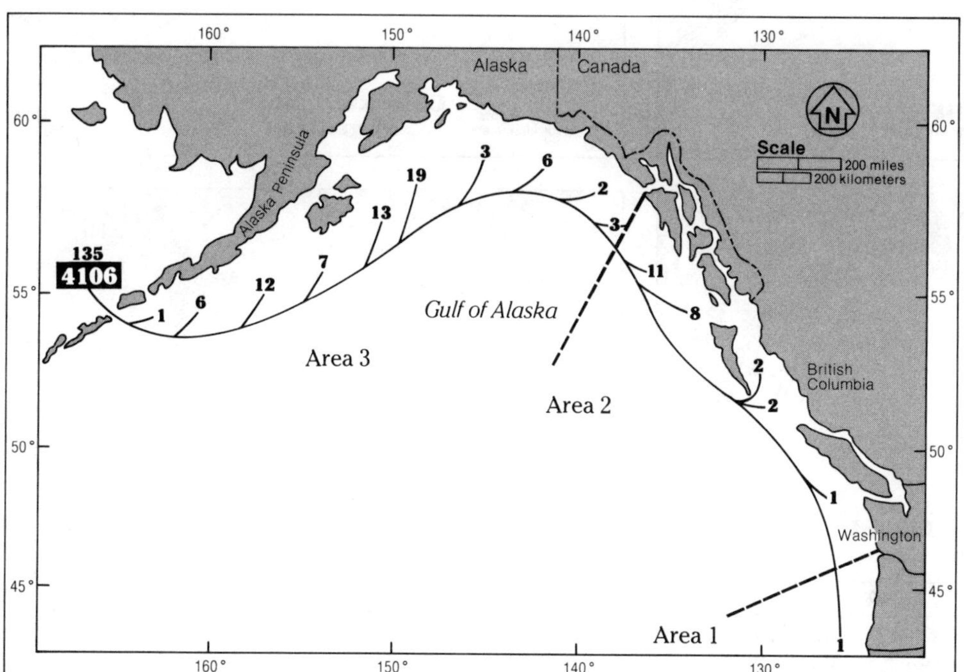

FIGURE 118. Extensive eastward migration of the halibut tagged in the Bering Sea.
International Pacific Halibut Commission

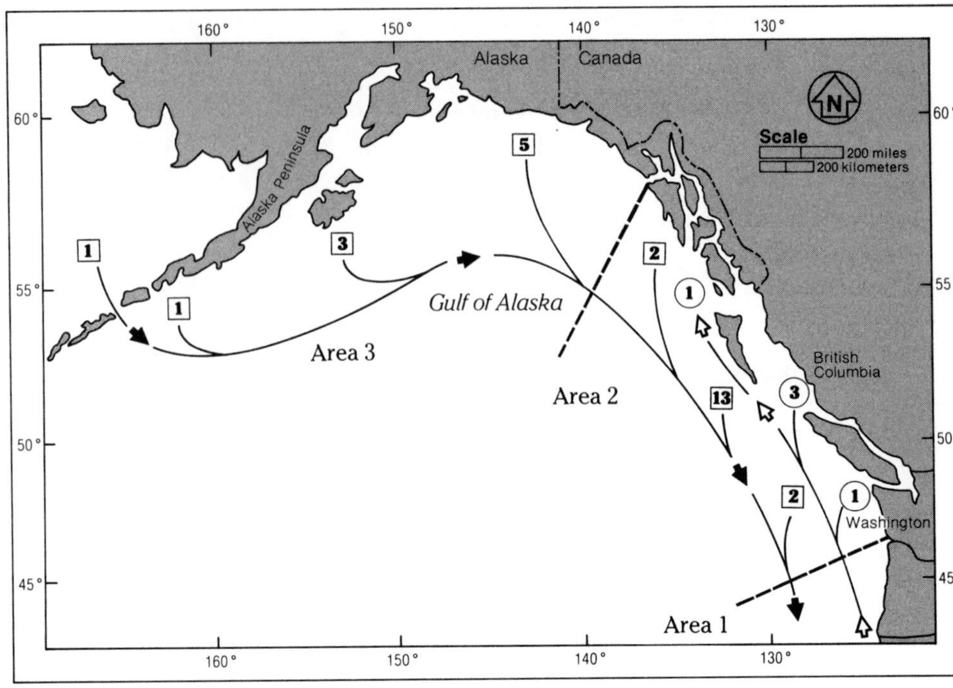

FIGURE 119. Tagging experiments demonstrated that halibut from the southern extremities of the species' range intermingled with halibut populations from British Columbia and as distant as the Bering Sea. Twenty-seven tagged halibut released from various locations in Areas 2 and 3 were recovered in Area 1. Boxed numbers show locations of release and numbers recovered in Area 1. Five tagged halibut emigrated from Area 1 into Area 2. Circled numbers show location of recovery and amounts recovered.
International Pacific Halibut Commission

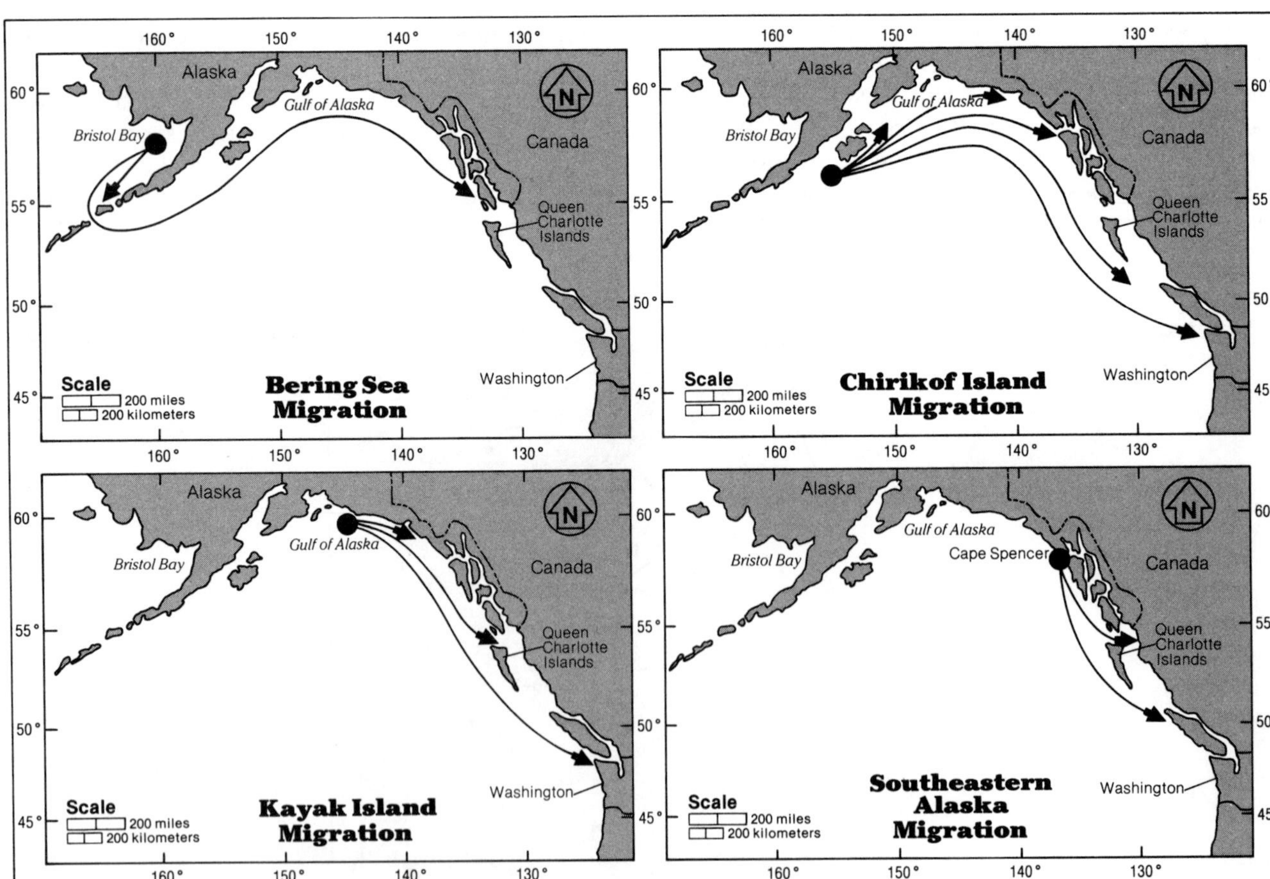

FIGURE 120. Migratory patterns of juvenile halibut from different tagging sites (described by Best, 1968).
International Pacific Halibut Commission

fleets to report tags taken, the pattern of movement has been consistent. Juveniles tagged in the southeastern Bering Sea have been recovered from the Gulf of Alaska and as far east as Southeastern Alaska. Those juveniles tagged in the Gulf of Alaska have been taken as far south as the coast of British Columbia (Figure 120).

The foregoing movements and migrations of the species at all stages of the life history (eggs, larvae, juveniles and adults) are adapted to the general ecology and movements of the water masses of the North Pacific and the Bering Sea. Although in a 1977 report the Halibut Commission claimed to have developed a "new thesis" regarding such movements and relationships, all the general concepts involved had been reported previously in earlier reports of the commission. Refinement of such general concepts is under study by the commission.

Dunlop et al (1964) described the compensatory eastward movement of adults to spawning grounds that countered to the passive westward drift of the eggs and larvae. The movements of individuals marked in spawning grounds in Areas 2 and 3 contrasted with those marked in other seasons and have been discussed and evaluated in many of the early reports of the commission, commencing with that of Thompson and Herrington (1930). The prevailing eastward movement of juvenile halibut from the Bering Sea to the Gulf of Alaska and from both the latter regions to the distant Southeastern Alaska and British Columbia coasts were described by Best (1970).

ADULT STOCK ASSESSMENT

Stock size

Since it is not practical to make direct enumerations of halibut populations only relative measures of their size are available such as provided by the catch per unit of fishing effort (CPUE). Such estimates have been widely used to judge stock condition by fishery investigators and theoreticians in all parts of the world. The CPUE used in the Pacific halibut fishery up to recent years has been the catch in pounds, avoirdupois, heads-off, eviscerated of the "standard skate." The standard skate was defined as six 50-fathom lines spliced together to form a ground line approximately 300 fathoms (548.6 meters) long with 120 hooks. From many thousands of interviews and numerous actual counts of hooks on commercial gear when hung in winter storage lofts, it was ascertained that each of the six lines would *on the average* have 20 hooks when spaced at intervals of 13 feet, more or less. The precise spacing depended upon the stretching or shrinking properties of the fiber being used in the construction of the ground line. Setline gear rigged otherwise was equated to this standard and indexed accordingly.

The use of the term skate in describing the unit of setline gear may have arisen from the fact that a piece of canvas used for the skate bottom has the rhomboidal appearance of a small skate (*Raja*). The bottom consists of an 18 inch by 18 inch square of canvas upon which the ground line is coiled as it comes off the gurdy or deck winch. The four straps at each corner of the canvas consisting of four lengths of one-half-inch rope, two about 6 feet long and two about 1-foot long, are used to bind the coil of retrieved gear. When laid on the deck preparatory to the coiling, the skate bottom and the attached four straps do possess some resemblance to a small skate (*Raja*).

The term skate in the Pacific halibut fishery contrasts with the expression "tub of trawl" used in Canadian and United States North Atlantic line fisheries wherein units of setline gear are in fact set from tubs and are recoiled therein. From time to time a few small boats in the Pacific halibut fishery have used such tub or "basket gear" in fishing for halibut. The snap-on gear which came into wide use in the 1970s is coiled on drums.

Prior to the early 1920s most of the Pacific halibut catch was taken by vessels fishing with dories and using lighter weight ground line and with hooks spaced at 9-foot intervals. During the 35-year period from about 1921 to about 1955 a very large proportion of the fleet used gear that approximated the standard skate, namely six lines of 13-foot gear, although in Southeastern Alaska this 13-foot gear began to be replaced with 18-foot spaced gear in the very late 1920s. By the early 1950s United States vessels fishing west of Cape Spencer and Canadian vessels by the late 1950s fishing in the same region began using 18-foot gear. Beginning about 1966 a few vessels fishing chiefly west of Cape Spencer began to experiment with 21-foot gear and by 1969 such gear represented a significant proportion of the gear used in the fishery. A growing proportion of even wider-spaced gear has been in use by some vessels after 1970.

When the 9-foot dory gear was being replaced by the 13-foot longline gear in the early 1920s, it was found for grounds west of Cape Spencer that as long as the same length of gear was being compared the catch was substantially the same for the 9-foot gear fished from dories as for the 13-foot longline gear despite the greater number of hooks on the former

TABLE 48. Comparison by Seasons of the CPUE of 9-foot Dory Gear and 13-foot Longline Gear Between Cape St. Elias and the South End of Kodiak Island (Sections 24-30) in 1926

	9-Foot Dory			
	Skates	Pounds	CPUE	CPUE*
April to October	27,143	2,715,800	100.0	91.8
November to March	4,654	685,850	146.6	134.0
	13-Foot Longline			
	Skates	Pounds	CPUE	
April to October	69,096	6,449,604	93.2	
November to March	17,631	2,192,272	124.0	

*Corrected to same length as the longline gear.

gear (Thompson, Dunlop, and Bell; 1931). This was based upon 1926 log data for the grounds between Cape St. Elias and the south end of Kodiak Island.

The data in Table 48 suggests that the CPUE of the dory gear with the greater number of hooks per unit length is higher in the winter months when fishing tended to be conducted on the more concentrated schools of spawning fish. Though cognizant of this situation, Thompson, Dunlop, and Bell did not believe it to be of a magnitude that would materially affect the average *annual* CPUE involving as it did such a large and preponderant volume of nonspawning season data in the annual averages.

Consequently, on the basis of these early comparisons of 9-foot dory and 13-foot longline gear, no adjustments were made in the CPUE for hook spacing until 1950. (IPHC Report No. 54 erroneously stated the year to be 1943.) It must be pointed out at this late date that, although Table 48 was based on a fairly large sample, it involved data for one year only, 1926. It is also probable that the changeover from dory fishing with 9-foot gear to longlining with 13-foot gear involved some unaccounted-for bias due to possible difference in the rate of changeover of the more, or of the less, efficient vessels such as occurred in late years with the shift from 13-foot longline to 18-foot longline.

By the late 1940s it became apparent that the early findings on hook spacing for grounds west of Cape Spencer involving 13-foot longline gear and 9-foot dory gear were not necessarily applicable to longline gear on Southeastern Alaska grounds where there had been a sharp increase in the use of 18-foot longline gear after 1930.

The annual CPUE of the 13- and 18-foot longline gear of the same length of ground line for United States vessels fishing Southeastern Alaska is compared in Figure 121 for the 20-year period, 1930 to 1949. These years included periods of *both high and low stock densities.* During that period the CPUE of each type of gear was, in general, *directly proportional to the counted numbers of hooks on each rig.* There are indications in Figure 121 that at the low stock levels of the early 1930s the 18-foot gear was slightly more effective per hook than the more closely spaced 13-foot gear. The converse was indicated at the higher densities in the late 1940s, namely the 13-foot was more effective per hook than the wider-spaced 18-foot gear. However, caution should be exercised in concluding that the difference in hook effectiveness is real and is not an artifact. It could have been that in the early 1930s, when the transition to 18-foot gear was rapid, the inherently more effective vessels tended to make the changeover to 18-foot gear first. There were indications of this, but the data were insufficient to confirm this tendency. This could then account for the apparently higher effectiveness per hook of the 18-foot gear from 1930 to about 1940 and the lower hook effectiveness of the 13-foot gear in the second 10 years when the inherently less effective vessels were the last to shift to the wider-spaced gear.

The ratios between the CPUE of the 18- and 13-foot gear for each year as well as the ratio between the average hook counts on the two types of longline gear are also shown in Figure 121.

During the late 1950s and through the 1960s the shift to longline gear with wider-spaced hooks was kept under close observation. It was found that the inherently more effective vessels were tending to shift to the wider-spaced gear first. Also, as recounted above for Southeastern Alaska, the wider-spaced gear was more effective per hook on the sparser stocks of fish.

Superimposed upon the problem of hook spacing was the abrupt replacement of herring as bait by octopus, particularly after the first trip in 1958. This sharply increased the effectiveness of the gear. There were some indications that the bait effect differed with the hook spacing and with the size of fish on the grounds. During this period the gear was being indexed for the number of hooks and equated to the 6-line, 120-hook standard unit. It was also being corrected for the use of octopus bait.

In 1972 a report by the Halibut Commission, No. 54 (Skud, 1972), endeavored to show that earlier

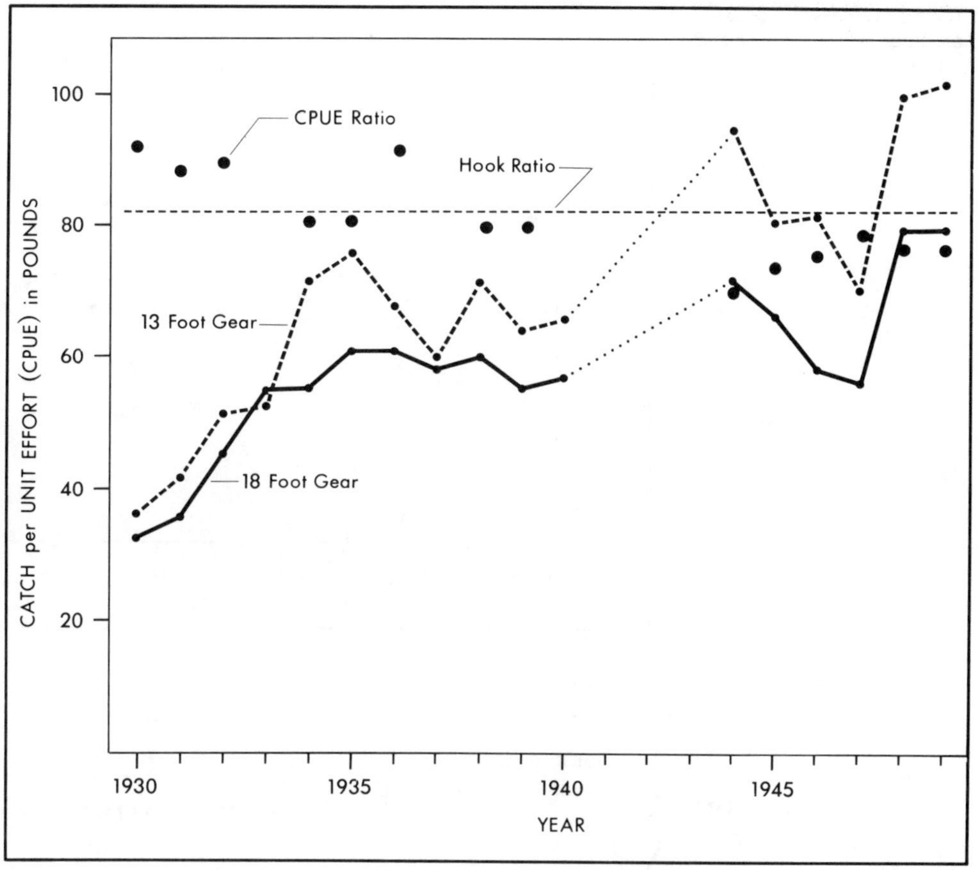

FIGURE 121. CPUE of 13-and 18-foot gear, Dixon Entrance to Cape Spencer, 1930 to 1949, and a comparison of CPUE ratios of the two types of gear and the average ratio of their hook counts.
F.H. Bell

workers, by not making allowance for hook spacing had caused an over-estimation of CPUE. Substantially the same erroneous statement was repeated in various commission reports for the next six years.

In Report 54 of the commission the catch per hook of the 13-foot gear for the 17-year period, 1954 to 1970, was shown to average 1.07 pounds and for the 18-foot gear, 1.28 pounds. Even more unreliable averages were given for still wider-spaced gear. No provision was made in any of these averages for the effects of changes in stock densities during the 17-year period, nor were any allowances made for the fact that the inherently more effective vessels tended to have been among the first to have shifted to wider-spaced gear. If in fact the differences between the catches per hook of the 13- and 18-foot gear had been due to hook spacing, the ratios between the two averages *should have been approximately the same from year to year.* This was not the case as can be seen in Figure 122. Instead there was a radical change in the ratio over the 17-year period changing from about 1.25 to about 1.0 during the period.

Equally misleading are the data in Figure 123 taken from the same 1972 report. They again suggest that the "old CPUE" used prior to 1970 underestimates the decline in the stock as portrayed by the trend of the 18-foot gear.

The old CPUE in Figure 123 is of a six-line (1,800 foot), *120*-hook standard skate and that of the 18-foot gear is of a six-line *100*-hook skate. By the definition given in Report 54, the latter unit is a 0.833 skate. Consequently, to plot the respective catches of the two gears on the same ordinate scale is incorrect. In fact, it is reverting to what Thompson, Dunlop and Bell (1931) did in comparing 13-foot longline gear and 9-foot dory gear in 1926. They had concluded that the length of the gear rather than the number of hooks was the critical element—a conclusion that was no longer found to apply when comparing 13-foot *longline* gear with 18-foot *longline* gear. The aforementioned plotting error becomes even more obvious when Figure 123 is compared with Figure 127.

A later report of the Halibut Commission, No. 56 (Skud 1975), claims that the effects of hook spacing, soak time, changes in the season of the fishery, the rate of decline of the resource from 1915 to 1930 and from 1958 to 1970, and the amount of fishing effort 1915 to 1930 and other parameters of the fishery had not been correctly evaluated by earlier workers. It

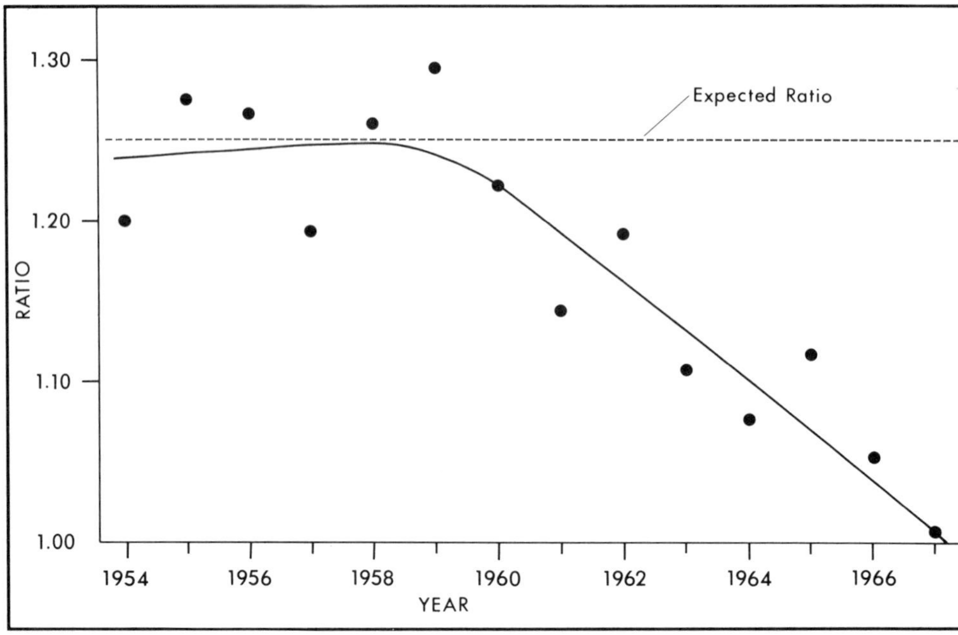

FIGURE 122. Ratios between catch per hook of 13- and 18-foot gear, 1954 to 1967.
F.H. Bell

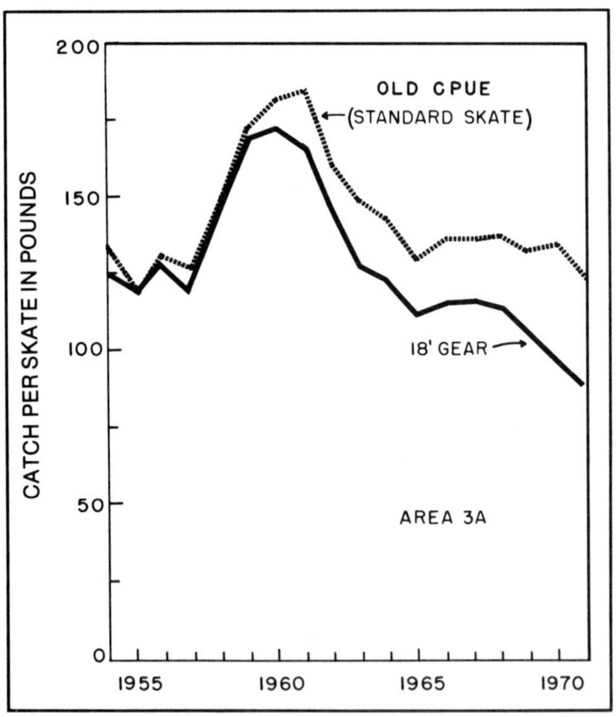

FIGURE 123. Comparison of CPUE based on the standard skate and CPUE of 18-foot gear.
International Pacific Halibut Commission

was also claimed that their assessment of the fishery was in error.

Although the comprehensive coverage of this book precludes refuting all the foregoing claims, one or two may be examined as examples. As to the contention that the *old* CPUE of earlier workers underestimated the decline of the stocks between the late 1950s and late 1960s, the fact is the old CPUE during that period shows a greater rate of decline than does the CPUE of Halibut Commission Technical Report No. 14, 1977, a 13.8% decline compared to 9.8%.

As for the rate of decline from 1915 to 1930, it was claimed that the revised rate was 38% less than that of earlier workers. However, if the data were shown as curves, the two series of CPUE would have to be smoothed by 2s, as necessitated by the paucity of the data in the early years, fitted by inspection, and each phased separately. The difference between the two declining legs would average 1.8%, far removed from the 38%.

The two curves would have different levels due to the fact that the revised data covers April to July only and those of Thompson are annual figures which are higher due to the inclusion of fishing records of the early spring and fall when fish are more concentrated and CPUE accordingly higher.

The following quotation from that report also appears to disregard the fact that the revised CPUE is only for April to July:

> Both the revised CPUE and Thompson's CPUE were divided into the catch to compare the two estimates of fishing effort. The results showed that Thompson's effort was consistently low, that the greatest percentage difference occurred in the early years, and that the difference was greater in Area 3 than in Area 2.

In dividing the annual catch by the CPUE for April to July, the months when the CPUE is at a seasonal low, the calculated effort will be higher than when the annual catch is divided by the annual CPUE which includes early spring and fall fishing on more concentrated stocks with a consequent higher CPUE. The difference between the two calculated efforts in earlier years and in Area 3 is greater for the simple reason a larger proportion of the fishing was conducted in the early spring and fall months of higher concentrations of fish.

There are many factors other than those discussed above that need to be considered as to whether they might significantly bias the use of CPUEs as a measure of the fishable stocks. Two will be briefly discussed at this time as examples.

In some areas the catch per unit effort can be materially affected by the tidal cycle. To reduce the impact of unfavorable tides upon operations of the fleet in such areas at the opening of the season, the commission usually set the opening date at a propitious phase of the tide. Since the annual fishing seasons have usually encompassed many complete tidal cycles they would buffer any effect tides might have upon the year-to-year comparison of CPUE. However, when the fishing seasons were very short, they were opened at about the same phase of the tidal cycle each year. This tended to minimize any bias that might be caused by differences from year to year in the number of low or high tidal periods during the fishing season.

The influence of tidal currents upon the CPUE is indicated in Figure 124. For the type of data involved and the circumstances of the observations, a coefficient of correlation of .70 can be regarded as particularly high.

While the tide effect is not generally discernible in all areas on all occasions, it is probably always present minimally or masked by other factors. One could speculate at length on the manner in which the tidal current might influence the CPUE. For example, stronger tidal flows may cause lightly hooked fish to be lost off the gear. Heavy tidal flows may also cause the natural prey of halibut such as sand lance, herring, and shrimp to move farther off-bottom, and thus attract the halibut away from the baited hooks on the bottom.

Another factor of great importance is whether available fishing records over the years have consistently represented those of the fleet as a whole. This was a matter of some concern to Thompson in 1950 and others.

It has been claimed that the lesser amount of fishing records, available for earlier years, would tend to be from the more intelligent and more effective fishermen and would so bias the halibut CPUE, compared to later years when there was a broader representation of the fleet in the records. It has been the belief and observation of this writer, that halibut fishing skill was, and still is, not necessarily associated with any superior capacity to keep or to provide governmental authorities their private fishing

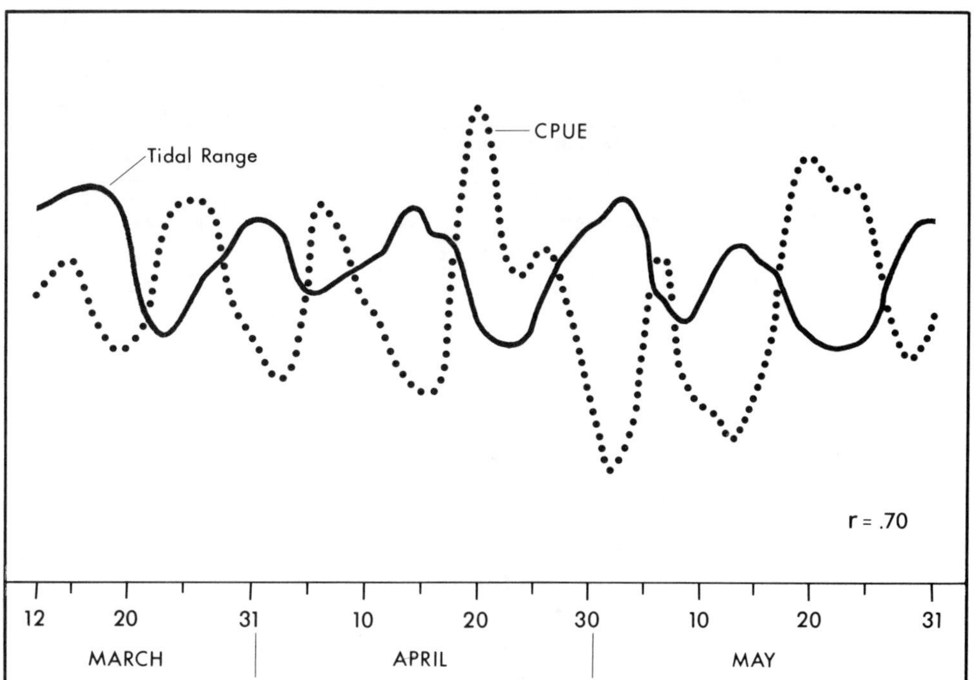

FIGURE 124. CPUE and daily tidal differences on Goose Islands grounds from March 12 to May 31, 1934.
F.H. Bell

records. This can be shown by comparing the fishing effectiveness of vessels who have not provided the Halibut Commission with log records, with those that did.

The CPUE of the commission from 1906 to 1915 was based on records of five Canadian steamers. Table 49 indicates that those vessels were on the average no more effective than similar United States vessels that were not represented in the CPUE. A total of 52,309,000 pounds and 4,935 dory-trips for a two-year period were represented in the comparison.

The same statistics were gathered in later years for other fleets as shown in Table 50, and the same conclusion can be drawn.

Stock Conditions and the Fishery 1910 to 1970

The fishery as represented by the annual catches and amounts of fishing effort expended and stock conditions as indicated by the CPUE are reflected in Figures 125 and 126. The data about catch or removals from the fishable stocks after 1960 are distorted by the relatively unknown amounts of adult halibut that have been removed and the large numbers of young recruits that have been destroyed by foreign fishing. It is known that both such losses are large and probably much larger than recent Halibut Commission reports would indicate.

To place the trends in Figures 125 and 126 in their proper perspective, they are plotted on a logarithmic scale that shows the relative size of the change from year to year, i.e., if the stocks are very low a 10-pound change in the CPUE is more significant than when the halibut are very abundant. To emphasize the trends the data have been smoothed by moving threes.

The trends are self-explanatory. After precipitous declines in the stocks in the unregulated fishery came the turn-around after 1930 due to the reduction of fishing because of economic conditions, followed by sustained improvements under the control measures initiated in 1932 by the Halibut Commission. By the late 1950s most segments of the population appeared to have reached levels of their maximum sustainable yields under the then prevailing environmental conditions. While such levels had been estimated by both empirical and model studies, they were only estimates, and as with all statistical data they possessed an upper and lower limit of confidence. To test the upper limits, the permitted removals were raised by the commission. As soon as the CPUE showed a decline, the domestic removals were reduced. However, about the same time losses to the foreign fleets began to escalate, which nullified such reductions made by the Halibut Commission. The stocks continued to decline. It would have been naive to believe that imposing greater and greater reductions on the domestic halibut fleet would have benefited any group other than the foreign fishery that was sharply increasing the drain on the stocks in Area 3 until 1970. After that time foreign fishing showed some decline, which could be real justification for further reducing the domestic removals. By 1977, under the combined effect of both reduced foreign

TABLE 49. Average Catch Per Dory Per Trip of Various Categories of Dory Vessels Fishing in Area 2

	Number Dory Trips		Total Catch (1,000's lbs.)		Average per Dory (1,000's lbs.)		
	1911	1912	1911	1912	1911	1912	1911-1912
Canadian Company 5-Steamer Fleet	408	536	4,717	4,550	11.6	8.5	9.8
U.S. Company Steamers	1,357	1,025	15,531	11,918	11.4	11.6	11.5
U.S. Independent Vessels	837	772	8,070	7,523	9.6	9.7	9.7
Total	2,602	2,333	28,318	23,991	10.9	10.3	10.6

TABLE 50. Catch Per-Man Per-Trip by Areas By Vessels Represented and Not Represented in the CPUE but of Similar Size

	1920		1925	
With (A) Without (B) Logs	A	B	A	B
South of Cape Spencer				
No. of Man-Trips	659	1,745	615	716
Poundage in 1,000's	1,892	4,790	1,390	1,804
Average lbs. per Man-Trip	2,871	2,745	2,260	2,520
West of Cape Spencer				
No. of Man-Trips	572	1,060	2,543	2,189
Poundage in 1,000's	2,100	4,021	7,562	7,307
Average lbs. per Man-Trip	3,672	3,793	2,974	3,338

and domestic removals, decline in the stocks ceased. However, stock recovery could be a very slow process unless there was a drastic reduction in the Bering Sea losses that continued to increase sharply deep into the 1970s.

Subsequent to 1970 after this author retired, changes were made in the tabulation of catch and fishing statistics preparatory to machine processing the historical and current data. The revised data in the new statistical report (Technical Report No. 14, 1977) show almost exactly the same basic trends of CPUE as was evident in Figures 125 and 126 which were based on the data compiled by this author from 1929 to 1970. Comparison of the two series for Area 3 are shown in Figure 127.

With respect to the commission's Technical Report No. 14 a few further comments might be made as it will probably be used for some time to come as a basic reference on the statistics of the fishery. Because data is on computer tape and no longer requires tedious and expensive manual analysis, adjustments should have been made for any possible bias introduced by the increased representation after 1950 of data from the Canadian fleet. Also, differential effects caused by standardizing to the 18-foot wider-spaced gear which favors CPUE when the stocks are sparse and vice versa might have been explored. Weighing the CPUE by the total catch on each section of the coast rather than by the poundage in the usable log records was a useful contribution although test patterns from time to time down through the years had indicated that it had no significant effect. What differences were present were far overshadowed by other problems that can not be evaluated and corrected.

With regard to the changing representation of Canadian log data in the CPUE from 1950 through 1967, there was a consistent and significant difference between the relative effectiveness of the United States and Canadian vessels during that period. This also varied from one section of the coast to another. For example on the grounds between

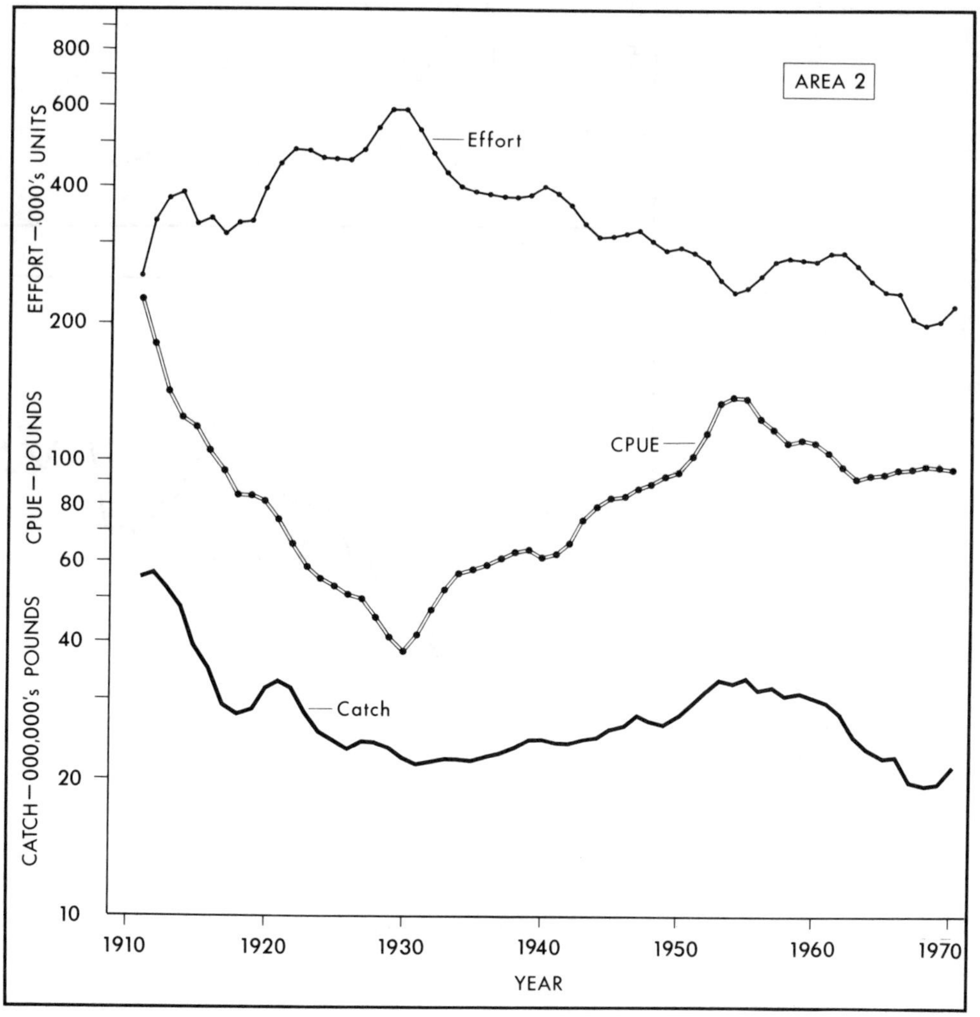

FIGURE 125. Catch, fishing effort and CPUE, south of Cape Spencer (Area 2) from 1910 to 1970 (Log 10 scale smoothed by moving 3's).
F.H. Bell

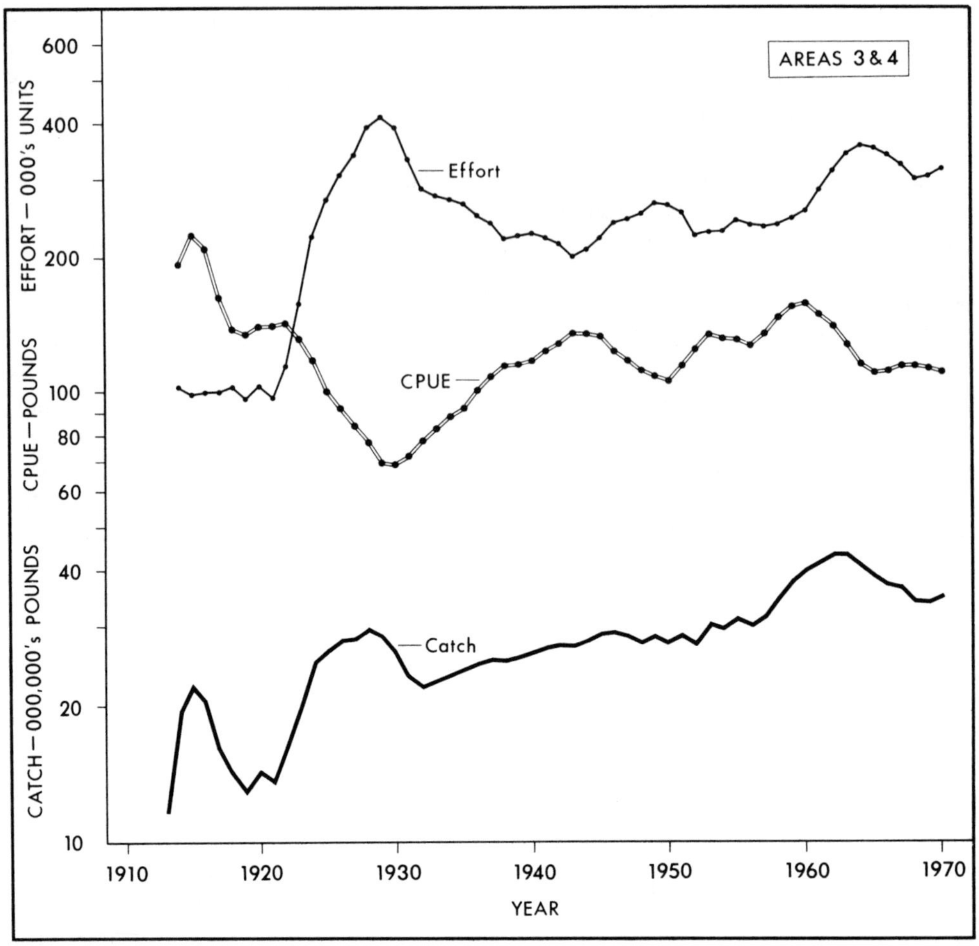

FIGURE 126. Catch, fishing effort and CPUE west of Cape Spencer from 1913-1970 (Log 10 scale smoothed by 3's).
F.H. Bell

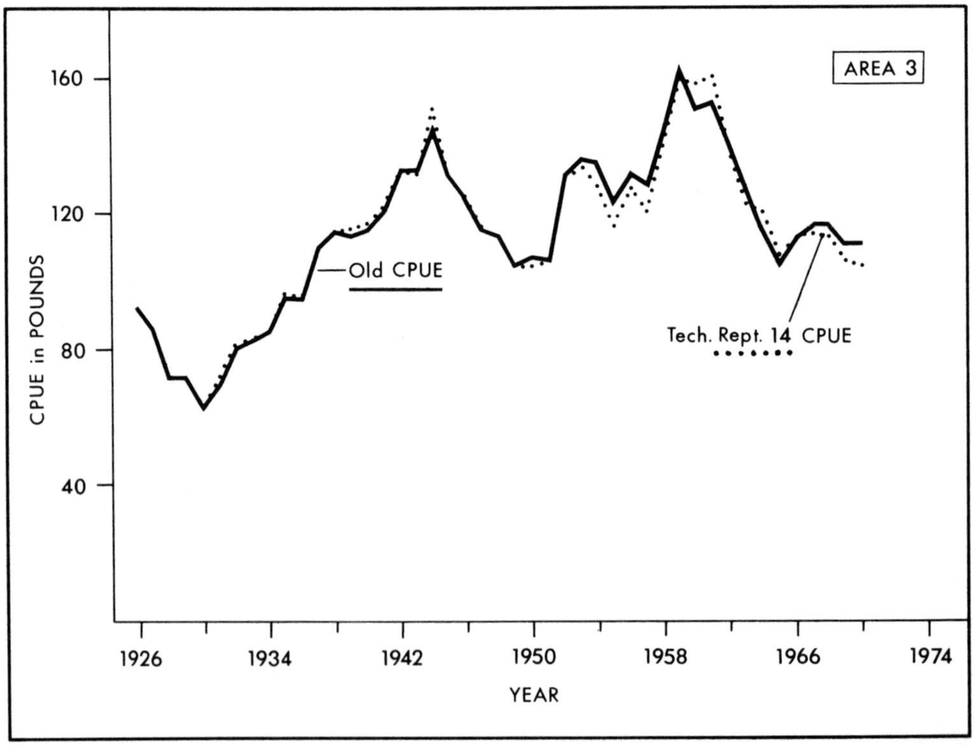

FIGURE 127. Comparison of Old CPUE and that shown in IPHC Tech. Report No. 14.
F.H. Bell

Cape Spencer and Cape St. Elias the CPUE of United States vessels exceeded those of the Canadian during the above 18-year period by an average of 12%. Between Cape St. Elias and Kodiak Island the United States vessels' CPUE exceeded the Canadian in 16 of the 18 years by an average of 7% and in the 2 years the Canadian exceeded the United States by 1%. West of Kodiak there was insufficient data until 1960 for stable averages. However, in that region from 1960 to 1967, United States vessels' CPUE exceeded that of the Canadian vessels by an average of 7%. Because of the aforesaid changing representation of Canadian data in the CPUE during this 18-year period, the relative efficiency of the two fleets may be of importance in establishing reliable CPUE for the fleet as a whole. It is doubtful, however, that it would change the basic trends of the CPUE.

Not including in the catch statistics in Technical Report No. 14 commission estimates of poundage documented to have been taken in contravention to the regulations is without justification. The poundage involved in earlier years was considerable. It was great enough to have influenced the year-to-year changes in the CPUE. Relatively small increases or decreases in the removals were usually followed by an inverse change in the catch per unit effort in the following year. This remarkably close relationship between the removals and CPUE in the succeeding year is evident in Figure 128. Seven of the 11 exceptions in the 42 years covered in Figure 108 occurred when there was relatively little change in the removals from one year to the next. This close relationship is all the more noteworthy when the inherent variability of fishery data is considered. The possible effects of the entry of particularly strong (or weak) year-classes or sudden changes in the availability of the fish to the gear would also impair if not destroy the relationship. In this example the relationship did not seem to have been suppressed by the fact that the long-term trends of the two parameters were increasing or decreasing at the same time.

The existence of the above relationship is also useful in indicating that the parameters involved, CPUE and total removals, were reasonably reliable measures from year to year at least in the above area, Southeastern Alaska. In most other areas and times where there have been unknown and variable amounts of halibut removed by the foreign fisheries the above relationship would of course probably not be evident.

RACIAL STUDIES

To understand what had occurred in the fishery and to develop programs that might lead to the

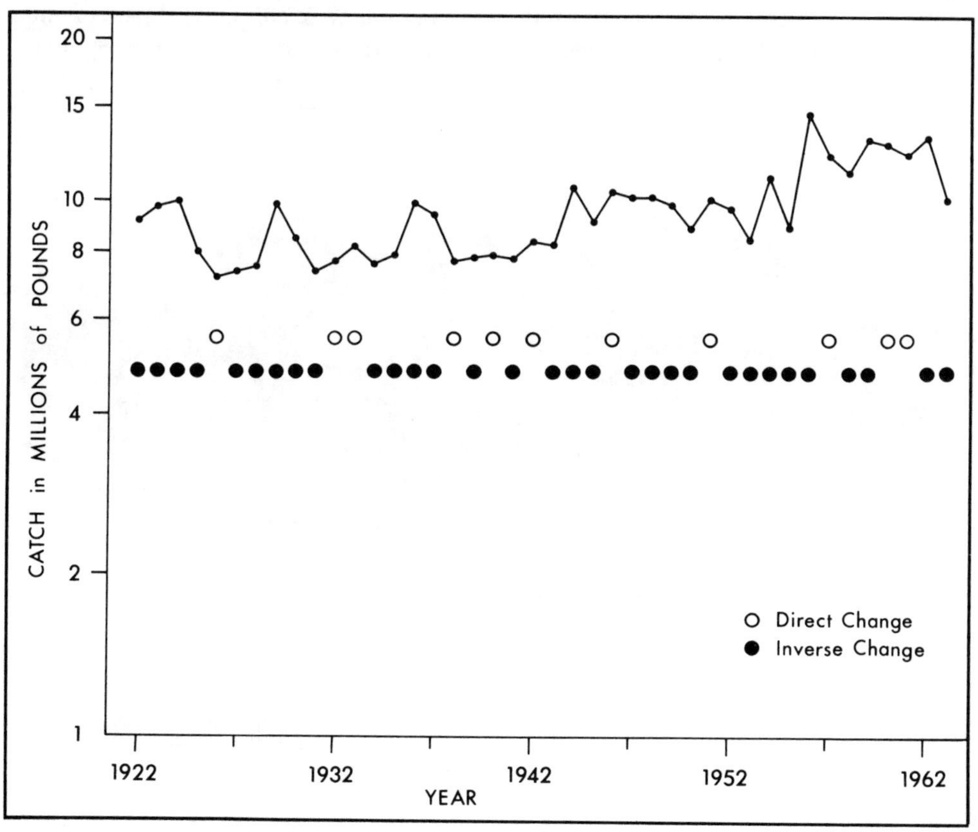

FIGURE 128. Annual catches of halibut in Southeastern Alaska from 1922 to 1963 and changes in CPUE for each succeeding year.
F.H. Bell

rehabilitation of the resource, it was necessary to ascertain the nature of the halibut population. Did the adults intermingle rapidly throughout the whole geographical range of the species so that the species could be regulated as a single unit or were there components each requiring different management measures?

Tagging, the most direct approach to the problem, indicated that the adult halibut west of Cape Spencer did not as adults intermingle freely with those south of Cape Spencer. This was also supported by difference in the rate of growth, age composition and the stock levels between the two regions. In addition the commission conducted an extensive four-year investigation of the morphometric or "racial" characteristics of the various populations of Pacific halibut. This study was the writer's exclusive concern, and some of the salient unpublished features will be reviewed. Also, incidental to such studies the academic question of the applicability of "Collet's Law" with respect to flatfish (Heterosomata) will be examined.

To secure undistorted body or morphometric measurements as well as to ascertain the sex of halibut, which have no external secondary sexual characters, all sampling had to be conducted on board the fishing vessel where the fish were eviscerated as caught. To secure each sample required from three to five weeks at sea on commercial or chartered halibut vessels. Securing dimensional characters singlehandedly, such as head length, width, and fin insertions, from such large fish required the development of a special carriage that could be moved on side rails of a cradle or trough in which the fish was secured (Figure 129). The carriage designed by the author had devices provided with sighting hairs and millimeter scales that permitted the vertical projection of body dimensions on both the longitudinal and transverse axes of the fish. In heavy weather it was necessary to strap the fish in the cradle with the head impaled on fixed spikes in the trough of the latter.

Numerical or meristic characters such as ray counts of the median fins, gill raker and vertebral counts were also made. Due to the relatively high market value of halibut, vertebral counts were made chiefly of undersize individuals or those culled at the time of landing.

Samples were secured from all major fishing grounds from the Oregon coast to the Shumagin Islands as indicated in Figure 130 as counts of meristic characters were taken from Bering Sea samples. In all, the writer measured about 28,000 individuals in the study. A few of the findings will be recounted here along with notes upon ambicoloration, left handedness and other abnormal, teratologic features.

One of the more useful dimensional characters was the length of the head. The fishermen had long reported that halibut west of Cape Spencer had longer heads than those south of Cape Spencer.

FIGURE 129. Measuring cradle
F.H. Bell

Figure 131 expresses head lengths as a percentage of the total length for each size class of fish.

It will be observed that the relative size of head changes with length. This is quite general with fish and frequently applies to other body proportions as well. The early generalization that growth of fish was indeterminate as opposed to the determinate growth in mammals has not proven to be valid. However, one cannot rule out the possibility that the change in the proportion of the head of halibut with size may in part be due to sampling selection rather than only to differential or heterogonous growth. Since the relative length of head is also associated with growth rate, the question is obviously one of no little com-

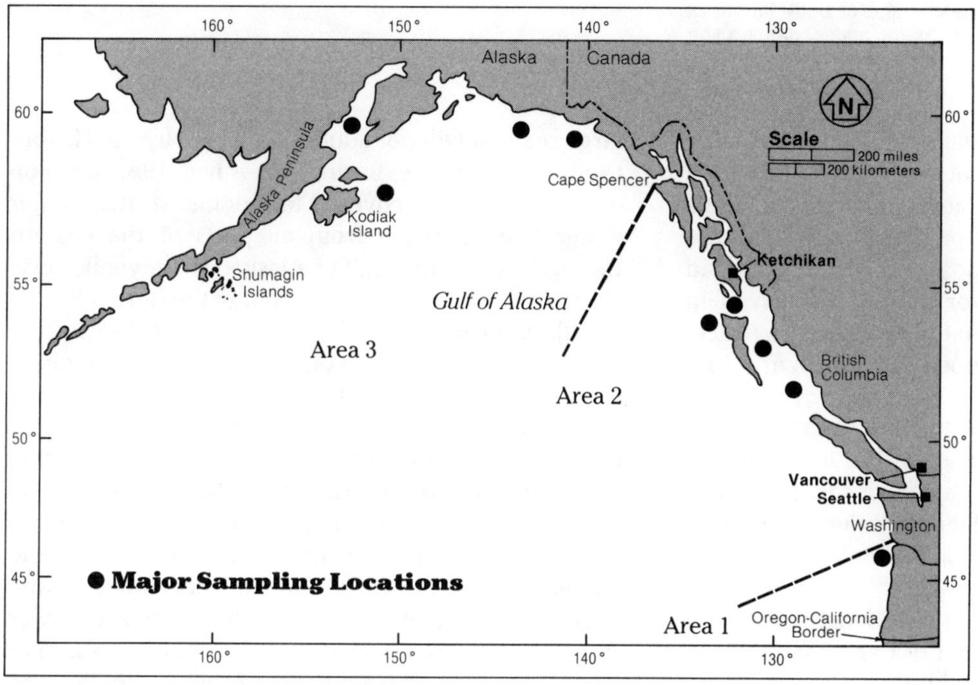

FIGURE 130. Sampling locations used in racial studies.
F.H. Bell

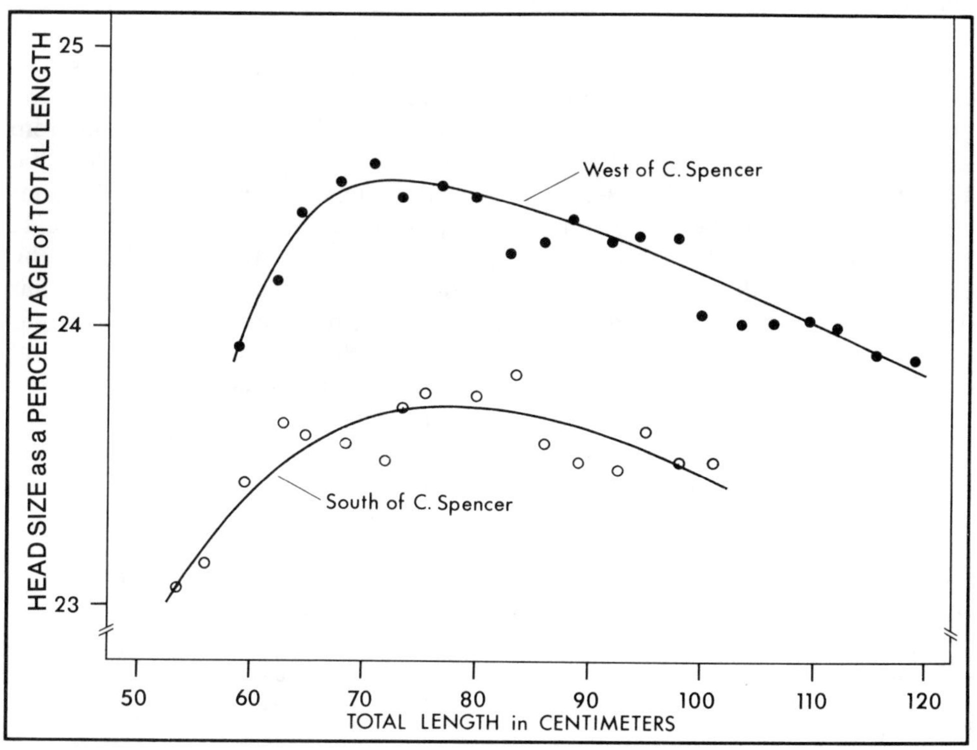

FIGURE 131. Head lengths as percentages of total length for samples from west of Cape Spencer and for those from south of Cape Spencer.
F.H. Bell

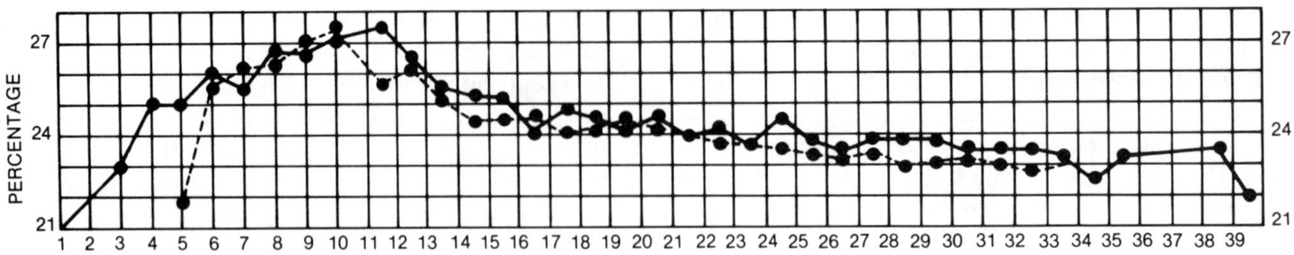

FIGURE 132. Head length as a percentage of body length of 588 plaice from Helgoland. Solid line is for males, dashed line for females.
Keilhoek

plexity. Nevertheless, regardless of what the biological implications may be, there were observed regional differences in average head lengths, and this contributed to the practical problem of defining the stocks for management purposes. The trends of head lengths with body length (or age) as portrayed in Figure 131 are similar to what prevails in the plaice, *Pleuronectes platessa* (Keilhoek, 1913) (Figure 132) and in other flatfish.

The head-length values for the individual locations shown in Figure 130 cluster about the trend for the combined data from west of Cape Spencer. There was a similar clustering of head-length values of various locations from south of Cape Spencer about the trend for the region as a whole.

The general resemblance in head lengths between the samples from grounds west of Cape Spencer is in conformity with that indicated by tagging, i.e., there is a relatively extensive interchange between the halibut in that region. The contrast between the head lengths on grounds south of Cape Spencer with those west of Cape Spencer suggests a lesser interchange of adults between those two large regions.

Initially, no head lengths or other racial measurements were available from Bering Sea as there was no fishery to sample in that area in those early years. More recently a large amount of data on vertebrae and fin-ray counts have been collected in both the Bering Sea and the Gulf of Alaska, chiefly from small halibut. As may be seen in Table 51, there was no significant difference between the counts of such meristic characters in the two regions. This lack of difference might be expected from the results of the large scale tagging of adults conducted by the Halibut Commission in the Bering Sea since 1956 (Dunlop et al, 1964). The recovered tags demonstrate a major emigration of adults from all parts of the eastern Bering Sea into the Gulf of Alaska. The juveniles also show an emigration eastward out of the Bering Sea.

This intermingling of stocks between the Bering Sea and the Gulf of Alaska, coupled with strong indications that the halibut in the eastern Bering Sea were fully utilized, provided the evidence required by the International North Pacific Convention for Japan to abstain from fishing halibut in the area from 1958 to 1962. Despite continued strengthening of the evidence of full utilization of the stocks, Canada and United States in 1962 suddenly capitulated and decided that the evidence of such full utilization was insufficient. This resulted in Japan being no longer required to abstain from halibut fishing in the eastern Bering Sea after 1962.

Concurrently, data were prepared by Japan concerning the intermingling between the halibut stocks in Bering Sea and in the Gulf of Alaska. Head and width measurements on 43 adult halibut ranging from 30 to 60 centimeters in length and taken from the Bering Sea were compared in a 1962 Fisheries Agency of Japan study (based on halibut data obtained by biological investigations during the winter operations of the *Chichibu Maru*) with similar measurements on specimens caught in other areas (Figure 133). The latter specimens consisted of photographs of three 2-cm. postlarvae determined to be from Area 3 by Thompson and Van Cleve in 1936 and one tagged halibut recovered by the Japanese

TABLE 51. Meristic Characters in Juvenile Halibut

Locality	No. of Fish	Ray Counts		Vertebrae
		Anal	Dorsal	
Idaho Inlet				
SE Alaska	66	74.23	99.26	50.162
Cape Fairweather	66	74.12	99.20	50.111
Cape St. Elias	99	74.36	99.62	50.217
Kodiak Island	199	74.94	99.87	50.151
SE Bering Sea	833	74.43	99.64	50.281

and assumed by them to have been tagged in the Gulf of Alaska by the Halibut Commission.

Comparing the body dimensions of postlarvae in one area with those of adult halibut in another to distinguish stocks, manifests ignorance of larval development in general and of flatfish in particular. Body proportions change radically during development in most organisms. Also the assumption that the recovered tagged halibut had been tagged in Area 3 was in error. The fish had been tagged by the Halibut Commission in the Bering Sea within 30 miles of where it had been recovered by Japan. On the other hand the relative widths of the 43 fish measured by Japan in the Bering Sea are close to what prevails in Area 2.

There are similar defects in the Japanese data on halibut head length, as was the case with the width data (Figure 133). The head length of the recovered tagged fish showed a 7% *contraction* in three years. However, the total length of the individual *increased* 36.5% in the three years between release and recovery. The fact is, Figure 134 shows that the relative head lengths of the Bering Sea halibut are similar to those found in Areas 2 and 3.

With respect to vertebral counts, it was claimed from information in Table 52 that the Bering Sea individuals might have originated in more northerly waters than those examined by Van Cleve and Bell. It is noteworthy that 592 vertebral counts of Bell were taken on grounds off British Columbia and Alaska. The counts for some of the individual samples differed more from one another than did the composite sample from the Bering Sea counts in the Japanese study. For example one sample in the Gulf of Alaska had a mean count of 50.41, statistically, no different than that of the Japanese sample from the Bering Sea.

It is evident from the above that the morphometric and meristic materials in the 1962 Japanese study lack substance.

The variations in meristic or numerical characters from various grounds are shown in Figure 135. Generally, they support the findings from tagging. Vertebrae counts from west of Cape Spencer are higher than those from south of Cape Spencer, and while the number of rays of the median fins differ in the two regions, they appear to vary inversely to the vertebrae counts. This inverse variation was contrary to the generally held principle that meristic characters such as median fin rays, vertebrae, scale counts and pores of the lateral line should *a priori* vary directly if not proportionately by reason of their common embryological metameric origin.

Because of the inherent difficulties of counting rays and also, in some instances, vertebrae, it was necessary to ascertain whether the violation of the long-held principle may have been caused by errors of determination by the writer. However, the high correlation between dorsal and anal ray counts, the close correspondence with the independent larval vertebrae counts by Van Cleve (Thompson and Van Cleve, 1936) added to the validity of this writer's determinations. Also median ray counts of Thompson made in 1914 (manuscript notes) showed the same differences for areas west and south of Cape Spencer as those of this writer. However, they were at a lower level as Thompson used different criteria in interpreting the split rays. It will also be shown that in other species of Heterosomata (flatfish) an inverse or at least a nonproportionate relationship between median fin ray and vertebral counts is widespread.

TABLE 52. Frequency Distribution of Vertebral Counts

Number of Vertebrae	Van Cleve: Larvae & Post Larvae	F.H. Bell: Adults	Japanese: Young*
49	10	22	4
50	91	392	60
51	55	177	54
52	0	1	2
Mean	50.29	50.27	50.45
Mean Probable Error	0.010	0.005	

Source: Fisheries Agency of Japan. 1962. Halibut data obtained by biological investigations during the winter operations of *Chichibu Maru*. Fisheries Agency of Japan (1-32). (Document submitted to the Annual Meeting of the International North Pacific Fisheries Commission.)

*Young halibut from winter operations in the southeastern Bering Sea.

Collet's Law

Collet (1880) is usually credited with the generalization of positive correlation between the numbers of vertebrae and ray counts of the median fins. He stated that "increased number of fin rays (in median fins) involves a proportionate augmentation in number of scales, vertebrae and pores in the lateral line." However, his conclusions, spread over a number of papers, were based on very few individuals from each area studied. Hubbs (1922) gave wider currency to what some have named "Collet's Law." He demonstrated that it existed in several non-flatfish species, and also developed some fundamental concepts regarding the phenomenon in a number of important papers.

There are three levels from which one may view correlation between meristic characters in flatfish. However, it is not always clear from which level some authors may be viewing the matter. Collet (1880),

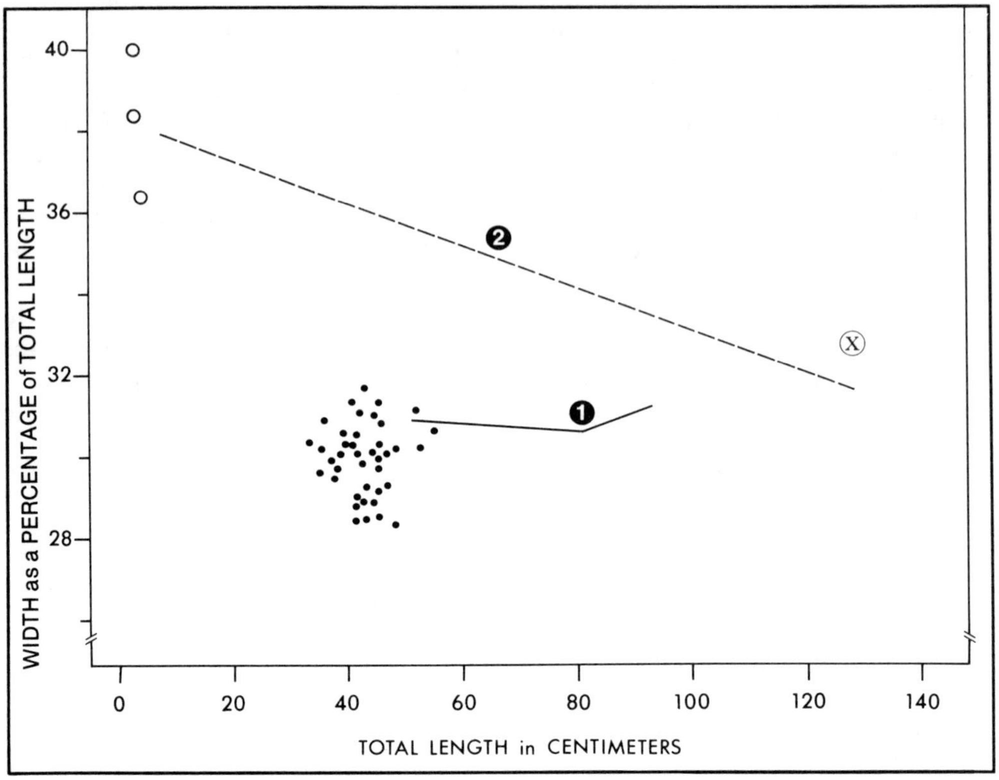

FIGURE 133. Changes in the relative height of body — width: *circled one,* regression line of 467 individuals from Area 2 (Bell, IPHC); *circled two,* regression line as in the 1962 Fisheries Agency of Japan study; *open circle,* 3 postlarvae from Area 3 (from photographs, Thompson and Van Cleve, 1936); *small dots,* 43 small adults from the Bering Sea (Japan); *circled X,* tagged halibut No. 91166.
F.H. Bell

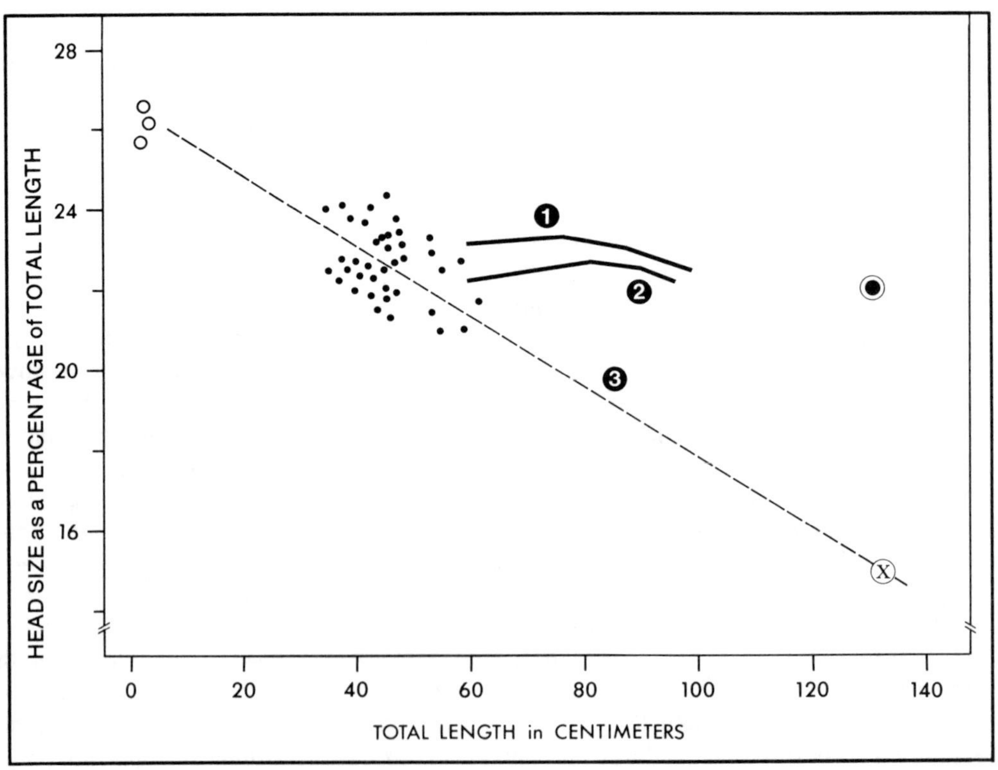

FIGURE 134. Changes in relative head length with total length: *circled one,* regression line of 510 individuals from Area 3 (Bell, IPHC); *circled two,* regression line of 326 individuals from Area 2 (Bell, IPHC); *circled three,* regression line as in the 1962 Fisheries Agency of Japan study; *open circle,* 3 postlarvae from Area 3 (from photographs, Thompson and Van Cleve, 1936); *small dots,* 43 small halibut from the Bering Sea (Japan); *circled X,* tagged halibut No. 91166; *circled dot,* tagged halibut No. 91166 with corrected length.
F.H. Bell

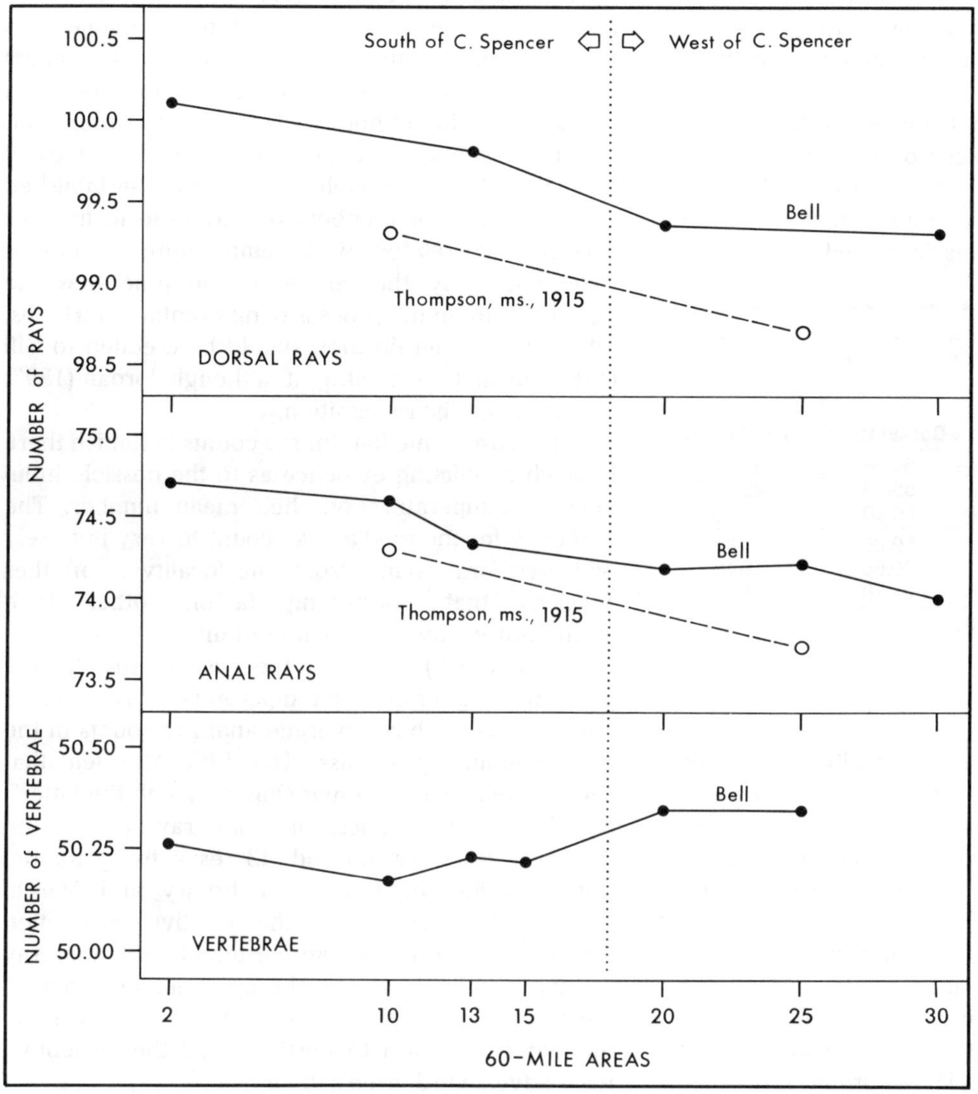

FIGURE 135. Variation of median fin rays and vertebrae south and west of Cape Spencer.
F.H. Bell

who is credited with the generalization that numbers of vertebrae and ray counts of the median fins are positively correlated, was referring to subpopulations of the same species. Others have viewed the question from the standpoint of groups of species. For example, Townsend (1936) showed a high positive correlation between median ray counts and vertebrae in 18 species of North Pacific flatfish. This finding has been supported by those of others. The third point of view is directed to the individual fish. The writer has not encountered any study of flatfish that has demonstrated a significant inverse nor direct correlation between the median ray counts and vertebrae of individual flatfish. Some correlation is indicated to be present, but it is not demonstrable statistically. One of the chief difficulties seems to lie in securing a large enough number of vertebrae counts.

Norman (1934) supports Collet's Law with respect to subpopulations or the regional variation of the meristic characters, vertebrae and median fin rays of flatfish. Norman (p. 85) in support of such views, quoted the observations of Collet and the conclusions of Jordon and Hubbs (1925) with respect to *Paralichthys olivaceous.* The latter authors observed that the vertebrae and median fin rays of *P. olivaceous* in Japan both increased in numbers from south to north. However, close examination of Jordan and Hubbs' data shows that while the vertebrae did increase from south to north, the rays actually *decreased.* Jordan and Hubbs, apparently subscribing to Collet's Law felt this defection from the rule was merely fortuitous and believed that further study would rectify the situation.

Norman's other supportive evidence was an indicated positive correlation between the vertebrae and median rays for subpopulations of *Hippoglosoides platessoides limanoides* from several localities in Europe. The data are unconvincing as they involved no more than from one to six individuals from each of the regions except one.

The inverse relationship between vertebral and median fin ray counts as observed in Pacific halibut is paralleled for a European flatfish in Table 53. As in the case for halibut the vertebrae increased from south to north, and the fin rays decreased.

TABLE 53. Comparison of Vertebral and Median Fin Ray Counts for *Pleuronectes flesus* for several European Localities From North to South

Locality	Vertebrae	Dorsal Rays	Anal Rays
White Sea	35.41	54.38	37.83
Barents Sea	35.17	55.76	39.44
Baltic Sea	35.40	56.90	39.80
SE North Sea	35.00	59.60	41.60
S. North Sea	35.00	60.50	42.40
Plymouth	NA	61.70	43.60

Note: After Ssytsch-Awerinzewa, 1930.

The inverse correlation for halibut and for *Pleuronectes flesus* receives support from the experiments of Taning (1952) and Molander (1957). They show that the numbers of vertebrae and rays in the median fins of flatfish vary inversely in larvae raised under controlled temperature conditions. However, these experiments do tend to nullify the hypothesis that the inverse relationship arose from the difference in the time of formation of the characters in the embryo and their consequent exposure to different conditions as the developing larvae drifted into other habitats.

There are also numerous instances where changes in vertebral counts were not necessarily proportionate to changes in median fin ray counts in flatfish.

Devold (1942) found in the Norwegian plaice that from one locality to another the change in vertebral count was not proportionate to the change in median ray count and that the two characters varied independently from one locality to another.

Smith (1936) recorded that for *Parophrys vetulus* there was a graduated series of median ray counts with significant differences from locality to locality in the Puget Sound while vertebral counts remained practically constant throughout the sound.

Examination of Townsend's (1936) material on *Lepidopsetta bilineata* and *Platichthys stellatus* reveals significant differences in median ray counts between Alaska and Puget Sound samples but practically no difference between vertebral counts.

Accordingly, from the evidence at hand it does appear that Collet's Law is no longer tenable with respect to the counts of median fin rays and numbers of vertebrae from one subpopulation to another in flatfish. This should not be miscontrued as ruling out the applicability of Collet's Law to other groups of fish. Also, it does no violence to the well-established principle that the numbers of vertebrae in fish are inversely correlated with temperature conditions experienced by the larvae at the time that the vertebrae are in the process of differentiation (Hubbs, 1922). This generalization should be credited to Gill (1863) for first formulating it, although Jordan (1893) is often given the recognition.

In the case of median fin ray counts in flatfish there is much conflicting evidence as to the possible influence of temperature on their mean number. The tendency for the median ray count to vary inversely with vertebral counts from one locality to another suggests that controlling factors other than temperature may be more important.

Jensen (1939) found in the case of the Jutland plaice that higher bottom temperatures in some years coincided with above average anal ray counts in the corresponding year-class. Devold (1942) felt that Jensen's data was unconvincing, although the latter's correlative "r" values of anal ray count and temperatures were .40 and .43 respectively for the critical period of January, February and March, respectively. Devold felt the positive correlation found between the mean vertebral and anal ray counts in individuals and the apparent tendency of the ray counts of the Norway plaice to increase in number from south to north, as did the vertebrae, were adverse to Jensen's theory.

Examination of Devold's data on this latter point shows it to be even less convincing than Jensen's data may have appeared to Devold, particularly in view of the fact that the effects of varying salinity or other environmental factors upon ray counts in the various localities sampled was not taken into consideration by Devold.

It should be remembered that pelagic larvae may change their environment at various stages of their development. With the time of fixation of one character fixed, the change in habitat may subject other characters to a different set of circumstances.

It may be said that the apparent defection of halibut from the hitherto accepted generalization known as Collet's Law regarding median fin ray and vertebral counts occurs in other Heterosomata and that, as far as this large group of fishes is concerned, Collet's conception is not an acceptable generalization. In some instances it is indeed contrary to fact.

Ambicoloration

Ambicoloration, where pigment is found on the blind, and normally white side, is relatively common in Pacific halibut. This writer observed or handled, over a period of nearly 50 years, several thousand halibut that exhibited varying degrees of ambicoloration. Whether ambicoloration is a teratological return to symmetry is an unresolved biological question (Hubbs and Hubbs, 1945).

Although no extensive study has been made of the incidence of ambicoloration, the frequency of the phenomenon was estimated by this writer from time to time. When sampling the catches of six vessels landing in Seattle in May, 1936, ambicoloration was present on 18 of the 3,941 individuals handled. Based on the writer's observations at other times and in other places, this proportion is reasonably representative.

While ambicoloration sometimes gives a rather bizarre appearance to the fish and sometimes confers upon them the designation of "circus fish" on the East Coast, the pigmented areas tend to exhibit certain patterns. Figure 136 shows the more frequently occurring distributions of coloration.

1. There is often a dearth of pigment along the end of the caudal (a).
2. Pigment is more generally found on the posterior part of the fish (b).
3. Pigment tends to be found along the base of and sometimes on the unpaired fins (c).
4. Pigment also appears to move anteriorly along the lateral line of the fish (d).
5. Pigment frequently appears either on or near the base of the pectoral fin (e).

FIGURE 136. Typical patterns of ambicoloration. There is often a dearth of pigment along the end of the caudal (a). Pigment is more generally found on the posterior part of the fish (b), and tends to be found along the base of, or sometimes on, the unpaired fins (c). Pigment appears to move anteriorly along the lateral line of the fish (d), and frequently appears either on or near the base of the pectoral fin (e).
F.H. Bell

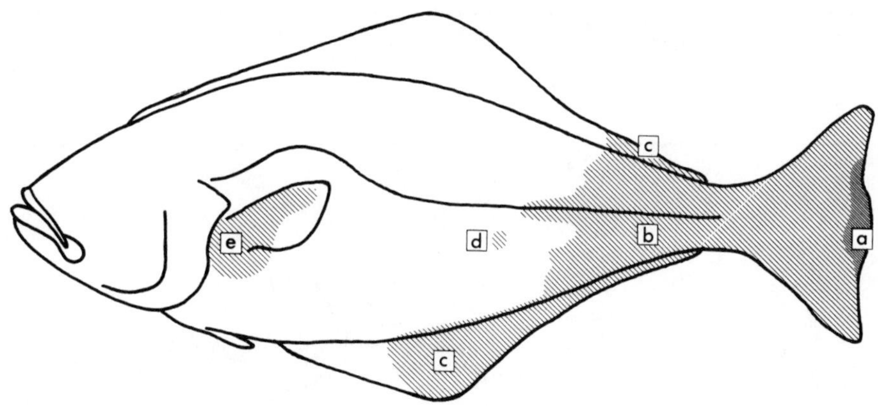

An individual fish might exhibit one or more of these patterns and less frequent ones, too. Whether such patterns bear any relation to an embryonic and characteristic distribution of pigment, such as can be identified on the normally dark side of the fish, has not been determined from any study conducted to date.

Although quite rare, the normally white side may be almost totally pigmented except for some very small areas. The abnormal notch at the anterior end of the dorsal fin is common in such cases as indicated in Figures 137 and 138. In Figure 139 the pigmentation covers about half the normally white side. Figure 140 shows a very unusual condition where a normally rare left-handed halibut has pigment on the blind and normally white side. Figure 141 portrays a normal right-handed halibut with reduced pigment on the normally dark right side.

Gray Halibut

Graying is quite different from the pigmentation found in ambicolored individuals. It has none of the sharply defined patterns and none of the adaptive coloration possessed by the ambicolored halibut. It tends to be found mainly on large fish, and the graying is often more concentrated along the base of the unpaired fins. In the fish trade such fish are called "grays" and command a lesser price per pound. In the process of eviscerating many thousands of halibut at sea, the writer observed that there might be some correlation between the degree or extent of the grayness on the blind side, with the incidence of roundworms on the viscera and particularly on the liver.

FIGURE 137. Ambicolorate individual showing left-side pigmentation.
F.H. Bell

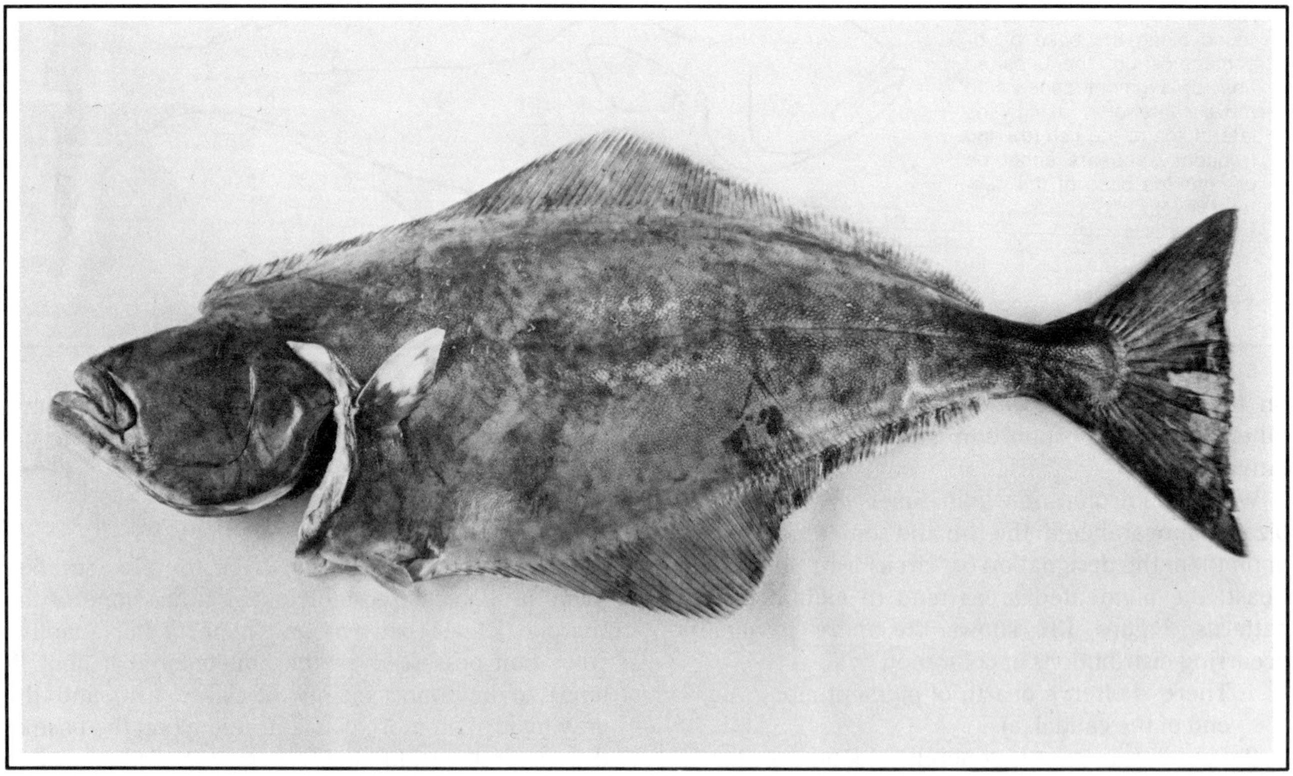

FIGURE 138. Ambicolorate individual
F.H. Bell

FIGURE 139. Pigmentation covers posterior half of halibut's normally white side.
F.H. Bell

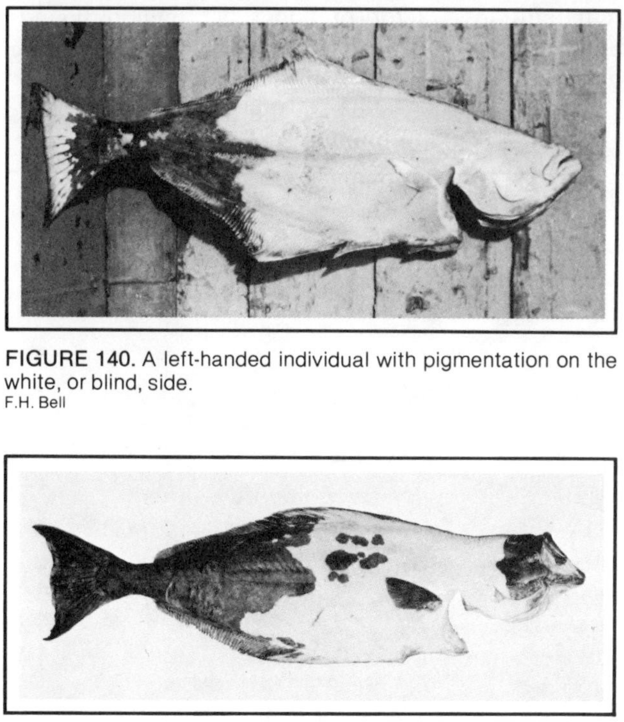

FIGURE 140. A left-handed individual with pigmentation on the white, or blind, side.
F.H. Bell

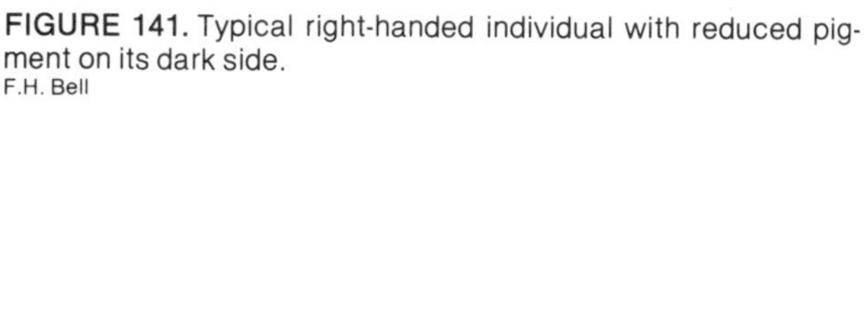

FIGURE 141. Typical right-handed individual with reduced pigment on its dark side.
F.H. Bell

Albinism

Albinism or lack of pigment on most of the ocular or eye side of halibut seems to be very rare. Over a period of nearly 50 years only one individual was observed by the author that would qualify as an albino. There was only a very small amount of pigment on the anterior portion of the head about three-eighths inch in diameter. There have been a few individuals lacking pigment on some small area of the ocular side, but they could not be described as albino.

Reversed halibut

Halibut are normally right-handed or dextral with both eyes on the right and pigmented side. Unlike some flounders, reversal in halibut is infrequent, if not rare. Of the several hundreds of thousand individuals handled by this writer, about one in 20,000 was "left-handed." Goode (1884) reported that for the North-east Atlantic, left-handed or sinistral individuals were infrequently observed at "perhaps one out of 5,000." Norman (1934) stated that only "very occasionally a sinistral individual is recorded." Gudger and Firth (1937) reports that Gudger (1935) could find only "three brief records" of such sinistral individuals in literature on both Atlantic and Pacific halibut. Two such fish were observed by the Halibut Commission during their field work in 1978.

Predator-Prey Relationships

The commission did not conduct extensive studies upon the predator-prey relationships of the Pacific halibut. Investigation of this complex aspect would have been extremely costly. Although of great academic value, little or nothing could be done to modify or mitigate any undesirable conditions that such a study might have revealed.

There were two matters about which the commission did collect some information that indirectly bear on this category of concern: namely, food preferences and the effects of sea lion predation on halibut fleet catches.

In discussing the impact of Steller's sea lions on the halibut fleets, it must be remembered that this account is based on the damage caused by sea lions upon halibut that are "on the gear" or that have been discarded overboard in varying states of viability. It is the opinion of the writer that *in nature* halibut may not be a significant food source for the sea lion for two reasons. First, the two creatures frequent different levels under normal conditions; sea lions, the surface and subsurface, and halibut, the greater depths. Second, because the halibut are extremely powerful and rapid swimmers due to the almost equal and massive development of the musculature of both the ocular and blind sides, sea lions present little threat to a free-swimming halibut.

The matter of the Steller's sea lion predation was one of longstanding concern to the halibut fleets of both Canada and United States. While the commission could not engage in any control measures, it repeatedly used its good offices to attempt to persuade the governments of Canada and United States to initiate some measures to alleviate the situation. The fleets claimed that on many grounds, particularly those west of Cape Spencer, sea lions interfered with fishing and destroyed large quantities of halibut when on the gear.

The commission made a number of surveys incidental to its primary concerns, the results of which fully sustained the fleets' claims. One study was conducted in 1958 to 1960 using about 40 selected halibut vessels as volunteer observers. The grounds covered ranged from British Columbia to the Shumagin Islands in Western Alaska.

There were 1,518 vessel-days upon which both sea lions and severely damaged halibut were carefully counted and recorded. The average sea lion count was about 4.3 animals. About 8% by weight of the halibut catch on those days was damaged and unsalable. The observations and area covered suggest a minimum loss factor of about 3-4 million pounds annually on the coast as a whole. The foregoing values are the average of three years of study.

The relationship between the number of sea lions

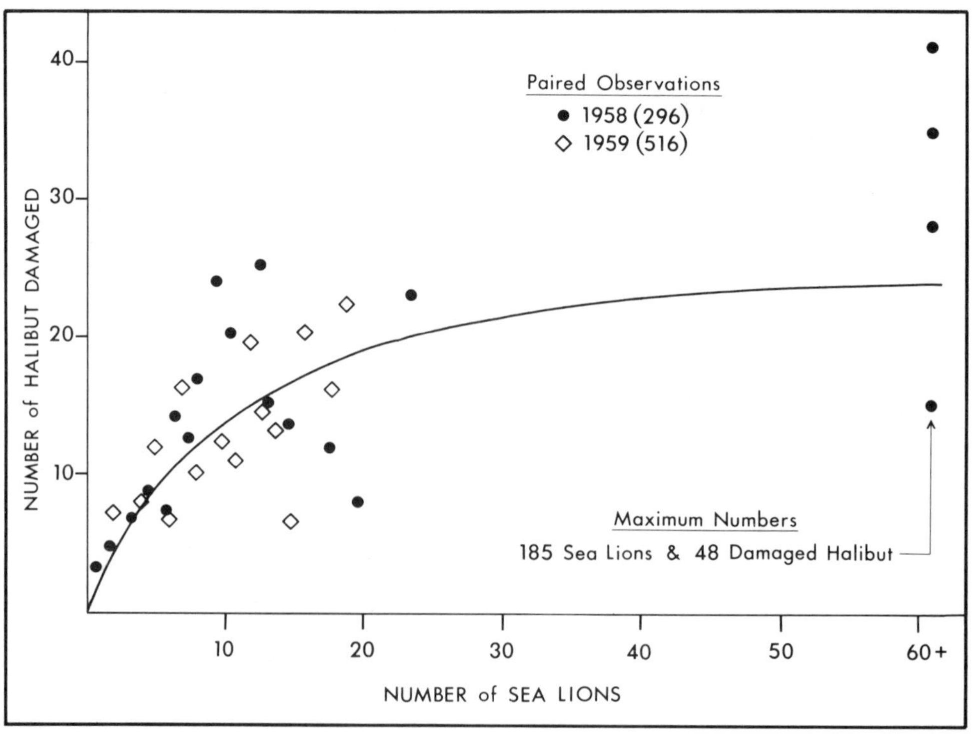

FIGURE 142. Sea lion damage. Paired observations of number of sea lions and number of damaged halibut. F.H. Bell

and number of damaged halibut is shown in Figure 142 for observations reported in the years 1958 to 1960. The consistency of the relationships from year to year confirms the conclusion that the damage was in fact caused by sea lions and that the recorded observations of the fleet are reliable.

The data also suggested that damage was caused chiefly by solitary individuals or relatively small groups of animals and that the herds when foraging were widely dispersed. No more than a few animals established their territory about any one vessel. The observations also indicated that the damage from sea lions seemed to be proportionate to the estimated size of the sea lion population on the contiguous coast. Yet when fishing was good the fleet could tolerate the damage up to a certain level, beyond which it probably became necessary to shift fishing operations.

On a number of occasions, the commission conveyed both its findings and the concerns of the fleet to the two governments usually in the following or like terms:

> The commission recommends that the responsible agencies of the contracting parties continue the research in progress to indicate the importance of sea lions as predators of marine food fishes, and furthermore, that immediate steps be taken within their respective jurisdictions to reduce substantially the numbers of sea lions on rookery and hauling grounds adjacent to halibut grounds to prevent the losses and damage reported by the fleet.

Effective action was never taken as the control of predator in its biosystem is a sensitive problem politically, with complications and repercussions. Such actions should be entered into with caution.

The losses in yield that resulted should be credited in the population equation to "other losses" rather than natural mortality, for it is only the hooked or rejected-at-the-rail halibut that seem vulnerable to damage or destruction by the sea lions.

In the course of making estimates of sea lions, it was observed that a published minimal estimate of 175,000 individuals was indicated for the entire North American stock of the Steller's sea lions. This is probably about half of a more probable minimum. The published estimate was very defective as it omitted counts for the coast from Biorka Island in Southeast Alaska to Cape Hinchinbrook. Certain rookeries were not visited in the year in question, too, nor was allowance made for animals foraging offshore or in transit from one region to another.

What the probable losses to sea lions have been to the present Canadian and United States setline fleets is not known. With the development of the massive foreign trawling operations in the Gulf of Alaska and the Bering Sea, it is noted that halibut that might be discarded by such vessels are heavily set upon by sea lions due to their weakened state from having been taken in the trawl gear. Even though foreign trawlers may be required under 200 mile laws to discard all halibut, the excessive losses of halibut to such foreign fishing will probably not diminish unless all trawling is restricted to enforceable off-bottom fishing.

Food Preferences

Observations on the food preferences of adult halibut since those of Thompson (1916) have been scattered and are not comprehensive. What there is tells that the adults are omnivorous, eating anything that is available. The mouth is large, and the fish a large and powerful animal. The diet of the juveniles consists of small crustacea and fish. Novikov (1964) made a detailed tabulation of the various forms in the stomachs of halibut of various size groups in the Bering Sea. He noted that the diet changes with age, season and area. From his observations it can again be said that halibut seem to consume anything that is available within the boreal or temperate seas, including both bottom- and upper-level life forms. To make a more definitive statement would require a large amount of research, and no matter what the findings, nothing could be done about it.

Parasites

While halibut have the normal parasites and commensals of various types, as do all animals, they are not a significant problem from the standpoint of the industry or the consumer. They offer no public health problem. Many have been described by parasitologists, and the frequency with which such descriptions occur in literature far exceeds their importance.

Being a marine species, halibut are free of a number of parasites such as the tapeworms that trouble freshwater fish. Also, being a wide ranging, offshore species, halibut seem to have fewer parasites than most marine fish. There are a number of roundworms and a few flatworms that have been identified in halibut, but one rarely finds an infestation of the flesh. Their presence is important to the functioning and survival of the biological system, as in the case of man's normal bacterial flora, without which one could not survive.

There are other conditions that affect the marketability of halibut, and they also are of no health hazard to humans. They result in unfavorable textures and are variously described as chalky, white-meated, milky or mushy. The milky or mushy are caused by a small unicellular sporozoan, and its incidence was greatest in the primeval stocks first fished, resulting in a high "cull" of the catch at the unloading dock.

The most widespread condition causes chalkiness or "white-meated" individuals. In contrast to the shiny translucent and sometimes opalescent appearance of the raw flesh, the flesh of chalky halibut is dull, opaque and has a cooked appearance. It is softer and the myotomes seem to separate from each other more readily than the normal uncooked fish. The condition develops after death and appears to be related to the pH arising from the glycogen-lactic acid balance of the feeding process.

As a consequence of the foregoing physiological processes, halibut caught in the setline fishery have a very much lower incidence of chalkiness than those caught in the bottom trawl-net fishery. The difference based on research by several agencies in Canada and United States is so considerable that it should probably be considered in the management of the resource.

15 THE SPORT FISHERY FOR PACIFIC HALIBUT

With the great interest in sportfishing, both fresh- and salt-water, in the United States and Canada it is understandable that halibut, a large and very active fish might be sought after. Some reach such a large size as to become a most desirable trophy fish, and as such, are sometimes given wide publicity in the press and sport magazines.

Since halibut are often found in close association with salmon on most sections of the Pacific coast, a halibut is regarded as an alternative rather than a primary target of the sportsman's efforts. Many sport fishermen who have been successful in having made a good or legal-limit catch of salmon regard the taking of a good-sized halibut as a bonus for their efforts, a "frosting on the cake" as it were. Some of the less successful salmon anglers will capture a halibut as a consolation prize, a means of not returning home empty-handed!

Halibut also provides a welcome variation to a steady diet of salmon, and there are a few regions on the Pacific coast where halibut is often the primary target of the sportsman's effort. Up to the present these areas are in Alaska only.

CALIFORNIA AND OREGON

Off the coast of California, the southern portion of the species range, the very few halibut that are taken by marine anglers are from waters in the northern part of that state. Along the Oregon coast halibut are landed in the marine sport fishery, chiefly out of Coos Bay, and occasionally at Umpqua, Newport and other Oregon centers of marine angling for salmon. Their occurrence in the sport catch, though only occasional, seems to be of sufficient interest for some Oregon coastal resorts and marinas to mention the fact in their advertising.

WASHINGTON

In the State of Washington halibut are taken in connection with the marine sport fishery for salmon both on the outside ocean coast as well as in the inside waters of the Strait of Juan de Fuca and Puget Sound. Although halibut is only an incidental objective of marine anglers it possesses a position of fairly high "prestige." Any halibut over 10 pounds seems to be of mentionable value, and the local newspapers and sport magazines publish many statements and photographs of such catches, and the bigger the halibut the bigger the story!

The incidence of halibut caught by salmon anglers varies widely in the state, based on an analysis of state angling statistics. In the southern Puget Sound the taking of a halibut in the salmon sport fishery is a rare occurrence. In the northern Puget Sound, on the other hand, halibut are taken quite frequently at Possession Bar, Point No Point, Fort Casey and Port Townsend. They occur chiefly in the spring and early summer months, and May is the peak month in most areas. The incidence of halibut in the bags of the salmon anglers seems to vary from year to year, being more frequent in some than others. In the Point No Point-Possession Bar region halibut are found in one in every hundred daily catches of salmon. In the Port Townsend-Fort Casey area the ratio may be as high as one in twenty. In Washington Sound and in the eastern portion of the Strait of Juan de Fuca there are a number of spots that a few sport fishermen actually seek out in the spring and early summer months where at least 5% of the sport catches of salmon are halibut. (Washington Department of Fisheries files.)

Halibut are taken with some frequency by the salmon anglers fishing in the western section of the Strait of Juan de Fuca, off Sekiu and particularly in the vicinity of Neah Bay. Some anglers may catch five or six halibut a day. The average frequency of halibut in the anglers' bags is usually about 1 in 50.

While the above ratios of occurrence per daily bag may seem low the total number of angler-days in the inside waters of Washington number several hundred thousands in a year. Consequently, halibut is far from being a rarity in the marine sport catches in Puget Sound. By the same token, however, the total annual volume of sport-caught halibut in the Strait of Juan de Fuca and Puget Sound of say, 2,000 fish or so, is inconsequential compared to the total commercial catch. In recent years the state has compiled statistics of the numbers reported, usually under 1,000 annually. The total is probably understated because of the large proportion of individually-owned vessels and many launching points compared to the large amount of marina fishing in which bags are more visible. In either case the sport catch remains inconsequential compared to that taken in the commercial fishery.

On the outside coast of Washington halibut are also taken in the salmon sport fisheries of Willapa Bay, La Push, Grays Harbor and Ilwaco. The frequency of halibut in the daily catches of the anglers varies from place to place, probably averaging about one in a hundred, although usually higher in the La Push region where the salmon sport fishery is on a scale more modest than at Westport or Grays Harbor.

Halibut fishing is a long-established sport in Washington and is on the increase. From a management standpoint the activity has not created any significant problems for the Halibut Commission nor the state fisheries agency. The total catch is extremely small in relation to the yields of the coastwise stocks. Until 1973 Washington placed a daily catch and possession limit on the species for anglers. State regulations of halibut and of salmon were enforced by the same fisheries agents.

BRITISH COLUMBIA

In British Columbia most marine angling is concentrated in the Strait of Georgia where relatively few halibut are now found. However, in spite of their relative infrequency, a large specimen of halibut taken by a fisherman while saltwater angling for salmon carries considerable prestige value and will often be noted in the media or sport press. In recent years, sport catches have been reported from the west coast of Vancouver Island and in the vicinity of Prince Rupert.

In both British Columbia and Washington, an occasional halibut was taken while sport fishing before or after the open season set by the Halibut Commission. Down through the years one or two individuals have been apprehended by a diligent officer for retaining sport-caught halibut out of season, but in each case the local authorities failed to prosecute or the local magistrate failed to convict. There have been no indications of any commercialization of sport fishing for halibut in the closed season. As in Washington the reported numbers of sport-caught halibut are relatively few compared to the commercial catch and have little management significance.

ALASKA

In some places in Alaska, unlike in Washington and British Columbia, the catching of halibut has been, in many instances, the primary objective of the marine angler. Competitions known as "Halibut Derbies" have been conducted on a number of occasions with a variety of prizes for the winning contestants. Such contests were held regularly in Kodiak from 1958 to 1964. In the 1959 contest, for which detailed data are at hand, there were 117 entrants, and 61 halibut were caught. The largest of the 61 halibut taken was 106 pounds, and the smallest 11.5 ounces. Total prizes were worth over $1,000.

Also in Alaska, unlike elsewhere, some of the so-called sport fishing for halibut is in fact subsistence fishing. It usually involves the use of gear that would not qualify within any reasonable definition of angling or sport gear. In many instances the catches are sold, which very clearly removes them from the sport classification. Such sales were often identifiable in the records of the various fish buyers. In one year 554 sales were noted involving about 111,000 pounds. While this is a considerable quantity of fish to be taken in the commercial-sport (comsport) fishery, it is but a negligible fraction of what is taken in the legitimate commercial setline halibut fishery, or in the illegitimate retention of halibut by Japanese trawlers off the coasts of Alaska.

The Alaska Department of Fisheries has reported that as much as 200,000 pounds of sport-caught halibut have been taken in some years since 1973. This also is an extremely small amount compared to what is taken illegally and probably destroyed by foreign trawlers off Alaska. Many of the sport-caught halibut in Alaska are caught with jig gear. While it technically may qualify as sport gear, it converts the "sportsman" into a subsistence fisherman.

TROPHY FISH

In recognition of the interest in angling, the Alaska

Department of Fish and Game has, since 1963, maintained a Trophy Fish Award Program for all species of sport fish. To qualify, the fish must meet a minimum weight, 110 pounds for halibut, and it must be taken subject to certain rules and affidavits. Three types of parchment certificates are awarded: 1) Trophy Certificates of Record for any fish complying with the minimum weight for the species; 2) Annual Certificates of Record for the largest of each species each calendar year; and 3) Champion Certificate of Record for any fish that exceeds the previous record for Alaska.

A record halibut of the trophy program was 340 pounds, 90 inches long, and was caught in Thomas Bay near Petersburg in 1971 by George Rice. A slightly larger specimen, 346 pounds, was taken in 1969 in Portage Bay near Petersburg by rod gear using a 30-pound test leader and 40-pound test line. In 1979, 10-year-old Dennis Smedley II caught a 361-pound halibut with rod gear, the largest sport-caught halibut registered with the Alaska Department of Fish & Game for that year (Figure 143).

While other sections of the Pacific Coast do not maintain trophy award programs, a number of halibut over 100 pounds in weight are reported caught by anglers each year. A 36-pound halibut caught in Puget Sound on July 31, 1969, had the distinction of setting the then world record for the Saltwater Fly Rodders of America in the 10-pound test tippet class.

FIGURE 143. Dennis Smedley II, age 10 of Juneau, shows his 361-pound sport-caught halibut taken August 11, 1979.
Alaska Department of Fish & Game, reprinted from *ALASKA®* magazine

Until 1973 the Halibut Commission made no distinction in the regulations for halibut taken while marine angling, and the control measures were applied equally to commercial and sport fishing. There were no problems of any major proportions. Occasionally an enforcement agent was required to warn sportsmen against retaining any halibut caught in an area closed to halibut fishing. On very infrequent occasions, zealous officers would apprehend violators. In these cases the officers believed that the suspect was selling "comsport" halibut.

Washington had long found it useful to limit the taking of halibut in all state and contiguous waters, the limit being "two halibut per day or in possession at any one time." No other state or province had found it necessary to place a restriction upon the possession of sport-caught halibut, provided the fish were caught in accord with the regulations of the Halibut Commission.

During the 1960s a United States federal agency repeatedly recommended to the Halibut Commission that it include in its regulations control measures for halibut sport fishing including bag limits per person or per vessel and a stated length of sport season. However, repeated surveys continued to indicate that the total amount of halibut involved was of no significance from the standpoint of management. The proper implementation and enforcement of such controls on a uniform basis for all sections of the coast would have added materially to the cost of fisheries enforcement. Thus, there seemed to be no merit in adding more regulations to those that were already being inadequately enforced, including the depredations on the halibut by the massive fleets of foreign nations. There was the further and even more critical consideration as to whether the 1953 convention allowed the commission to implement such class action as would be involved in establishing sport-fishing regulations. Many years previously both governments had definitely declared that the commission had no power nor would be granted such power to limit the amount of halibut that an individual person or vessel could catch. This would preclude any bag limit.

The 1953 convention does not mention sport fishing season or any other class of season except open and closed. The convention does permit the commission to establish any number of open or closed seasons. An open or closed season in the context of the Halibut Convention is a period of time when halibut may or may not be caught for any purpose—be it sport, commercial, bait, subsistence, reduction, fertilizer—provided such fishing is conducive to attaining the maximum sustainable yield. The plural "seasons" mentioned in the convention refer to multiple open or closed seasons—a numeric matter. It could not be construed as referring to kinds of seasons—a generic matter.

Indeed the primary reason why Canada and the United States negotiated the 1953 convention was to allow the Halibut Commission to establish more than one fishing season in any area each year, not to introduce kinds of seasons. This purpose is clear and unequivocal. To infer that the applicable provision in the 1953 convention refers to kinds rather than numbers of fishing seasons is also to say that *only* under the 1953 convention could the commission have had a sport fishing season, for prior conventions only provided for a single season in any area.

For the commission to deviate from this and assume authority that was neither stated, implied, nor intended in the convention is to court disaster. Its past close adherence to the directives of the various conventions has brought respect to the commission's decisions, even though they may not always have been popular nor enthusiastically accepted by other governmental agencies.

Furthermore, any *class* action not provided for under the convention such as providing a season for sport fishermen would make it nearly impossible to deny requests for a Makah Tribal season, a king crab bait season, a Kake Tribal season, a troll-caught halibut season and seasons for other class groups that have heretofore petitioned the commission for such consideration. Other things being equal, a trawl fishing season for halibut would be difficult to deny, creating management anarchy.

On the question of class action, to paraphrase a United States Federal Court decision dealing with the salmon fishery (United States vs. Oregon, 1968), were the commission to recognize separate sport and commercial fishing seasons, it would create *separate statuses* for the two, and it is obvious that more types or *classes* of seasons may be added. It would be constitutionally and politically impossible to resist such demands.

There has been no evidence presented during the past decade that would indicate any significant *public demand* for a sport fishing season for halibut. Thus, from a practical standpoint, there is little need for a special sport season. The present open halibut season now extends over a period of about five to six months during which halibut may be caught, with other than net gear, and used for bait, sport, subsistence, freezing, reduction, or any purpose one may choose. In fact, the season during which one may sport fish for halibut is longer than that allowed for many other species of sport fish.

The claim is often made that with an established sport fishing season, statistics could be compiled of the sport catch. This is an old and over-used argument of special pleaders. The administrative task involved in implementing a statistical system would be expensive and futile.

SUBSISTENCE FISHING

A separate season for subsistence halibut fishing is unneeded. Except for a few tribal groups there is no significant subsistence fishing and, furthermore, no need for it in the present economic climate along the Pacific Coast. The availability of low-cost home-freezing facilities makes it feasible for anyone so inclined to store an ample 12-month supply during the long open seasons.

Although the commission may be provided with advisory interpretations of its authority under the convention, the judgment of the validity of such interpretations and the actions that may be taken thereunder are within the commission's exclusive competence. The authority of the commission is derived from the convention, and its members in the United States are responsible to the president, or his specific designates. As such the commission depends *only* on the Department of Justice for its convention authority. Directives are not received from other administrative agencies (Truman, 1946).

In addition to any interpretation that might be placed upon any provision of the convention, there remains the paramount matter of intent. To provide a season for sport or subsistence fishing or any other class of fishing is contrary to the intent of the convention which is the sole guideline for the commission.

The language of treaties, conventions and even agreements between sovereign powers has been stated to be of "high dignity and should be construed and utilized to achieve the purpose or purposes sought by the signatory powers at the time of consummation of the treaty, convention or agreement."

The purposes of the 1953 convention are clearly expressed in the document as well as the large file of official reports to governments, minutes of public hearings and messages within and between governments that developed during the seven years that the 1953 convention was under consideration by the governments. At the time of the consummation of the convention there was no mention of a sport or subsistence fishing season or of any kind or class of season, only the numbers of seasons in any one year.

It should be evident that to attempt to provide, under the current convention, for the isolated and occasional instances of halibut being taken by sport fishermen usually while fishing for salmon when the halibut season is closed, would create more problems than would be solved. It would require a shadowy interpretation of the terms of the convention and a flat denial of their intent.

The repeated requests during the 1960s for a sport or personal-use season originated from a federal agency in Alaska. No similar request had ever been *initiated* by government agencies in British Columbia, Washington, Oregon, or California.

While some departmental domains might have been enhanced by having the responsibility for the surveillance of a sport or subsistence halibut season, there was little justification for the latter considering so many surveillance problems relative to the halibut fishery remained unresolved.

Moreover, there was clearly no evidence of any groundswell or public demand for either a sport or a subsistence season for halibut. Except in Alaska, halibut does not enjoy any widespread popularity as a sport fish. For the commission to take some questionable position to satisfy a local demand, or the unsupported request of some local enforcement group, would not be in the best interests of those it serves. Furthermore, to advocate and initiate such a season in order to stimulate the interest or cultivate the support of sport fishermen does not seem appropriate.

Furthermore, were a sport or subsistence season established there is no conceivable manner by which the commission could provide any type of individual bag limit *under the terms of the present convention* as well as under an earlier decision of the United States Department of Justice. Without some limit on the amount of catch that might be retained for sport or subsistence purposes by an individual during the closed halibut season, illegal traffic would be rampant and the control program a travesty.

While a sport fishery for halibut has many social and recreational values both for the individual and the community, it does not warrant the subversion of the purposes and intent of the 1953 convention. If there is strong public demand for a sport fishery and if the benefits of such a fishery outweigh the disadvantages, the convention should be amended, not subverted.

Notwithstanding the foregoing, the Halibut Commission in 1973, after the retirement of this writer, made provision for the retention of sport-caught halibut other than during the open halibut fishing season with the gear limited to a hand-held rod or line. It also provided a bag limit but no size nor possession limit.

It was also claimed that action was justified on the grounds that the sport catch was increasing and the halibut fishery declining. Estimates made in the years since 1973 indicate that the sport catch is still relatively small. Much more is eaten as "cook fish" by crews of commercial vessels that catch halibut. Only in Alaska was the estimated catch of any significance, about 200,000 pounds in 1976. How much of the total should have been in the subsistence or comsport category is not indicated. The total is not much higher than was estimated to have been taken many years ago.

It was claimed in the commission's 1973 annual report that there was an untenable enforcement situation because heretofore the commission did not "encourage attempts by authorized federal agents to apprehend sport fishermen who violated the (halibut) regulations."

It should be made clear by one who was intimately familiar with enforcement problems in both countries over a period of 40 years that the commission never encouraged nor discouraged any agency from carrying out its duties under the convention. It would have been most inappropriate and presumptuous for the Halibut Commission to have so meddled in the exclusive responsibilities of the respective countries. Also, such a statement as quoted above from the commission's 1973 annual report had no basis in fact. However, when requested the commission always assisted the respective enforcement agencies in their difficult responsibilities. It could not, and did not however, accede to each and every request of local agencies to place this or that restriction upon those engaged in the halibut fishery. This was particularly true when the restrictions had negligible conservation value or did not further the stated objectives of the convention. Some proposals were regarded as redundant and harrassing.

16 OVERVIEW

Along the 4,000-Mile Arc of the Northeastern Pacific Ocean and the Eastern Bering Sea

Every section of the coast from the Mexico-United States border to Attu Island has made some distinctive contribution to the history of the Pacific halibut fishery. In this condensed geographical overview, some notable instances will be mentioned as one moves along this 4,000-mile coastal margin of the Northeastern Pacific Ocean and the eastern Bering Sea. The variety of the features chosen may interest different readers.

A SOUTHERN RECORD!

The southern limit of the halibut story is within 10 miles of the United States border with Mexico. At 32°45′ north latitude, the jig boat *Albatross,* fishing south out of San Diego, California, in 1942 caught a 42-pound halibut, *Hippoglossus.* This is the southernmost recorded occurrence for the species.

A SHORTLIVED BONANZA

North 1,100 miles lies the Lightship Ground, a shoal region off the mouth of the mighty Columbia River. In 1916 through an inebriated crew member in a Seattle saloon, it came to be known that the Puget Sound dory vessel *Tom and Al* had in late April encountered extremely heavy halibut fishing near the Columbia River Lightship. By May, practically the entire Puget Sound halibut fleet and some Canadian vessels from Vancouver were fishing the hitherto secret ground. That month alone, 3 million pounds of halibut were taken from the relatively small area that came to be known as the Lightship Ground (Figure 144). One of the vessels was the S.S. *Chicago*, a large 149-net-ton steamer owned by Chlopeck Fish Co. of Seattle, which with 12 dories caught 320,000 pounds—the second largest single fare of halibut ever taken in the Pacific halibut fishery! This record catch which was landed in Seattle, was caught in less than two days fishing time on the Lightship Ground.

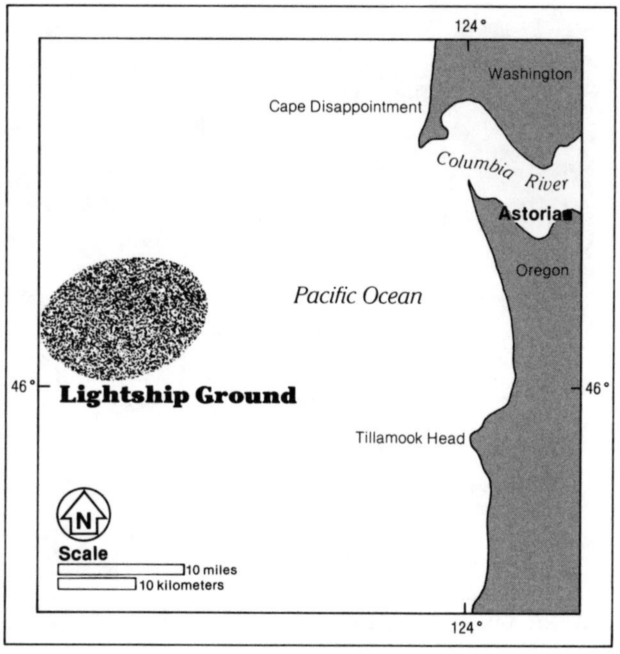

FIGURE 144. Chart of Lightship Ground off the Columbia River.
F.H. Bell

Notwithstanding the phenomenal initial productivity, the "Ground" was literally cleaned out in a few months. Despite a practically complete cessation of fishing in the region during the subsequent 50 years, there has been no recuperation. On September 16, 1966, a halibut vessel under charter to the Halibut Commission fishing 28 skates (3,360 hooks) over the identical ground, caught but four halibut with a total weight of 82 pounds.

RIGHTS OF RED MEN

Farther north Cape Flattery in Washington brings to mind the *Oscar and Hattie*, the Makah and many other names important in the early halibut fishery. Long before the schooner *Oscar and Hattie* from New England inaugurated in 1888 the modern vessel fishery off Cape Flattery, an Indian tribe, the Makah, had been at the "Cape." This tribe, along with the Haida of British Columbia and Tlingit of Southeastern Alaska, were the original master halibut fishermen of the Pacific Coast. Halibut, fresh, dried, or smoked was a staple food of the Makah and some were used in trade with interior Indians. The 1856 Treaty between the Makah and the United States, dealing with fishing and other traditional rights, has been the subject of continuing tribal petitions, claims, litigation, and negotiations with the United States government for the past 100 years.

HALIBUT FISHING IN SUBURBIA!

Proceeding through the Strait of Juan de Fuca, one realizes that the fishing for halibut in early times was conducted practically in the suburbs of what are now great metropolises. In the 1890s small sailing sloops fished halibut at the south end of Whidbey Island and at Point No Point, a few miles out of Seattle harbor. Even today, some halibut are still taken on these grounds in the spring months by sportsmen fishing primarily for salmon. Race Rocks, within sight of Victoria, the capital city of British Columbia, was also an early favored location for small boats operated chiefly by Italians fishing halibut for the city market. English Bay, now the outer harbor of Vancouver was in the 1880s a fishing location for a few small boats

FIGURE 145. A Tlingit halibut crest in Saxman, Alaska, on Revillagigedo Island. Conversion of the building from a community house to a boat house damaged the Indian artwork.
F.H. Bell

catching halibut for the local trade of Gastown, as Vancouver was earlier called. Some were taken in Burrard's Inlet (now Burrard).

BIRTHDAY OF A DEEP-SEA FISHERY — SEPTEMBER 20, 1888

One cannot leave this Northwest region, certain to become one of the world's great centers of population, without recalling that it was in Tacoma, in what was then the Washington Territory, that the modern commercial halibut fishery was born. On September 20, 1888, express car number 14044 of the Northern Pacific Railroad laden with iced fresh halibut highballed for the East Coast. The fish had been landed that day by the *Oscar and Hattie* in the Cape Flattery Grounds—the first carload of fresh Pacific halibut shipped east.

PACIFIC RAILHEADS

Three railroads that were destined to play most important roles in the development of the halibut fishery established their termini in this region. The Canadian Pacific Railway was located in Port Moody in 1885 with a branch line to Vancouver and Kitsilano Beach. The Northern Pacific Railroad was located in Tacoma in 1887 with a branch to Seattle, and the Great Northern Railroad was in Everett in 1893 with a spur to Seattle.

"BULL OF THE WOODS" AND A BACKFIRE

One departs from this populous region to again proceed northward, leaving much unsaid about halibut. On the west coast of Vancouver Island, there is a secure habor called Sydney Inlet, off which there once existed a small but extremely productive halibut ground. It was discovered in a rather whimsical manner. Captain Ben Joyce, who was the skipper of the S.S. *Edith* out of Tacoma, the first large Puget Sound halibut steamer, was asked by an inexperienced crew on the halibut vessel *Queen of the Pacific* for a good place to fish for halibut. Captain Joyce, not wishing to feel responsible for any mishap that might befall the rather poorly built vessel in the event of bad weather, told the greenhorn crew to fish close by what he knew to be a good harbor. The *Queen of the Pacific* was almost universally known to the men of the halibut fleet as the "Bull of the Woods" because it was prone to spend a great deal of time in harbor. Much to the surprise of everyone (including Captain Joyce) the *Queen of the Pacific* took on board, in half a day, nearly 100,000 pounds of halibut just outside the anchorage! And thus was the Sydney Inlet ground discovered!

A CONSERVATION DISASTER

The Goose Islands grounds are located at the north end of Vancouver Island in Queen Charlotte Sound (Figure 146). These grounds have produced over 140 million pounds of halibut during the 50 years since 1928 and probably over 500 million since the time they were first fished in 1903. The S.S. *New England*, a United States vessel fishing out of Vancouver is credited with their discovery in 1903 as well as with their subsequent early exploitation along with other large steamers.

In the 1970s the annual catches from Goose Islands grounds have averaged below 1 million pounds compared to a recent high of 7.5 million in 1954. The growing domestic trawl fishery and many night forays in the 1960s of large foreign trawlers on these grounds from the deeper water perch grounds contributed to the decline. There may also have been a decline in recruits from such distant regions as the Bering Sea where the young halibut have been decimated by foreign trawlers. However, one cannot rule out the possibility of some natural decline in recruits as well.

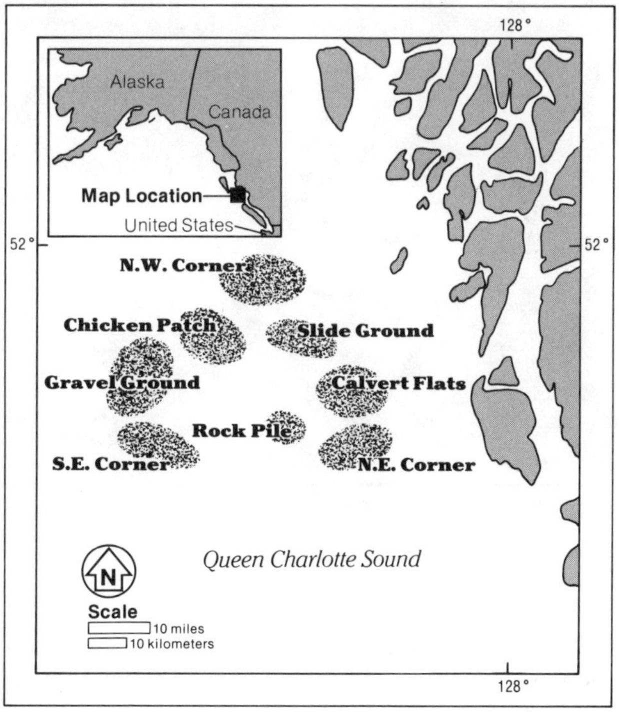

FIGURE 146. Chart of Goose Islands grounds
F.H. Bell

FIGURE 147. The crew of the S.S. *New England* with halibut on deck.
International Pacific Halibut Commission

A DAY'S CATCH OF HALIBUT

Off British Columbia in Hecate Strait there are many places that have a long history and a present importance to the fishery. One could linger long in this region recounting the many significant events that have occurred over the years, but only one will be cited. On August 31, 1904, the S.S. *New England*, under Captain A. Freeman, fishing 12 dories 10 miles southwest of Dundas Island caught 188,000 pounds of halibut, the largest single-day catch on record during the 90-year history of the Pacific halibut fishery.

A RESILIENT FISHING GROUND

At the northern end of Hecate Strait are the Rose Spit and Two Peaks grounds, a region of high halibut productivity. These grounds were first fished by the *Oscar and Hattie* in 1888 and were the source of many large trips by the steamers, both United States and British, in the early part of the twentieth century. Despite heavy fishing pressure, the stocks possessed great resilience and the region is still moderately productive in the early 1980s. The Seattle seven-dory schooner *Republic*, under Captain William B. Hegge, took over a million pounds from Rose Spit in 1915. Thirty-nine years later, in 1954 the Prince Rupert

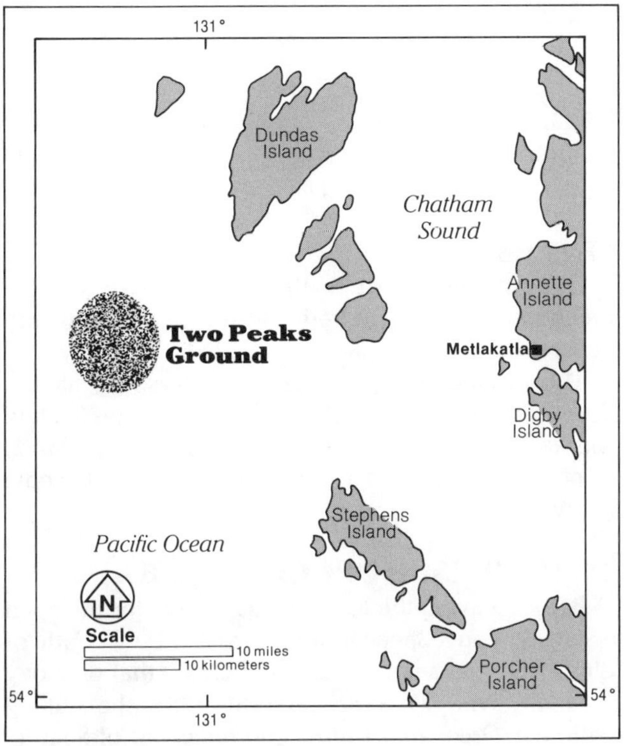

FIGURE 148. The Two Peaks grounds
F.H. Bell

vessel B.C. *Producer,* under Captain Fred Kohse, with a six-man crew on the Two Peaks ground caught 145,000 pounds in an eight-day special August season including the off-loading of two portions of the catch in nearby Prince Rupert. The vessel well deserved its name!

THE HALIBUT CAPITAL OF THE WORLD

Prince Rupert is also well-deserved of its designation, the "Halibut Capital of the World." Since becoming the Pacific terminus in 1908 of the Grand Trunk Pacific, later called the Canadian National, more halibut has been landed in Prince Rupert than in any other place in the world—about 1.1 billion pounds from 1908 to 1975, taking over the capital title from Seattle in 1915. By the mid-1970s the reputation of Prince Rupert as the "Halibut Capital of the World" was closely contested by Kodiak in Western Alaska.

FIGURE 149. Prince Rupert, British Columbia
Eugene Q. Mah

THE BROTHERS GROUND

As in Hecate Strait, the extensive channels and the outside waters of the Alexander Archipelago that make up Southeastern Alaska can contribute

FIGURE 150. The S.S. *Independent*
International Pacific Halibut Commission

innumerable interesting items and anecdotes to the halibut saga. The initial production from some of the primeval stocks of halibut on some locations in the Inside Passage was phenomenal. One relatively small ground near the Brothers, a group of small islets at the south entrance to Stephens Passage, witnessed some very productive fishing. The 12-dory steamer *Independent* (Figure 150) alone in 1912 caught 2.3 million pounds on 13 trips to the Brothers. Also, unlike the bonanza fishing at the previously mentioned Columbia River Lightship Grounds, the Brothers Ground continues to this day to provide moderate yields of halibut.

FIGURE 151. Petersburg, Alaska
Sharon Paul, reprinted from *The MILEPOST®*

PETERSBURG

One cannot depart Southeastern Alaska without mention of the town of Petersburg with a long history of participation of its people in the halibut fishery.

The production record of the Petersburg halibut fleet has been the most consistent of any fleet in Alaska. Because a large proportion of the population is of Norwegian extraction, the town has become known as "Little Norway."

TO REMINISCE OFF GLACIER BAY

Before leaving the Inside Passage to enter the open ocean at Cape Spencer one goes by the mouth of Glacier Bay. It is not difficult to imagine that on a day in June over 90 years ago in 1888 the sailing schooner *Oscar and Hattie,* 800 miles out of Seattle, could be seen coming out of Glacier Bay laden with glacier ice and bound for fishing grounds in Hecate Strait.

ON THE OPEN OCEAN

Once outside Cape Spencer, conditions change. While the fishery in this region started as late as 1913, its history has been eventful and often tragic. With

FIGURE 152. A glacier enters the sea at Muir Inlet, Glacier Bay, in Southeast Alaska.
William D. Boehm, reprinted from *ALASKA GEOGRAPHIC®*

FIGURE 153. Halibut vessel in heavy sea
International Pacific Halibut Commission

most of the fishing conducted in offshore open ocean waters, the vessels used were larger than those fishing south of Cape Spencer, for the hazards of the sea were greater and the weather more rigorous, particularly during the winter fishing season. There have been many tragic losses.

CATASTROPHE!

On a frigid stormy November day in 1916, the Canadian halibut steamer *Onward Ho* out of New Westminster with 36 men was last seen in the failing light of early afternoon by the United States halibut vessel S.S. *Manhatten*. The *Onward Ho* was heading into an icy gale from the northeast sweeping down off the 150-mile-wide Malaspina Glacier.

The *Onward Ho* was loaded with halibut and probably attempting to make Cape Spencer 100 miles to the east. To the S.S. *Manhatten*, which lay hove to in the cold gale, the *Onward Ho* appeared iced down and heavy in the water. The *Onward Ho* was never seen again—not even a trace of the wreckage or the 36 brave men.

FIGURE 154. An iced-down halibut vessel
International Pacific Halibut Commission

FIGURE 155. The M.V. *Idaho,* which later became the *Eider* of the United States Bureau of Fisheries.
International Pacific Halibut Commission

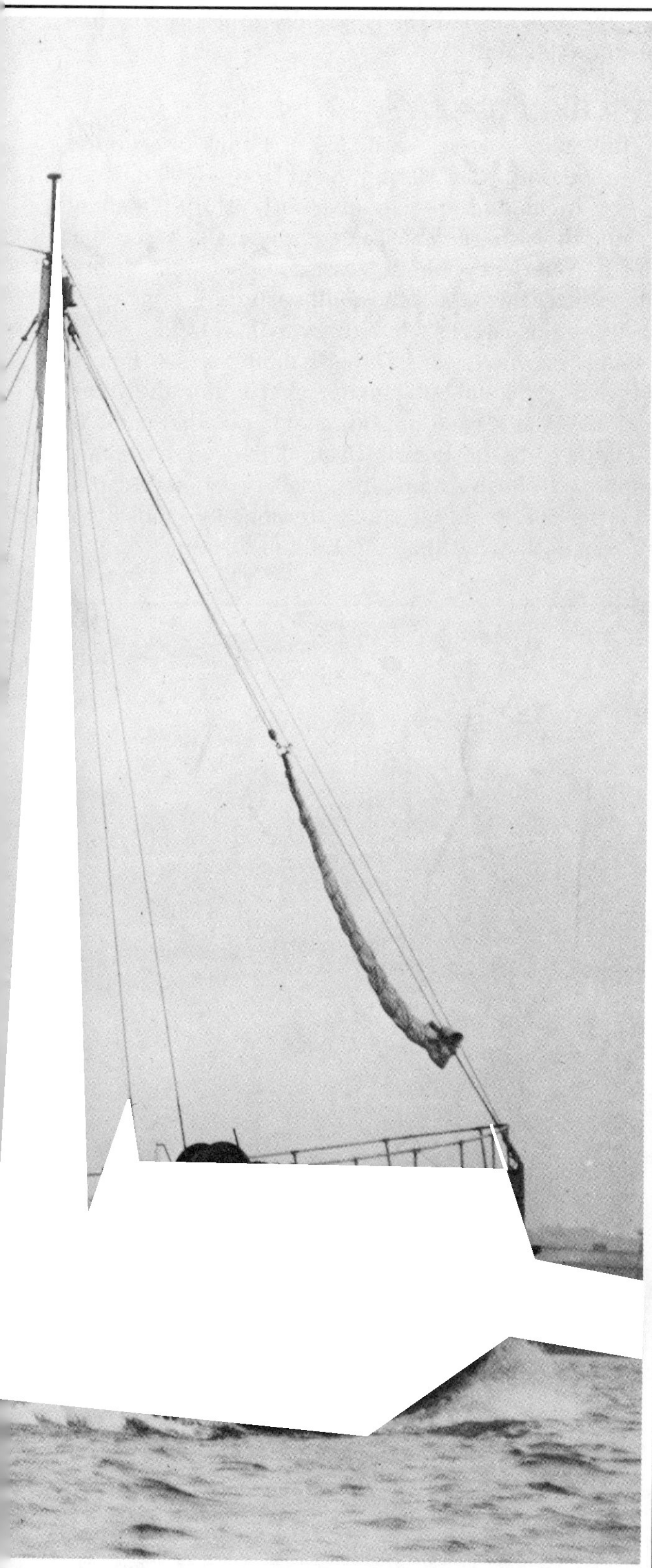

A SPAWNING CONGREGATION — YAKUTAT SPIT

Three years before the *Onward Ho* disaster, the Seattle motor schooner *Idaho*, under Captain T.P. Quinn, fished off Yakutat Bay on the continental edge in water between 175 to 225 fathoms where the halibut were congregated in extremely dense schools for spawning from November to February. This new ground came to be known as Yakutat Spit. The phenomenal fishing encountered by the *Idaho* soon became generally known, and most of the more seaworthy vessels in the halibut fleet ventured into the stormy winter waters of the open ocean beyond Cape Spencer to share in the bounty.

DISCOVERING THE "W" GROUNDS

Proceeding west towards Cape St. Elias, one recalls the discovery of another famous fishing ground, the "W," where halibut congregate for winter spawning. This ground was first sounded out in the fall of 1913 by Captain John Daniels on the *Tom and Al* out of Seattle. The stocks were dense. On one trip that fall, the *Tom and Al* caught 155,000 pounds on 63 skates of seven-line dory gear—over 2,000 pounds per skate, which is extremely productive fishing. The name of the ground is derived from the "W" that can be seen on the face of a mountain in the Chugach-St. Elias range from a point 35 miles east of Cape St. Elias.

Many events of greater or lesser importance in the history of the fishery have occurred in the Gulf of Alaska. Vessels were lost and many dorymen and others died in the rigors of the open ocean fishery, conducted to a large extent on the edge of the continental shelf.

DEEP LIE THE HALIBUT

Prince William Sound is located at the northern arc of the gulf. Although the sound was not a productive area for halibut, a few vessels in the 1930s conducted a very shortlived fishery in a deep fjord and encountered catches of halibut at a depth of 600 fathoms, a record depth for commercial catches of the species on this coast.

THE LITTLE HALIBUT CAPITAL OF THE WORLD

Near Kodiak Island the continental shelf is very wide, forming what is called Portlock Bank, the largest of all the North Pacific halibut fishing grounds. With Kodiak Island practically surrounded by productive halibut fishing grounds, it is no small wonder that the town of Kodiak was, in the 1970s, fast gaining a reputation as the "Little Halibut Capital of the World."

DEEP WATER FISHING OFF TRINITY ISLANDS

Near the south end of Kodiak Island lie the Trinity Islands, Sitkinak, Tugidak and Aiaktalik. Good fishing was encountered off these islands in the spring of 1926 at 550 fathoms, unusually deep for productive halibut fishing in the Northeastern Pacific.

IT HAPPENED AT SHUMAGIN IN 1778 AND 1916

At Shumagin Islands on a summer Sunday, about 200 years ago, Captain Cook wrote of replenishing his ship's stores with 100 halibut caught in the space of one hour by the crew while at anchor.

Very much later, in 1916 the S.S. *Starr,* a large halibut fishing steamer under Captain Keough and owned by the San Juan Fishing and Packing Co., Seattle, first prospected the Shumagin Grounds for halibut. By the mid-1920s, a substantial fishery had been established.

THE BERING SEA

Still farther westward lies Unimak Pass, the entrance into the Bering Sea. Several significant events highlight the relatively short history of halibut fishing in this large shallow and generally arctic sea. North American halibut vessels occasionally fished those distant waters in the southeastern Bering Sea as early as the late 1920s. However, it was the heavy fishing encountered by the halibut vessel *Polaris* under a 1956 halibut tagging charter for the International Pacific Halibut Commission on what came to be known as the Polaris Ground that gave a sharp impetus to North American vessels to fish in the area. By the early 1960s all "reasonably indicative evidence" showed that the Halibut Commission had

FIGURE 156. The S.S. *Starr*
Webster and Stevens

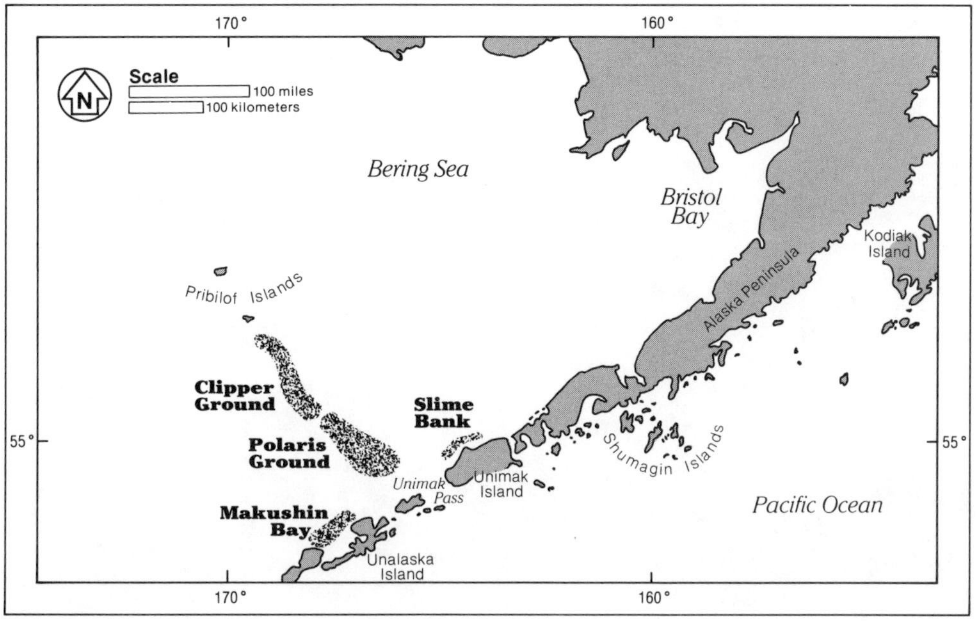

FIGURE 157. Southeastern Bering Sea halibut grounds
F.H. Bell

been successful in developing a fishery in the area that was fully utilizing the stock of halibut in the region.

A SURRENDER TO "SUBTLETY"

In 1962, the United States and Canada, in what was to be a "subtle" means of convincing Japan to curb its pressure upon the salmon stocks, agreed after four years to the contrary, that proof of full utilization of the halibut in the eastern Bering Sea by Canada and United States was lacking and the abstention provision of the Tripartite Convention with Japan was no longer applicable. The fact that the convention only required reasonably indicative evidence, a vastly different criteria than proof, was disregarded. In 1963, the Tripartite Commission, which had assumed control of the fishery, established on its own initiative, a completely unrealistic catch limit of 11 million pounds for a small fishing ground in the region. Needless to say the halibut stocks were almost destroyed in one season by the combined setline fleets of Canada, United States and Japan.

DISASTER!

Consequently, the following year, 1964, brought heavy economic losses for many North American halibut vessels in Bering Sea. The Halibut Commission, on its own initiative, recommended complete closure or at the most a token fishery in the area for 1965. While there was a subsequent slight recovery, any significant response has been eliminated by losses accruing to the halibut stocks from the massive foreign bottom trawl and Danish seine fishery developing in the region.

A RECORD TRIP

The Bering Sea may be pleasantly remembered for the all-time record halibut catch made anywhere by a halibut vessel with a total fishing crew of only six men including the captain. The *Eclipse,* under Captain Adolph Samuelson, was on a tagging charter to the Halibut Commission in 1967. The single trip of 180,700 pounds was taken in 10 days' fishing on grounds in close proximity to St. Matthew Island at 60°30′ north latitude. While it was a seasonal concentration, this is the farthest north that a full fare of Pacific halibut had ever been taken. The accumulation was quickly cleaned out, reminiscent on a small scale of what occurred 50 years previously on the Columbia River Lightship Ground off Oregon.

THE END OF THE ROAD

The Bering Sea, at the end of a long road from San Diego where this overview commenced, probably once possessed an important key to the continued existence of the Pacific halibut fishery. Prior to the multi-billion-pound foreign fishery in the region, the southeastern Bering Sea was populated by 65 million young two- to four-year-old halibut from spawning grounds in the Gulf of Alaska and off British Columbia. In turn these young fish were an important source of recruitment to the fishable stocks supporting the Canadian and United States fishery as far south as the coast of California.

17

BIOGRAPHICAL SKETCHES

Among the thousands of vessel captains, fishermen, dealers, union officials, administrators and scientists who have contributed to the development and management of the Pacific halibut resource during the past 85 years, there are a few whose contributions have been particularly notable.

The Public Sector. Edward W. Allen, John P. Babcock, Henry A. Dunlop and William F. Thompson are citizens and public servants who by their deeds and writings made lasting imprints upon the events of the fishery.

The Private Sector. William C. Calvert Sr., Absalom Freeman, Miller Freeman, Alvah L. Hager, E. Aemilius Jarvis, Harold E. Lokken, Mattias Madsen, Eugene A. Ruthford, Grier Starratt, and Roy M. Winslow worked to create a sound industrial structure for the halibut fishery and consistently supported the conservation of the resource.

The Scientific Sector. Fedor I. Baranov, William E. Ricker and Michael and Georg Ossian Sars were scientists whose basic and pioneering studies had a substantial influence upon Halibut Commission research.

There are many others whose contributions have been notable, but space precludes their naming.

EDWARD WEBER ALLEN 1885-1976

A distinguished citizen of the United States, the State of Washington and the City of Seattle with 58 total years of public service on three international fisheries commissions between 1932 and 1973.

FIGURE 158. Edward Weber Allen
International Pacific Halibut Commission

Edward Weber Allen was born in 1885 in Oshkosh, Wisconsin. He attended Oshkosh State Normal School and the University of Chicago. Allen received his legal education at the University of Washington (LL.B. 1909) and was admitted to the American Bar Association the same year.

In 1913 he was the deputy prosecuting attorney, Pacific County, Washington, and from 1914 to 1916 was the assistant attorney general of Washington. After that time he was engaged in general civil practice with a considerable number of clients associated with Alaska, including the fisheries.

In 1947, Allen was chosen Alumnus Cum Laude of his alma mater, the University of Washington. He was author of an authoritative study of the North Pacific explorations of LaPerouse drawn from original documents in France. His book, *North Pacific*, reviews the fisheries of Alaska and other developments in that region including his detention by Japan when suspected of being in a prohibited area prior to Pearl Harbor. He published other works of a lighter vein, both of prose and poetry, and wrote many articles on international law for various law journals.

Appointed by President Herbert C. Hoover, he served on the International Fisheries (Halibut) Commission 23 years from 1932 to 1955, twice as long as any commissioner in either Canada or United States. He was appointed to the International Pacific Salmon Fisheries (Sockeye) Commission by President Franklin D. Roosevelt and served from 1937 to 1953. In 1954 Allen was appointed by President Dwight D. Eisenhower to the International North Pacific Fisheries Commission and served until resigning to President Richard M. Nixon on December 14, 1973.

In addition to the above gratuitous service to the United States, Allen was involved in Secretary of State Cordell Hull's negotiations with Japan in the 1930s pursuant to Japanese salmon fishing in Bristol Bay. He participated in the FAO Conference in Baguio, Philippine Islands, in 1948; was a member of a three-man advisory committee to General Douglas MacArthur in occupied Japan in 1949; and was a member of the United States Delegation that negotiated the International North Pacific Fisheries Convention with Japan and Canada in 1951.

It will be difficult to ever equal Edward Weber Allen's 58 total years of public service on three international fisheries commissions, his service to the State of Washington and his broad and prolonged participation in the business and cultural life of Seattle.

JOHN PEASE BABCOCK 1855-1936

During his lifetime of public service and devotion to scientific fisheries administration, Babcock initiated and assured the consummation of both the Halibut and Sockeye Salmon conventions.

FIGURE 159. John Pease Babcock
F.H. Bell

John Pease Babcock, LL.D., was born in St. Paul, Minnesota, in 1855. On leaving college he went to San Francisco with the California Fish and Game Commission. His work on the salmon of the Sacramento River came to the attention of the British Columbia government, and he was appointed fisheries commissioner in 1901. With the elevation of the commission to a department, he became deputy commissioner.

After serving on the Boat Rating Commission of 1910 he went back to California but returned to British Columbia in 1912 to become assistant to the commissioner of fisheries, a position he held until 1934 when he retired due to ill health. He had become a Canadian citizen in 1912.

His education and early experience had convinced him that trained experts must be employed full-time to solve the complex biological problems involved in the management of renewable resources. Although the Department of Fisheries did not have a scientific staff of its own, Dr. Babcock recognized that the problems of the commercial fisheries required the services of persons other than just university professors who were being employed intermittently during their summer escapes from academia. Such casual employment had been the rule in North America up to that time. It had solved few problems for the fisheries.

Babcock sponsored the work of Dr. Charles H.

Gilbert of Stanford University on the Fraser River sockeye salmon and its continuation by the doctors Clemens, husband and wife, from the University of Toronto. The work became an early classic in the field of salmon research. The preliminary studies of W.F. Thompson on halibut, that ultimately led to the establishment of the Halibut Commission, were sponsored and published by the Provincial Fisheries Department. Later work of such specialists as Dr. John L. Hart and Dr. D.B. Quayle on pilchards and shellfish, respectively, were subsidized by the department. As California led all other states in the Union in furthering applied fishery research, British Columbia led all other provinces in Canada.

Dr. Babcock, who had received a Doctorate of Letters from the University of British Columbia, had the full confidence of the industry, although the management of the fisheries was a federal responsibility. Dr. Babcock always recognized the contributions of the younger fishery biologist.

FEDOR ILICH BARANOV 1886-1966

First to formularize in mathematical terms the basic principles of population dynamics.

FIGURE 160. Fedor Ilich Baranov
International Pacific Halibut Commission

Fedor I. Baranov was born in Orel, Russia, in 1886. He attended St. Petersburg Polytechnic Institute and graduated in 1909 with a thesis on the effects of a ship's roll on a gyroscope.

In 1911 he secured employment in the Fisheries Division of the Department of Agriculture and in 1915 became a professor in the Department of Commercial Fisheries of the Moscow Agricultural Academy. In 1930 he joined the Moscow Fishery Technical Institute where he remained until he retired in 1959. In 1950 the institute was moved to Kaliningrad but after his retirement, Baranov returned to Moscow where he continued to write for fishery journals until his death in 1966.

Baranov's scientific contributions followed two paths: fishing gear and population dynamics. His textbook on the theory and construction of fishing apparatus went through many editions from 1933 to 1960, and he is recognized in the USSR chiefly for his work on fishing gear.

In the field of fishery biology, Baranov was the first person to formulate in mathematical terms the basic principles of population dynamics. In 1914 he wrote a short paper comparing fishery management to forestry, and in 1916 he presented the classic and definitive study "On the Question of the Biological Basis of Fisheries." It was published in 1918. This is a landmark contribution.

Although Baranov engaged in controversy within Russia's scientific community, most of his precepts are to a large extent still applicable to conditions prevailing today in many countries, including the USSR. In his 1916 treatise, he deplored the existence of "so much obscurity in understanding the term overfishing." He felt the lack of reliability of statistical material (sample sizes) was due to the fact that "the energy of investigators is being dissipated on very diverse problems instead of being applied to the one that leads directly to the goal." He also believed that "a systematic study of biological statistics provides a means (and the only one) for establishing control over the condition of a fishery and such control should be one of the principal problems for existing fishery laboratories."

Such criticisms did not endear him to the scientific hierarchy in Russia. Among other things, Baranov's use of mathematics in the formulation of the principles of population dynamics and, more importantly, his demonstration that fishing could have a major impact upon the stocks of exploited species were not favorably received. His critics seem to believe that in some vague and mystic manner the removals by man were somehow exempted from being part of the total mortality.

This philosophy which still prevails was a popular one in a country whose fishery policy still underscores the uncontrolled exploitation of distant water fisheries off the coasts of other countries.

Due to these circumstances Baranov was assigned after 1930 to work in the field of fishing gear for the rest of his life rather than being permitted to work in the field of population studies. However, he persisted in espousing his views in a forceful presentation of them at an all-Russian conference in 1951. They were again summarily rejected. Baranov passed away in 1966 at 80 years of age.

Baranov gave favorable notice to the applied fisheries research that was carried on so extensively in the United States and Canada and referred particularly to that on halibut and Pacific salmon.

WILLIAM C. CALVERT SR. 1868-1933

A successful industrialist, banker and strong believer in the need for restraint in using the sea's renewable resources, Calvert was a major contributor to the successful conclusion in 1923 of the initial Halibut Convention.

FIGURE 161. William C. Calvert Sr.
International Pacific Halibut Commission

William C. Calvert was born in Montreal, Quebec, June 24, 1868. His family, after moving to North Dakota, experienced the usual physical hardships of those pioneering times. They also underwent severe economic adversity. At the age of 29, Calvert decided to "go west." Arriving in Seattle in 1897 with his brother James, Calvert engaged in the stationery and printing business. Failing in this venture, he was tempted to try his fortunes in the gold fields of the Klondike. Instead, he recognized early that all that glitters is not gold and apparently concluded that more lasting values were to be found in the renewable resources of the fisheries.

To this end he and his brother, undaunted by previous business failure and in company with Edwin Ripley, bought the Seattle Fish Co.*, a division of the bankrupt Pacific Packing and Navigation Co., an overly promoted and overly capitalized fisheries conglomerate in Bellingham. In 1903 the Calverts and Ripley incorporated their enterprise under the name of the San Juan Fishing and Packing Co.

From a modest beginning of supplying the local Seattle market with fish and ice from a small dock at the foot of Pike Street, the company, known to most as "San Juan," within five years was able to build a large cold storage and fish processing plant at Pier 31, from which to ship its products (including fresh and frozen halibut) to all parts of the United States. Under the imaginative direction of William, both brother James and Edwin Ripley having withdrawn from the company, San Juan expanded into all sections of the fisheries in Oregon, Washington, British Columbia and Alaska. It soon became a major producer and distributor of halibut and in 1904 built the large halibut fishing steamer S.S. *Starr* and later the S.S. *San Juan* which was launched in January 1905. These events made Seattle the premier halibut port of Puget Sound, superseding Tacoma.

By 1907 the S.S. *Grant* was added to the vessels fishing halibut for the company. As the independently-owned fleet developed and the halibut fishery extended to Alaska, the company also kept pace. In 1917 San Juan built and operated a cold storage plant at Seward where it could service the halibut fleet fishing the Gulf of Alaska. This was a limited operation at the time and was terminated in 1931. It might have interested Calvert to know that 50 years later Seward would become a major halibut landing port of the Pacific Coast.

Although the company became very active in the salmon fishery in Alaska and in other divisions of the fishing industry, the distribution of halibut remained an important phase of the business both in the United States and Canada. Calvert was a major force in furthering the 1923 Halibut Convention in the United States and a strong supporter of the Halibut Commission thereafter.

Calvert not only created a sound corporate structure in San Juan, he endowed it with three sons, Laurence, Starr, and William, who continued the family traditions in the business and greatly expanded the scope of the enterprise. In addition, Calvert

*This was the successor to Ainsworth and Dunn, the pioneer fish dealing firm in the Pacific Northwest, dating back into the 1880s.

attracted to the company able men of integrity, such as Harry J. Tillman and Eugene R. Ruthford, both of whom "made their mark" on the fishing industry.

The years passed and San Juan Fishing and Packing Co. in 1962 merged with the expanding structure of the New England Fish Co. Although the words *San Juan* may have now lost much of their import in the fisheries, William Calvert's achievements and personal integrity are indelibly recorded in the business and fishery history of the Pacific Northwest and Alaska. Calvert passed away June 28, 1933.

HENRY ADAM DUNLOP 1898-1966

Second director of investigations, holding that position for 23 years to retirement.

FIGURE 162. Henry Adam Dunlop
International Pacific Halibut Commission

Henry (Harry) A. Dunlop was born in Dunrea, Manitoba, on July 8, 1898. He began his education in Manitoba prior to coming to Vancouver where he graduated from Britannia High School and entered the University of British Columbia. He received his B.A. in 1919 and M.A. in 1922 in zoology under Dr. C. McLean Fraser, a prominent fishery-oriented marine biologist. He was instructor in zoology from 1920 to 1924 at the university. He continued his graduate studies at the University of Toronto and the University of Washington working on the embryology of salmonids.

In 1925 he was appointed assistant director of investigation for the Halibut Commission and was acting director in 1937. In 1940, he became director and held that position for the next 23 years until retirement in 1963. During his tenure of the position the rehabilitation of the resource continued, and by the early 1960s, the stocks generally had attained their optimum size. This was accomplished under adverse and trying circumstances.

Maintaining a research program during World War II was most difficult with a scientific staff at one time reduced to three persons including the director. Also, funds allotted for research and management were steadily eroded by inflation. In the postwar period prolonged and frustrating efforts over a period of seven years were required to secure treaty authority that would permit measures to alleviate the drastic reduction that had occurred during the fishing seasons. In the 1950s, further demands were placed on the staff to "prove in" certain commitments made by Canada and the United States with Japan, culminating in the eastern Bering Sea debacle in 1962 when the two countries abdicated their valid exclusive claims to the halibut stocks in that region.

Dunlop was deeply concerned over the threat the fishing fleets of Japan and the USSR posed for the Pacific halibut stocks, and he feared for the survival of the Canadian and United States halibut fishery.

Dunlop passed away on May 3, 1966, at the age of 68. He had returned to Vancouver upon retirement in 1963, having lived in Seattle during his service with the Halibut Commission. As director for 23 years, he had the longest tenure of any incumbent.

ABSALOM FREEMAN 1863-1930

A man of broad interests, Absalom Freeman was a master fisherman and mariner, historian, cartographer, entrepreneur and conservationist.

FIGURE 163. Absalom Freeman
Canadian Fishing Company, Ltd.

Captain Absalom Freeman was born at Trinity Bay, Newfoundland, in 1863. At the age of eight he went to sea on his father's schooner, fishing herring off Labrador and cod on the Grand Banks. Sealing voyages out of St. John's, Newfoundland, were also part of the young man's indoctrination into the fisheries and maritime life.

Having mastered the many skills of his profession, Absalom Freeman came west arriving in Vancouver, British Columbia, in 1890. At that time the consolidation of the Burrard's Inlet Towing Company with the New Zealand-based Union Steamship Co. resulted in a fleet of coastal freighters and passenger vessels in Vancouver. Absalom Freeman served on their 324-ton passenger vessel *Cutch*, the 76-foot tug *Skidegate* and the S.S. *Glide*. He became first mate of the company's 200-passenger vessel S.S. *Comox,* which had been brought out from Scotland in sections with the S.S. *Coquitlam* and S.S. *Capilano.*

In November, 1893, the New Westminster Fish Co. chartered the *Capilano* from the Union Steamship Co. and Captain Freeman was in command. He established a halibut shore station on Banks Island producing about 200,000 pounds by March, 1894.

In the fall of 1894 Captain Freeman operated the *Capilano* out of Vancouver for the New England Fish Co. In conjunction with Captain Hanson B. Joyce, another experienced fisherman and "high liner" from the Atlantic, Freeman and Joyce developed a successful halibut operation where others had failed before. From experience gained over several years, the same team designed the first steamer built specifically for Pacific Coast halibut fishing, the S.S. *New England*, a vessel that was destined to be eminently successful.

On April 30, 1906, Captain Freeman, Captain Johnson and three Vancouver pharmacy owners formed the Canadian Fishing Co., which operated its own vessels. The group sold their production to the halibut distributing companies including the New England Fish Co. and the B.C. Packers. In 1908 the Canadian Fish Co. became a subsidiary of New England Fish Co., and its vessels the *Celestial Empire* and *Flamingo* became part of the New England Fish Co. fleet.

Captain Freeman's record as a successful operator of large dory vessels is almost a Pacific Coast legend. He discovered many famous fishing grounds off the British Columbia coast, and made reconnaisance charts of many portions of the coast, some of which were used for many years pending completion of official surveys. Freeman's detailed records, supplemented by a "picture" memory, made it possible to preserve an important segment of Northwest history.

His seafaring skill and knowledge of the sea are evident in the fact that for the many thousands of launchings of dories made by him whether in westerlies or dry southeasters of the summer or in bitter winter storms, no man nor vessel under him was ever lost.

We find his name commemorated in Freeman's Pass into Kitkatla Inlet, Porcher Island, on the British Columbia coast. It was there that the S.S. *New England* made the largest single day's catch of halibut on record: 212,000 pounds.

His first-hand knowledge of the reaction of the resource to overfishing was primarily responsible for the New England Fish Co. being in the forefront in Canada's promotion of the international control of the fishery.

LEGH MILLER FREEMAN 1875-1955

Miller Freeman was a self-educated, self-made, industrial-journal publisher, public-spirited citizen and conservationist. He was also a leader in furthering international management of the Pacific halibut and Fraser River sockeye salmon. He founded and published the *Pacific Fisherman,* the most prestigious chronicle of the Pacific fishery events for a period of 65 years.

FIGURE 164. Legh Miller Freeman
International Pacific Halibut Commission

Legh Miller Freeman was born in Ogden, Utah, in 1875. At the age of four after the accidental death of his mother on the "wagon trail westward" he was entrusted to the firm but benevolent discipline and teaching of Catholic sisters, for his father, the crusading publisher of the *Frontier Index* newspaper, moved from place to place. When Miller was eight, his father settled in Yakima, Washington, and Miller lived nearby with a farm family. He was able to attend the Ahtanum Creek one-room school for one three-month winter session. He then went to work for his father, spending twelve hours a day six days a week as a printer's devil and other arduous duties. Although his formal education ceased before the age of 10, later in life he was awarded an honorary degree from Whitman University for his achievements and public service.

In 1897 Freeman moved to Seattle. In 1903 Miller Freeman entered the industrial publishing field with the *Pacific Fisherman* and later in 1908, with the *Pacific Motorboat.* In the following years a growing list of other national industrial journals were produced.

From the outset of his career, Freeman played a leading and active role in many projects of great importance to the community, the state and the nation. His contributions to the fisheries of the Pacific Coast are a lasting monument to the man. He was dedicated to the wise use of all natural resources and placed on high priority the protection of the fisheries from depredations by foreign fleets.

As early as 1915, along with William Calvert and the Fishing Vessel Owner's Association of Seattle, Freeman advocated international control of the Pacific halibut fishery. Subsequent to the ratification of the 1923 Halibut Convention between Canada and United States on October 21, 1924, Freeman was appointed by President Calvin Coolidge to be a member of the International Fisheries (Halibut) Commission established under the 1923 convention to investigate the halibut resource.

The commission came into being at its first meeting in Seattle on November 19, 1924. Freeman was the first person from the private sector in both Canada and United States to be appointed to an international fishery commission. He served with distinction until October 1932.

The final ratification in 1937 of the 1930 convention between Canada and United States for joint control of the Fraser River sockeye salmon fishery was in no small measure due to the influential and persistent efforts of Freeman over a period in excess of 30 years.

His close association with President Suzzalo of the University of Washington led to the establishment of the School of Fisheries at the university, the first such institute in the western world.

Freeman was by 1910 acutely aware of the threat to the fisheries posed by the expansionist and destructive fishery policies of Japan and other countries. He subsequently marshalled the fishery forces of the Pacific Coast and carried forward many programs designed to thwart such destruction.

ALVAH L. HAGER 1877-1948

Hager projected the New England Fish Co. on a path destined to make it one of the largest fish producers in the western world. The first Halibut Convention in 1923 owed so much to the efforts of Alvah L. Hager that it came to be known as the "Hager Halibut Treaty."

FIGURE 165. Alvah L. Hager
Canadian Fishing Company, Ltd.

Alvah L. Hager was born of a large family in the farm community of Olewein, Iowa, November 19, 1877. Experiencing adversity, the family moved on to Mitchell, South Dakota. However, conditions did not improve and they moved to Chicago. During these difficult times Hager's mother Mary was the rock upon which the family depended while father Hager was making successive trips to the West Coast in search of better opportunities. The lessons of diligence, thrift, and resourcefulness learned during this hard period made their mark on Alvah Hager.

Unable to take advantage of an athletic scholarship to Northwestern University, he took employment as a junior librarian at the Chicago Public Library. Not satisfied, he went on to Boston to enter the fish business, probably recalling the stories about fish brought back to the farm from the West Coast by his father.

Although the New England Fish Co. controlled most of the East Coast market, this did not deter the resourceful Alvah Hager at the age of 24 from successfully engaging in the distribution of Pacific Coast salmon to Boston and New York dealers each of whom invested $500 in the project in the form of an advance. He soon acquired a half interest in the Doty Fish Co. owned by Spreckels in Kalama, Washington, on the Columbia River. A firm supply of fish thus assured, hard work and perseverence did the rest.

The salmon distributorship of Alvah Hager, which in 1903 had been incorporated as the Northwestern Fisheries Co., was finally acquired in 1908 by the New England Fish Co. after it had made several increasingly attractive offers both for the company and the services of Alvah L. Hager. As New England Fish Co. did not expand into the salmon trade at that time, it seems that the abilities of Alvah Hager were the prime reason for the acquisition.

Located in Vancouver in 1908, Hager, first as general manager of New England Fish Co.'s Pacific operations and later in 1931 as president of the company, was able to apply his business acumen and energy to raise New England Fish Co. into a preeminent position as a fully diversified operation in the North American fishing industry.

In October, 1909, New England Fish Co. purchased the Canadian Fishing Co., a vessel-owning company that had been supplying fish to New England Fish Co. This added another large steamer to the New England Fish Co. halibut fleet and also secured the services of Captain Absalom Freeman, the master halibut fisherman who had been with New England Fish Co. prior to forming the Canadian Fishing Co.

By making the Canadian Fishing Co. the dominant corporate entity in Canada rather than New England Fish Co., Hager, who himself became a Canadian in 1921, was able to dissipate much of the politically-inspired and demogogic hostility against foreign, particularly American, companies in Canada. The Canadian Fishing Co. developed a distinctly Canadian character under Hager, and practically all of its executives, staff and workers were Canadian.

His understanding of men and events were good for both the Canadian Fish Co. and New England Fish Co. In making corporate acquisitions he often placed a higher value on the people in a company than on the physical assets, probably not unmindful of his own experience. This was particularly evident when the Nootka-Banfield Co. was acquired in 1945 by Canadian Fishing Co. It was headed by Sydney M. Rosenberg, an astute and able executive. By 1947 Rosenberg was general manager and in 1950 was president of Canadian Fishing Co. He later moved to Seattle where he became president of New England Fish Co.

At the end of World War I and during the Great Depression of the 1930s, the possession of a strong capital structure permitted New England Fish Co. and Canadian Fishing Co. to acquire on favorable terms a large number of salmon canneries in British Columbia and Alaska. In this manner, New England Fish Co. in the United States and Canadian Fishing Co. in Canada became major factors in the canned salmon industry practically overnight. They also continued their dominance in the distribution of Pacific halibut.

Hager's dedication to conservation was evident early in his career. Even before World War I he voiced concern for the decline in the stocks of halibut. And, long before many, he saw the great benefits that could accrue to the people of Canada and United States by uniting to rehabilitate the Pacific halibut and the sockeye salmon of the Fraser River.

EDWARD AEMILIUS JARVIS 1860-1940

Jarvis was a financier and investor of risk capital so essential to resource development. He was a director and chairman of the British Columbia Packers, Ltd. for 31 years.

E. Aemilius Jarvis was born in Bonshaw, York County, Upper Canada, on April 25, 1860. He passed away in 1940 at the age of 80.

Upon graduation from Upper Canada College at the age of 17, he went into the banking business. At 32 he formed his own firm of investment banking, Aemilius Jarvis and Co. in Toronto. The company was successful from the outset, and in 1902 at the age of 42 he executed a most difficult corporate merger in the British Columbia salmon industry that others in the past had failed to accomplish. He effected a combination of 42 salmon canneries to form what is now the British Columbia Packers, Ltd. While he was on the board of directors and later chairman, Jarvis possessed the business sagacity to leave the operations of the company for most of its existence in the hands of some of the most competent people in the salmon industry.

Jarvis maintained active participation in the affairs of the B.C. Packers until he retired from the board on May 8, 1933, after 31 years of service. Although primarily salmon oriented, the firm played a major role in the Pacific halibut industry.

HAROLD ELMER LOKKEN 1905-

A staunch advocate of conservation and of the interests of the setline halibut fleets, both owners and fishermen, for over half a century, Lokken has been an implacable adversary of all who deny the just claims of those who underwent long years of restraint to rehabilitate the Pacific halibut stocks.

FIGURE 166. Harold Elmer Lokken
International Pacific Halibut Commission

Harold E. Lokken was born in Seattle, Washington, on June 20, 1905. He attended the Seattle public schools and Lincoln High School. Upon graduation in 1922, he began a long and productive career in the fishing industry. As the opportunity presented itself, Harold Lokken was able to greatly extend his academic training through extension courses at the University of Washington. Later, he was special lecturer in fisheries at the university for many years, speaking on various phases of the commercial fisheries.

Since 1924 he has represented the Fishing Vessel Owner's Association of Seattle in the capacity of general manager. Under his one-half century of leadership, this organization became the most effective fishing-vessel-owner group in North America. The many functions and activities of that group are discussed elsewhere in this book.

Harold Lokken has been a member of many standing committees dealing with national and international fishery matters, including among others:

Advisory Committee, United States Section, International North Pacific Fisheries Commission;

Conference Board, International Pacific Halibut Commission;
Ocean Affairs Advisory Committee, United States Department of State;
Sea Grant Advisory Panel, United States Department of Commerce;
National Advisory Committee on Ocean and Atmosphere.

He was appointed in 1961 commissioner for the State of Washington on the Pacific Marine Fisheries Commission by Governor Daniel J. Evans.

He has been a member of a number of United States delegations to international fishery conferences, including the United States negotiating delegation with respect to the International North Pacific Fisheries Convention in Tokyo, Japan, in 1951. He represented United States employers on the Committee of Experts at the International Labor Organization (ILO), Geneva, Switzerland, in 1954. He was also a member of the United States delegation to the Law of the Sea Conference, Geneva, Switzerland, in 1960.

Lokken is widely respected in the fishing industry. In 1963 he was chosen "Man of the Year" by the Northwest Fisheries Association, an organization of Puget Sound fish dealers. In 1971 he was recipient of the Golden Halibut Award presented annually by the Halibut Fishermen's Wives Association for outstanding service to the fishing industry.

Lokken is recognized by all sections of the fishing industry in both Canada and United States including vessel owners, fishermen, and dealers for his able advocacy of the welfare of the Pacific halibut fleet in particular and the fishing industry in general. He has been in the forefront in furthering conservation. Although always a strong supporter of the work of the Halibut Commission, he has also been a constructive critic of some of its actions as they may have affected the halibut fleets over the past 50 years.

In Tripartite negotiations dealing with the North Pacific fishery matters over the past 20 years, Lokken was usually the lone voice in support of the conservation of the Pacific halibut and of the interests of the setline fisheries of Canada and United States. While perfunctory support was sometimes visible in government circles, the overriding concerns of Canada and United States were always the anadromous salmon. The halibut fishery was probably regarded as expendable.

MATTIAS MADSEN, 1890-

A strong advocate of conservation, Madsen worked for the best and the wisest use of our resources.

Madsen was a master halibut fisherman and the first vessel-owner member of the Halibut Commission from the United States.

FIGURE 167. Mattias Madsen
International Pacific Halibut Commission

Mattias Madsen was born in Norway on September 15, 1890, in the community of Bjorneren. His schooling ended early when at 15 he became a line fisherman and also engaged in the winter Icelandic herring fishery. In 1909 he went to the Arctic for seal and collected bears off Jan Mayen for various zoos in Europe.

In 1911 Madsen and two friends left for the United States arriving at Ellis Island, New York, on May 21st. As many had done before, he gravitated to North Dakota where he worked on a farm for two summers and in the woods of Minnesota in the winters. He made his way to Seattle in 1913 where he had many and varied jobs, including making rail ties for the Great Northern and worked for $2.00 per 10-hour day on the locks of the Lake Washington Ship Canal then under construction.

In late spring he got his first "chance" as deck hand on the 10-dory steamer, *Weiding Bros.*, under Captain Holstrom. He soon advanced to doryman making two trips before the vessel was lost on Graham Island, British Columbia. The crew made their way in dories to Jedway where the passenger vessel, *Prince John,* took them to Prince Rupert.

Returning to Vancouver, Madsen shipped out on

the halibut vessel, *Carlotta G. Cox,* Captain George Selig. After two months of mediocre fishing, Madsen returned to Seattle to join the newly launched 12-dory steamer, *Chicago,* Captain Johannsen, making four trips in the winter of 1913-14. He then joined the M.V. *Kodiak* with Captain "Ole" Larsen. In the fall and winter of 1915-16 Madsen attended Everett College. In 1919 on account of one of the few strikes in the industry, Madsen and three others bought the *Eidsvold,* a 5-man 15-ton halibut vessel. He operated the vessel successfully until selling out in 1923 to become majority partner with Jacob (Jack) Engdahl in the M.V. *Eagle.* Disposing of his interest in the *Eagle* in 1927, he built and was sole owner of the M.V. *Ilene.*

From 1928 to 1963 under the ownership and skilled direction of Captain Madsen, the *Ilene* probably created the most consistent production record over a long period of years of any vessel in the halibut fleet. It continued in the halibut fishery under new ownership until it was lost off Kodiak in 1975.

Mattias Madsen was very active in the Fishing Vessel Owner's Association and served as president. Madsen also helped initiate the very successful vessel-owners insurance fund. When United States decided to appoint a vessel owner to the Halibut Commission, Madsen was the logical choice. He was appointed June 9, 1955, by President Dwight D. Eisenhower, resigning May 14, 1964, after nine years of service. Captain Madsen had an abiding interest in conservation and saw in the commission the instrument whereby the resource could be maintained and preserved for future generations.

WILLIAM EDWIN RICKER 1908-

A noted Canadian fisheries biologist whose many contributions to fishery theoretics and practice have had a major impact on the profession. No study of population dynamics is possible without some reference being made to the studies of William E. Ricker.

FIGURE 170. William Edwin Ricker
International Pacific Halibut Commission

Dr. William E. Ricker was born in Waterdown, Ontario, in 1908 and educated in the schools of North Bay. He attended the University of Toronto and received a B.A. in 1930, M.A. in 1931 and Ph.D. in 1936.

Dr. Ricker, after investigating the effectiveness of artificial propagation of sockeye salmon in British Columbia, was briefly on the staff of the International Pacific Salmon Fisheries Commission. He then spent 11 years in Indiana becoming professor of zoology at the University of Indiana and director of the Indiana Lake and Stream Survey.

Returning to Canada as editor-in-chief of the Fisheries Research Board Journal he made it one of the most prestigious scientific publications in the world. He was chief scientist of the fisheries Research Board of Canada until retiring in 1974.

His 150 scientific papers include many studies of the fundamentals of population dynamics, sampling and the basic concepts involved in the relationships between growth and mortality, spawners and progeny, and predators and prey. Many other subjects have engaged his attention including limnology and the taxonomy of Plecoptera.

Dr. Ricker has been the recipient of numerous honors including the first Award of Excellence of the

American Fisheries Society and the Flavelle Medal of the Royal Society of Canada of which he is a fellow.

From time to time he has been a consultant to Canadian members of the Halibut Commission, and he has maintained a continued interest in its work.

EUGENE ARTHUR RUTHFORD 1909-

An authority on marketing quality fishery products and a strong advocate of self-help in industry, Ruthford has been consistent supporter of resource management and the Halibut Commission.

FIGURE 168. Eugene Arthur Ruthford
International Pacific Halibut Commission

Eugene (Gene) A. Ruthford was born in Burlington, Washington, on October 31, 1909. Growing up, he lived in Bellingham, Anacortes and Everett and attended the Lake Stevens High School in the general vicinity of the latter city. After graduating from the University of Washington in 1931, he served with the San Juan Fishing and Packing Co., Seattle, from 1931 to 1962, becoming vice president in 1947. Here he was fortunate in having been associated with other self-made and successful men of integrity—William C. Calvert Sr., Harry J. Tillman and William C. Eardley Sr.

When San Juan was acquired by the New England Fish Co. in 1962 Ruthford continued on with that company as vice president. With a relatively new and imaginative entrant into the fishing industry of the 1970s, the rapidly growing Icicle Seafoods, Inc. of Alaska, it was no accident to find Ruthford its senior vice president, having joined that company in 1975. In this position he was well situated to draw upon a long and successful background in the marketing of a wide line of quality fishery products including halibut.

Ruthford was an early leader in the formation and development of the National Fisheries Institute, serving on many national committees. He was elected national president of the organization in 1965-66. In 1963 he received the honor of being chosen by the National Fisheries Institute as their "Man of the Year."

In 1960 Ruthford and S.M. Rosenberg of the New England Fish Co. were prime movers in the formation of the Halibut Association of North America (HANA), a Canadian and United States self-help organization whose primary purpose was to stabilize and expand the market for Pacific halibut through advertising and other promotions. Ruthford was first president of HANA and subsequently played a prominent role in the organization.

Ruthford has been a forceful advocate of self-reliance in the fishing industry. He believes it should initiate and underwrite its own advertising promotion, and be responsible for improvement in its plants and products. While government support at the appropriate times and places is to be sought, government controls and interference in the market place is to be resisted.

In the days of high volume halibut landings, Ruthford was regarded by other buyers as the "dean of the halibut market" and was depended upon to open the daily bidding on the Seattle Fish Exchange, and this had a major impact on all other daily halibut markets on the Pacific Coast.

Gene Ruthford's unfailing interest in resource conservation made him one of the strongest supporters of the work of the Halibut Commission. His volunteer service as a special lecturer in fisheries at the University of Washington is further evidence of his interest in young people and the development of fishery science and administration.

MICHAEL SARS 1805-1869 GEORG OSSIAN SARS 1837-DATE UNKNOWN

Father, Michael, and son, Georg Ossian, laid the foundations of modern marine fishery research.

FIGURE 171. Michael Sars
Aksjeselskapet J.W. Eides

FIGURE 172. Georg Ossian Sars
McMillan and Company, Ltd.

Michael Sars was born in Bergen, Norway, in 1805 and though trained as a theologian, he was destined to open up a new vista in the field of zoology. Assigned to the parish of Manger on the coast of Norway, he studied marine life in its natural environment and soon became one of the leading zoologists. His breadth of vision and understanding placed him alongside the giants of his time. In 1854, the chair of zoology was created for him at the University of Kristiania in what is now Oslo. This opened a new interest for many in the living resources of the oceans.

While residing in 1837 at his first pastorate in Floro, his son Georg Ossian was born. He became a man whose achievements in the fields of marine and fishery research would eclipse even those of his illustrious father.

In 1864, Georg Sars began his classic and painstaking studies of the Lofoten cod and of the effects of the environment on those stocks. Later Georg Sars extended his research to the herring, mackerel, lobster and other fisheries and led fishery expeditions to waters off Faeroes, Iceland, Spitsbergen and Jan Mayen, Norway, each of which greatly broadened the horizons of fishery knowledge. In 1870, Georg Sars became professor of zoology at the University of Kristiania but continued to head Norwegian Fishery Research until 1893 when Johan Hjort, another distinguished scientist, assumed the responsibility.

GRIER SAMUEL STARRATT 1867-1944

Starratt's managerial talents and construction capabilities laid the foundation for the two companies that have dominated the Canadian halibut industry in the twentieth century.

FIGURE 169. Grier Samuel Starratt
F.H. Bell

Grier Starratt was born in Lower Economy, near Truro, Nova Scotia, in 1867. Soon after completion of his education he came west and in 1891 found employment with a lumber mill, a short distance upriver from New Westminster

He was soon drawn to the fishing industry and in 1893 became manager of the office of the New Westminster Fish Co. upriver at Haney, B.C. The following year, 1894, he became Pacific Coast manager in Vancouver of the New England Fish Co. of Boston and chartered the S.S. *Coquitlan* to fish halibut for that Massachusetts company. During those pioneering and often very difficult years, he successfully established and expanded the Pacific Coast fortunes of the New England Fish Co. Other companies were less fortunate.

In 1907 the transcontinental Grand Trunk Pacific Railway was about to reach tidewater at Prince Rupert. A group of Canadian and English entrepreneurs enlisted the proven capabilities of Grier Starratt. They commissioned him to construct and later manage what was the largest fish cold storage in the world, the Canadian Fish and Cold Storage Co. at Prince Rupert. This company was later absorbed by the B.C. Packers in 1945.

By 1914, Grier Starratt, having established the Canadian Fish and Cold Storage Co. on a firm foundation, left the company. During the short space of but two decades the abilities and imagination of Grier Starratt had made a lasting imprint upon the Pacific halibut fishery.

In 1942 Grier Starratt represented United States interests in developing and organizing a $200,000 fish-freezing operation at Hay River on the southern shore of Great Slave Lake, Northwest Territories. During World War II the company planned to ship 500,000 pounds a month of frozen trout, whitefish, herring and inconnu over the Grimshaw winter road.

Grier Starratt died in Vancouver, British Columbia, on March 17, 1944.

WILLIAM FRANCIS THOMPSON 1888-1965

"One of the great figures in fisheries biology."—A.W.H. Needler, 1965.

FIGURE 173. William F. Thompson
International Pacific Halibut Commission

William F. Thompson was born in St. Cloud, Minnesota, April, 3 1888. His parents moved to Everett, Washington, in 1903. He attended the University of Washington from 1905 until 1909, studying zoology under Professor Trevor Kincaid. He became a research assistant at Stanford University under David Starr Jordan and received his B.A. in 1911. In 1930 he earned his Ph.D. from Stanford University.

Upon graduation in 1911 he made a summer survey of the shellfish beds in northern California for the state, and in the summers of 1912 and 1913 he made similar assessments in British Columbia for the provincial government. In the winters he collaborated with David Starr Jordan on the publication of ten studies on fish taxonomy.

During 1914 and 1915 he made an exhaustive one-man investigation of the Pacific halibut fishery on behalf of the Province of British Columbia. His detailed and comprehensive reports upon the halibut resource and fishery and their respective conditions, as well as their needs, represented a new departure in the field of fishery research. As Thompson often said, "I was going only where T. Wemyss Fulton, Heincke, Hjort, and C.G.J. Petersen had earlier pointed the way."

In 1917 Thompson organized and headed a comprehensive program of applied fishery research for the State of California, placing that state in the forefront of such research in North America. With the first Halibut Convention ratified, Thompson was in 1924 the unanimous choice of both countries to head the research for the commission established under the convention. Successful in turning the tide of decline in the halibut fishery, he moved on in 1938 to become the first director of investigations for the International Pacific Salmon Fisheries Commission. He demonstrated that the initial requirement for the rehabilitation of the Fraser River sockeye was to alleviate the river blockages and delays that the ascending spawning fish were experiencing. When remedial measures were well on their way to success, Thompson became in 1943 director of the School of Fisheries at the University of Washington and reorganized the main thrust of that institution both as to curriculum and the direction of research. In 1947, he organized the Fisheries Research Institute at the University of Washington and served as its director until his retirement in 1958 at age 70.

There were five program ingredients that William Francis Thompson stressed during his long and productive career. They were good staff, good library, adequate data, insight to identify the essentials, and the intellectual courage to initiate action upon the facts at hand.

ROY MAYWOOD WINSLOW 1886-1980

Known as "Mr. Halibut" in the fish trade of Canada, Roy Maywood Winslow had a long and unfailing interest in the work and accomplishments of the Halibut Commission.

FIGURE 174. Roy Maywood Winslow
International Pacific Halibut Commission

Roy M. Winslow was born in Davenport, Iowa, August 9, 1886, of Canadian parents. His mother and his father, a cooper, returned to Canada in 1887 with their 15-month-old son.

Raised in London, Ontario, Winslow later attended Guelph Agricultural College, the leading agricultural college in Canada.

In 1923 he became sales manager for the Canadian Fish and Cold Storage Co. of Prince Rupert, a major factor in the halibut trade. In 1929 he assumed the position of secretary of the canned salmon, meals, oils and saltfish sections of the Canadian Manufacturers Association located in Vancouver. In 1937 he returned to the Canadian Fish and Cold Storage Co. in Prince Rupert as the assistant manager.

When the company was purchased by the B.C. Packers in 1945, Winslow became the manager of its fresh and frozen fish division in Vancouver. He retired in 1958.

Roy Maywood Winslow, with his experience in the marketing of perishable or semiperishable land and sea products and his special aptitude for scholarly expression, was for many years a most articulate spokesman for the Pacific Coast halibut dealers at meetings with the Halibut Commission. He served his industry and his family well and had the respect of all.

Bibliography

The following Bibliography includes only those references noted in the book. It does, however, document in some detail the reports referred to in the discussion regarding the misuse of the term "Greenland halibut."

While there are no comprehensive treatises published dealing with halibut, reports of certain authors do present a large amount of useful information regarding the species in different parts of the world. The authors are Devold (1938); Jesperson (1917); McCracken (1958); McIntyre (1952); and Novikof (1964).

The major source for information on the Pacific halibut are the publications of the International Pacific Halibut Commission and its predecessor organization, the International Fisheries Commission.

Aleem, A.A.
1972 Proc. Sect. B. Roy. Soc. Edinburgh V. 73, 1972. Edinburgh.

American Fisheries Society
1948 Am. Fish. Soc. Spec. Pub. No. 1, p. 45. Wash., D.C.
1960 Am. Fish. Soc. Spec. Pub. No. 2, p. 102. Wash., D.C.

Anderson, A.W. and Peterson, C.E.
1952 to 1954 *Fish. Stat. of the U.S., 1949 to 1951,* F.W.S., U.S. Dept. Int. Wash., D.C.

Anderson, A.W. and Power, E.A.
1946 to 1951 *Fish. Stat. of the U.S., 1942 to 1948,* F.W.S., U.S. Dept. Int. Wash., D.C.
1955 to 1957 *Fish. Stat. of the U.S., 1952 to 1955,* F.W.S., U.S. Dept. Int. Wash., D.C.

Anon
1911 *Fisheries of the U.S., 1908.* Bur. of the Census, Dept. of Comm. and Labor, p. 324. Wash., D.C.
1931 *Rules and Regulations Respecting the Fisheries of Newfoundland.* Dept. Mar. and Fish. p. 122. St. John's, Newfoundland.
1932 *Annual Rept. of the Dept. of Mar. and Fish. for 1931.* p. 39. St. John's, Newfoundland.
1932 *First List of Fishes in the Newfoundland Fishing Area.* Rept. Nfld. Fish. Res. Comm. V. 1, No. 4: pp 107-110. St. John's, Newfoundland.
1934 *Third List of Fishes in the Newfoundland Fishing Area—1933.* Rept. Nfld. Fish. Res. Comm. V. 2, No. 2: pp 115-117. St. John's, Newfoundland.
1955 *Yearbook of Fish. Stat. Part I—Production and Craft.* FAO, p. 249. Rome, Italy.
1956 *The Commercial Fisheries of Canada.* Dept. Fish. Can., and Fish. Res. Bd. p. 191. Ottawa.
1957 Newfoundland Fisheries, 1956. Dept. Fish. Can., *Trade News*, V. 9, No. 9: p. 13. Ottawa.
1958 Newfoundland Fisheries, 1957. Dept. Fish. Can., *Trade News*, V. 10, No. 7: p. 8. Ottawa.
1964 U.S. B.C.F. Div. of Econ. M.N.S., Fish. Prod. Rept. B-236. Wash., D.C.
1964 to 1967 *Fish. Stat. Can., 1961 to 1964.* Dom. Bur. Stat. Ottawa.

Beard, Harry R.
1937 *Story of Canfisco;* The Canadian Fishing Co., Ltd. Vancouver.

Bell, F. Heward
1970 Am. Fish. Soc. Spec. Pub. No. 7. Wash., D.C.

Bell, F. Heward and Pruter, Alonzo T.
1958 J. Fish. Res. Bd. Canada. V. 15, No. 4: pp 625-683. Ottawa.

Best, Edgar A.
1963 Cal. Fish and Game. V. 49, No. 3: pp 213-214. Sacramento, CA.
1970 IPHC Tech. Rept. No. 7. Seattle.
1977 IPHC Rept. No. 62. Seattle.

Beverton, R.J.H. and Hodder, V.M.
1962 ICNAF: Supp. Ann. Proc. V. 2: Halifax.

Bigelow, Henry B. and Schroeder, W.C.
1953 *Fishes of the Gulf of Maine.* Bull. U.S. Bur. Fish. 74, p. 577. Wash., D.C.

Bigelow, Henry B. and Welsh, W.W.
1925 *Fishes of the Gulf of Maine.* Bull. U.S. Bur. Fish. 40, part 1, p. 567. Wash., D.C.

Breder, C.M. Jr.
1929 *Field book of marine fishes of the Atlantic Coast from Labrador to Texas.* p. 332. New York.

Brook, George
1886 Scot. Fish. BD. Ann. Rpt. V 4: 1885. Edinburgh.

Buchannan-Wollaston, H.J.
1926 Gt. Brit. Min. Agric. Fish, Invest. Ser., V. 9, No. 2. London.

Collet, Robert
1860 Fiske. Norske Nordhavs.-Exped., Zool. p 164.

Collins, J.W.
1883 U.S. Fish. Comm. Bull., 1882 V. 2: Wash., D.C.
1885 U.S. Fish. Comm. Bull., 1884 V. 5: p. 256. Wash., D.C.

Cornish, G.A.
1907 *Notes on the Fishes of Canso.* 1902-1905, Fur. Contri. Can. Biol. pp 81-90. Ottawa.

Devold, Finn
1938 Fisk. Skrift. Ser. Havunders., V. 5, No. 6. Bergen.
1942 Fisk. Skrift. Ser. Havunders., V. 7, No. 3. Bergen.

Dixon, George
1789 *A Voyage Round the World; But More Particularly to the Northwest Coast of America Performed in 1785, 1786, 1787 and 1788 in the King George and Queen Charlotte.* 2nd Ed. London

Dunbar, M.J. and Hildebrand, H.H.
1952 J. Fish. Res. Bd. Can. V. 19, No. 2: pp 83-128. Ottawa.

Dunlop, Henry A. et al
1964 IPHC. Rept. No. 35. Seattle.

Ehrenbaum, E.
1936 Bedeuting der Seefisch. Nordeuropas Bd. III. Handbuch.

Fabricius, O.
1780 *Fauna Groenlandica, Systematice Sistens, Animalia Groenlandiae occidentalis.* p. 452. Copenhagen & Leipzig.

Fiedler, R.H.
1930 to 1940 *Fish. Indust. of the U.S., 1928 to 1938.* Bur. Fish., Dept. of Comm. Wash., D.C.
1942 to 1945 *Fish. Stat. of the U.S., 1939 to 1941.*, F.W.S. Dept. Int. Wash., D.C.
Fulton, T. Wemyss
1904 Gt. Brit. Fish. Bd. for Scotland, part 3, S. Invest. Edinburgh.
1911 *The Sovereignty of the Sea.* p. 799. Edinburgh and London.
Furnivall, F.J.
1868 *Early English Meals and Manners,* Early Eng. Text Soc. Orig. Ser. No. 32.
Goode, G. Brown
1880 Bull. U.S. Nat. Mus. No. 18: p. 263. Wash., D.C.
1884 *Fisheries and Fishery Industries of U.S. Sect. 1,* U.S. Fish. Comm., pp 163-682. Wash., D.C.
1887 *American Fishes.* p. 496. Boston.
Goode, G. Brown and Bean, T.H.
1879 *A list of the fishes of Essex County.* Bull. Essex Inst. V. 11. Salem.
1895 Smithsonian Inst. Spec. Bull. Wash., D.C.
Gudger, E.W.
1935 J. Morph. V. 53, No. 1. Philadelphia.
Gudger, E.W. and Firth, F.E.
1937 Am. Mus. Nov. No. 925. New York.
Gunther, Albert
1868 *Catalogue of the Fishes in the British Museum. V. 7,* Physostomi. p. 512. London.
Hoag, Stephen H.
1971 IPHC. Rept. No. 53. Seattle.
1975 IPHC. Rept. No. 57. Seattle.
1978 IPHC. Rept. No. 61. Seattle.
Hoag, Stephen H. and French, R.R.
1976 IPHC. Rept. No. 76. Seattle.
Hoek, P.P.C.
1904 Pub. du Circon. No. 12, p. 90. Copenhagen.
Hubbs, Carl L.
1922 *Amer. Naturalist,* V. 61, pp 360-372. Chicago.
1945 Misc. Publ. Mus. Zool., Univ. Mich. Ann Arbor.
Hubbs, Carl L. and Hubbs, L.C.
1945 Pap. Mich. Acad. Sc. Avs. Lett. V. 30, pp 229-310. Ann Arbor.
Hubbs, Carl L. and Wilimovsky, N.J.
1964 J. Fish. Res. Bd. of Canada. V. 21, No. 5: pp 1129-1154. Ottawa.
Huntsman, A.G.
1918 Our Eastern Flat-fishes. *Can. Fisherman.* V. 5, No. 6: pp 788-790. Gardenvole.
1920 Standardization of the Trade Names of Fishes. *Can. Fisherman.* V. 7, No. 6: pp 115-121. Gardenvole.
International Pacific Halibut Commission (IPHC)
Reports:
Scientific and Annual No. 1, 1930 to present.
Annual, 1969 to present.
Technical No. 1, 1969 to present.
Jensen, A.J.C.
1939 Journ. du Conseil V. 14, No. 3. Copenhagen.
Jesperson, P.
1917 Medd. Komm. Havunders, Ser. Fiskeri, Vol 5, No. 5. Copenhagen.
Jordan, David S.
1887 Rept. U.S. Comm. Fish. V. 13: p. 921. Wash., D.C.
1893 Temperature and Vertebrae: A study in evolution. Wilder. Quarter-century Book. pp 13-26. Ithaca.
1903 *A guide to the study of fishes.* V. 2: p. 599. New York.
1923 Stanford Univ. Pub. Biol. Sci. V. 3, No. 2: pp 79-243. Stanford.
1925 *Fishes.* p. 773, New York.
Jordan, David S. and Evermann, B.S.
1898 *The Fishes of North and Middle America, Part III,* Bull. U.S. Nat. Mus. No. 47. Wash., D.C.
Jordan, David S., Evermann, B.W. and Clark, H.W.
1930 Rep. U.S. Comm. Fish., 1928, part 2: p. 670. Wash., D.C.
Jordan, David S. and Gilbert, C.H.
1882 Bull. U.S. Nat. Mus. No. 16: p. 1018. Wash., D.C.
Jordan, David S. and Goss, D.K.
1889 Rep. U.S. Comm. Fish. V. 14: pp 225-342. Wash., D.C.
Jordan, David S. and Hubbs, C.L.
1925 Mem. Carnegie Mus. V. 10: p. 297.
Jordan, David S. and Snyder, J.O.
1901 J. Coll. Sci., Imperial Univ., Tokyo. Vol. 15, part 2: pp 302-311. Tokyo.
Ketchen, K.S. and Forrester, C.R.
1966 Fish. Res. Bd. of Canada. Bull. 153. Ottawa.
Kohler, J.J.
1906 *Die altengulischen Fischnamen Ang Forschungen Heft 21.* Heidelberg.
La Perouse, J.F.G.
1798 *A voyage round the world in the years 1785, 1786, 1787 and 1788.* London.
Langley
1930 MSS. Report. Can. Embassy. Tokyo.
Leim, A.H. and Scott, W.B.
1966 Bull. No. 155, Fish. Res. Bd. Can. p. 485. Ottawa.
Lewis and Dryden
1895 E.W. Wright, ed. *Marine History of the Pacific Northwest.* p. 494. Portland.
Malm, A.W.
1877 *Goteborg Och Bohuslans,* Fauna Ryggradsdjuren. Goteburg.
Martin, W.R.
1963 *Trade News* Dept. Fish. Can. V. 16, No. 2: p. 8. Ottawa.
McAllister, D.E.
1960 *List of the Marine Fishes of Canada.* Bull. Nat. Mus. Can. No. 168: p. 76. Ottawa.
McCracken, F.D.
1958 Jour. Fish. Res. Bd. Can. V. 15, No. 6. Ottawa.
McIntyre, A.D.
1952 Statistics of the Scottish halibut fishery. Scottish Home Dept., Mar. Res. No. 1: p. 24, Edinburgh.
Meehan, W.E.
1918 *Can. Fisherman* V. 5, No. 7: pp 851-859. Gardenvole.
Moiseev, P.A.
1953 *Cod and Flounders of Far Eastern Waters.* Inst. Rybnoe, Khoziaistva Okeano. V. 40. Vladivostok.
Molander, A.R. and Molander-Swedmark, M.
1957 Rpts. Lysekil Inst. Mar. Res. (Biol) V. 77: p. 44.
Myhre, R.J.
1974 IPHC, Rept. No. 55. Seattle.

Needler, A.W.H.
1965 Prue. Twelfth Ann. Meet INPFC. Minutes 1st Plenary p. 16-17. Vancouver.
Nilsson, Sven
1855 *Skand. Fauna Fiske.* Fjerd Delen-Fiskarna. p. 768. Lund.
Norman, J.R.
1934 *A Systematic Monograph of the Flatfishes (Heterosomata)* V. 1: p. 459. London.
Novikov, N.P.
1960 *Bering Sea Halibuts.* Rybnoe Khoziaistvo. V. 36, No. 1: pp 12-15. Vladivostok.
1964 *Basic Elements of the Biology of the Pacific Halibut. H. hippoglossus stenolepis.* Proc. All Union Res. Inst. Mar. Fish. (TINRO) Part 2. Moscow.
Pacific Fisherman
Annual and Monthly issues. Seattle.
Quensel, Conrad.
1806 Svenska Vet. Akad. Nye Handl. V. 27.
Rae, B.R.
1959 Scot. Home. Dept. Mar. Res. No. 4. Edinburgh.
Regan, C.T.
1920 Ann. Durban Mus. V. 2: pp 205-222. Durban.
1929 *Fishes:* Article Encyclopaedia Britannica 14th Ed. V. 9.
Rendahl, H.
1931 Archiv. f. zool., Bd. V. 22, No. 18: pp 17-65. Stockholm.
Russel, John
1430 *Boke of Nurture* and *Kervynge of Fische.* Early English Text Society. England.
Schmidt, Johannes
1904 Denmark, Komm. f. havunders. Medd. Ser. Fisk. V. 3, No. 3. Copenhagen.
Schmidt, P.J.
1930 USSR Aead. Se Comptes-rend. Ser. A. V. 8. Moscow.
Schmidt, P.Y.
1950 Acad. of Sci. USSR, Trans. Pac. Comm., V. 6. Moscow-Leningrad.
Schroeder, W.C.
1955 Pap. Mar. Biol. and Oceanog., Deep-Sea Res. Supplement to V. 3: pp 358-372.
Smith, Richard T.
1936 *Rept. on Puget Sound Otter Trawl Invest.* Biol. Rept. 36B, Dpt. Fish., Wash. State.
Smitt, F.A.
1893 *A History of Scandinavian Fishes Part I.* London.
Stewart, Hillary
1977 *Indian Fishing: Early Methods on the Northwest Coast.* Seattle: Univ. of Wash. Press.
Stewart, T.N.
1956 Bull. No. 108, Fish. Res. Bd. Can. p. 26. Ottawa.
Swan, J.G.
1870. The Indians of Cape Flattery. *Smithsonian Contr. to Knowledge.* V. 26, Art. 2. Wash., D.C.
Taning, A.V.
1936 Medd. Comm. f. Danmarks. Fisk, Havunders. Bind X. Copenhagen.
1952 Biol. Revs. Cambridge Phil. Soc. V. 27, No. 2. Cambridge.
Templeman, W.H.
1966 Bull. No. 154, Fish. Res. Bd. Can. p. 170. Ottawa.
1966 *The Fisheries Situation in the Northwest Atlantic.* Fisheries of Canada. V. 19, No. 3: pp 3-10. Ottawa.
Templeman, W.H. and Fleming, A.M.
1956 Bull. No. 109, Fish. Res. Bd. Can. p. 55. Ottawa.
Templeman, W.H., Squires, H.J. and Fleming, A.M.
1957 J. Fish. Res. Bd. Can. V. 14, No. 6: pp 831-897. Ottawa.
Thompson, Harold
1930 Rep. Nfld. Fish. Res. Comm. V. 1, No. 1: p. 63. St. John's.
Thompson, W.F.
1915 Brit. Col. Fish. Dept. Rpt. (1914). pp 76-99. Victoria.
1916 Brit. Col. Fish. Dept. Rpt. (1915). pp 65-126. Victoria.
1950 Fish. Res. Instit., Univ. of Wash. p. 60. Seattle.
Thompson, W.F. and Van Cleve, R.
1936 IPHC Rept. No. 9. Seattle.
Thurston, C.E. and MacMasters, P.P.
1960 Food Res. V. 25, No. 2: pp 229-236. Chicago.
Tomlinson, N. et al
1964 *Chalkiness in Halibut in Relation to Muscle pH and Protein Denaturation.* J. Fish. Res. Bd. Can. V. 22, No. 3. Ottawa.
Townsend, L.D.
1936 IPHC Rept. No. 11. Seattle.
VanCleve, R. and Seymour, A.H.
1953 IFC Report No. 19. Seattle.
Vernidub, M.F.
1936 Leningrad Soc. Nat. Dept. Zool. V. 65, No. 2. Moscow-Leningrad.
Vladykov, V.D. and McKenzie, R.A.
1935 Proc. Nova Scotian Inst. Sci. V. 19, part 1: pp 17-113. Halifax.
Watt, B.K. and Merrill, A.L.
1963 *Agr. Handbook No. 8,* U.S. Dept. Agr. p. 190. Wash., D.C.
Wright, E.W.
1895 *Marine History of the Pacific Northwest.* Portland: Lewis and Dryden.

Index